The Design Warrior's Guide to FPGAs

Devices, Tools and Flows

The Design Warrior's Guide to FPGAs

Devices, Tools and Flows

Clive "Max" Maxfield

AMSTERDAM • BOSTON • HEIDELBERG • LONDON
NEW YORK • OXFORD • PARIS • SAN DIEGO
SAN FRANCISCO • SINGAPORE • SYDNEY • TOKYO

Newnes is an imprint of Elsevier

Newnes

Newnes is an imprint of Elsevier

200 Wheeler Road, Burlington, MA 01803, USA

Linacre House, Jordan Hill, Oxford OX2 8DP, UK

Illustrations by Clive "Max" Maxfield

 Recognizing the importance of preserving what has been written, Elsevier prints its
books on acid-free paper whenever possible.

Library of Congress Cataloging-in-Publication Data

Application submitted

ISBN-13: 978-0-7506-7604-5

ISBN-10: 0-7506-7604-3

British Library Cataloguing-in-Publication Data

A catalogue record for this book is available from the British Library.

For information on all Newnes publications

visit our Web site at www.newnespress.com

Transferred to Digital Printing, 2011

Printed in the United States of America

*To my wife Gina—the yummy-scrummy caramel, chocolate fudge,
and rainbow-colored sprinkles on the ice cream sundae of my life*

*Also, to my stepson Joseph and my grandchildren Willow, Gaige, Keegan, and Karma,
all of whom will be tickled pink to see their names in a real book!*

The **companion website** also contains a set of Microsoft® PowerPoint® files—one for each chapter and appendix—containing copies of the illustrations that are festooned throughout the book. This will be of particular interest for educators at colleges and universities when it comes to giving lectures or creating handouts based on *The Design Warrior's Guide to FPGAs*

Last but not least, the **companion website** contains a smorgasbord of datasheets, technical articles, and useful web links provided by Mentor and Xilinx.

Contents

Preface

This is something of a curious, atypical book for the technical genre (and as the author, I should know). I say this because this tome is intended to be of interest to an unusually broad and diverse readership. The primary audience comprises fully fledged engineers who are currently designing with *field programmable gate arrays (FPGAs)* or who are planning to do so in the not-so-distant future. Thus, *Section 2: Creating FPGA-Based Designs* introduces a wide range of different design flows, tools, and concepts with lots of juicy technical details that only an engineer could love. By comparison, other areas of the book—such as *Section 1: Fundamental Concepts*—cover a variety of topics at a relatively low technical level.

The reason for this dichotomy is that there is currently a tremendous amount of interest in FPGAs, especially from people who have never used or considered them before. The first FPGA devices were relatively limited in the number of equivalent logic gates they supported and the performance they offered, so any "serious" (large, complex, high-performance) designs were automatically implemented as *application-specific integrated circuits (ASICs)* or *application-specific standard parts (ASSPs)*. However, designing and building ASICs and ASSPs is an extremely time-consuming and expensive hobby, with the added disadvantage that the final design is "frozen in silicon" and cannot be easily modified without creating a new version of the device.

By comparison, the cost of creating an FPGA design is much lower than that for an ASIC or ASSP. At the same time, implementing design changes is much easier in FPGAs and the time-to-market for such designs is much faster. Of particular interest is the fact that new FPGA architectures containing millions of equivalent logic gates, embedded processors, and ultra-high-speed interfaces have recently become available. These devices allow FPGAs to be used for applications that would—until now—have been the purview only of ASICs and ASSPs.

With regard to those FPGA devices featuring embedded processors, such designs require the collaboration of hardware and software engineers. In many cases, the software engineers may not be particularly familiar with some of the nitty-gritty design considerations associated with the hardware aspects of these devices. Thus, in addition to hardware design engineers, this book is also intended to be of interest to those members of the software fraternity who are tasked with creating embedded applications for these devices.

Further intended audiences are electronics engineering students in colleges and universities; sales, marketing, and other folks working for EDA and FPGA companies; and analysts and magazine editors. Many of these readers will appreciate the lower technical level of the introductory material found in Section 1 and also in the "101-style" appendices.

Last but not least, I tend to write the sort of book that I myself would care to read. (At this moment in time, I would particularly like to read this book—upon which I'm poised to commence work—because then I would have some clue as to what I was going to write … if you see what I mean.) Truth to tell, I rarely read technical books myself anymore because they usually bore my socks off. For this reason, in my own works I prefer to mix complex topics with underlying fundamental concepts ("where did this come from" and "why do we do it this way") along with interesting nuggets of trivia. This has the added advantage that when my mind starts to wander in my autumn years, I will be able to amaze and entertain myself by rereading my own works (it's always nice to have something to look forward to <grin>).

Clive "Max" Maxfield, June 2003—January 2004

Acknowledgments

I've long wanted to write a book on FPGAs, so I was delighted when my publisher—Carol Lewis at Elsevier Science (which I'm informed is the largest English-language publisher in the world)—presented me with the opportunity to do so.

There was one slight problem, however, in that I've spent much of the last 10 years of my life slaving away the days at my real job, and then whiling away my evenings and weekends penning books. At some point it struck me that it would be nice to "get a life" and spend some time hanging out with my family and friends. Hence, I was delighted when the folks at Mentor Graphics and Xilinx offered to sponsor the creation of this tome, thereby allowing me to work on it in the days and to keep my evenings and weekends free.

Even better, being an engineer by trade, I hate picking up a book that purports to be technical in nature, but that somehow manages to mutate into a marketing diatribe while I'm not looking. So I was delighted when both sponsors made it clear that this book should not be Mentor-centric or Xilinx-centric, but should instead present any and all information I deemed to be useful without fear or favor.

You really can't write a book like this one in isolation, and I received tremendous amounts of help and advice from people too numerous to mention. I would, however, like to express my gratitude to all of the folks at Mentor and Xilinx who gave me so much of their time and information. Thanks also to Gary Smith and Daya Nadamuni from Gartner DataQuest and Richard Goering from *EETimes*, who always make the time to answer my e-mails with the dread subject line "Just one more little question ..."

I would also like to mention the fact that the folks at 0-In, AccelChip, Actel, Aldec, Altera, Altium, Axis, Cadence, Carbon, Celoxica, Elanix, InTime, Magma, picoChip, Quick-Logic, QuickSilver, Synopsys, Synplicity, The MathWorks, Hier Design, and Verisity were extremely helpful.[1] It also behooves me to mention that Tom Hawkins from Launchbird Design Systems went above and beyond the call of duty in giving me his sagacious observations into open-source design tools. Similarly, Dr. Eric Bogatin at GigaTest Labs was kind enough to share his insights into signal integrity effects at the circuit board level.

Last, but certainly not least, thanks go once again to my publisher—Carol Lewis at Elsevier Science—for allowing me to abstract the contents of appendix B from my book *Designus Maximus Unleashed* (ISBN 0-7506-9089-5) and also for allowing me to abstract the contents of appendix C from my book *Bebop to the Boolean Boogie (An Unconventional Guide to Electronics), Second Edition* (ISBN 0-7506-7543-8).

---

1. If I've forgotten anyone, I'm really sorry (let me know, and I'll add you into the book for the next production run).

Introduction

What are FPGAs?

Field programmable gate arrays (FPGAs) are digital *integrated circuits (ICs)* that contain configurable (programmable) blocks of logic along with configurable interconnects between these blocks. Design engineers can configure (program) such devices to perform a tremendous variety of tasks.

Depending on the way in which they are implemented, some FPGAs may only be programmed a single time, while others may be reprogrammed over and over again. Not surprisingly, a device that can be programmed only one time is referred to as *one-time programmable (OTP)*.

The "field programmable" portion of the FPGA's name refers to the fact that its programming takes place "in the field" (as opposed to devices whose internal functionality is hardwired by the manufacturer). This may mean that FPGAs are configured in the laboratory, or it may refer to modifying the function of a device resident in an electronic system that has already been deployed in the outside world. If a device is capable of being programmed while remaining resident in a higher-level system, it is referred to as being *in-system programmable (ISP)*.

Why are FPGAs of interest?

There are many different types of digital ICs, including "jelly-bean logic" (small components containing a few simple, fixed logical functions), memory devices, and *microprocessors (µPs)*. Of particular interest to us here, however, are *program-*

FPGA is pronounced by spelling it out as "F-P-G-A."

IC is pronounced by spelling it out as "I-C."

OTP is pronounced by spelling it out as "O-T-P."

ISP is pronounced by spelling it out as "I-S-P."

Pronounced "mu" to rhyme with "phew," the "µ" in "µP" comes from the Greek *micros*, meaning "small."

PLD is pronounced by spelling it out as "P-L-D."

SPLD is pronounced by spelling it out as "S-P-L-D."

CPLD is pronounced by spelling it out as "C-P-L-D."

ASIC is pronounced "A-SIC." That is, by spelling out the "A" to rhyme with "hay," followed by "SIC" to rhyme with "tick."

ASSP is pronounced by spelling it out as "A-S-S-P."

mable logic devices (PLDs), application-specific integrated circuits (ASICs), application-specific standard parts (ASSPs), and—of course—FPGAs.

For the purposes of this portion of our discussion, we shall consider the term *PLD* to encompass both *simple programmable logic devices (SPLDs)* and *complex programmable logic devices (CPLDs)*.

Various aspects of PLDs, ASICs, and ASSPs will be introduced in more detail in chapters 2 and 3. For the nonce, we need only be aware that PLDs are devices whose internal architecture is predetermined by the manufacturer, but which are created in such a way that they can be configured (programmed) by engineers in the field to perform a variety of different functions. In comparison to an FPGA, however, these devices contain a relatively limited number of logic gates, and the functions they can be used to implement are much smaller and simpler.

At the other end of the spectrum are ASICs and ASSPs, which can contain hundreds of millions of logic gates and can be used to create incredibly large and complex functions. ASICs and ASSPs are based on the same design processes and manufacturing technologies. Both are custom-designed to address a specific application, the only difference being that an ASIC is designed and built to order for use by a specific company, while an ASSP is marketed to multiple customers. (When we use the term ASIC henceforth, it may be assumed that we are also referring to ASSPs unless otherwise noted or where such interpretation is inconsistent with the context.)

Although ASICs offer the ultimate in size (number of transistors), complexity, and performance; designing and building one is an extremely time-consuming and expensive process, with the added disadvantage that the final design is "frozen in silicon" and cannot be modified without creating a new version of the device.

Thus, FPGAs occupy a middle ground between PLDs and ASICs because their functionality can be customized in the field like PLDs, but they can contain millions of logic gates[1] and be used to implement extremely large and complex functions that previously could be realized only using ASICs.

The cost of an FPGA design is much lower than that of an ASIC (although the ensuing ASIC components are much cheaper in large production runs). At the same time, implementing design changes is much easier in FPGAs, and the time-to-market for such designs is much faster. Thus, FPGAs make a lot of small, innovative design companies viable because—in addition to their use by large system design houses—FPGAs facilitate "Fred-in-the-shed"–type operations. This means they allow individual engineers or small groups of engineers to realize their hardware and software concepts on an FPGA-based test platform without having to incur the enormous *nonrecurring engineering (NRE)* costs or purchase the expensive toolsets associated with ASIC designs. Hence, there were estimated to be only 1,500 to 4,000 ASIC design starts[2] and 5,000 ASSP design starts in 2003 (these numbers are falling dramatically year by year), as opposed to an educated "guesstimate" of around 450,000 FPGA design starts[3] in the same year.

NRE is pronounced by spelling it out as "N-R-E."

---

[1] The concept of what actually comprises a "logic gate" becomes a little murky in the context of FPGAs. This topic will be investigated in excruciating detail in chapter 4.

[2] This number is pretty vague because it depends on whom you talk to (not surprisingly, FPGA vendors tend to proclaim the lowest possible estimate, while other sources range all over the place).

[3] Another reason these numbers are a little hard to pin down is that it's difficult to get everyone to agree what a "design start" actually is. In the case of an ASIC, for example, should we include designs that are canceled in the middle, or should we only consider designs that make it all the way to tape-out? Things become even fluffier when it comes to FPGAs due to their reconfigurability. Perhaps more telling is the fact that, after pointing me toward an FPGA-centric industry analyst's Web site, a representative from one FPGA vendor added, "But the values given there aren't very accurate." When I asked why, he replied with a sly grin, "Mainly because we don't provide him with very good data!"

What can FPGAs be used for?

When they first arrived on the scene in the mid-1980s, FPGAs were largely used to implement *glue logic*,[4] medium-complexity state machines, and relatively limited data processing tasks. During the early 1990s, as the size and sophistication of FPGAs started to increase, their big markets at that time were in the telecommunications and networking arenas, both of which involved processing large blocks of data and pushing that data around. Later, toward the end of the 1990s, the use of FPGAs in consumer, automotive, and industrial applications underwent a humongous growth spurt.

FPGAs are often used to prototype ASIC designs or to provide a hardware platform on which to verify the physical implementation of new algorithms. However, their low development cost and short time-to-market mean that they are increasingly finding their way into final products (some of the major FPGA vendors actually have devices that they specifically market as competing directly against ASICs).

By the early-2000s, high-performance FPGAs containing millions of gates had become available. Some of these devices feature embedded microprocessor cores, high-speed *input/output (I/O)* interfaces, and the like. The end result is that today's FPGAs can be used to implement just about anything, including communications devices and software-defined radios; radar, image, and other *digital signal processing (DSP)* applications; all the way up to *system-on-chip (SoC)*[5] components that contain both hardware and software elements.

I/O is pronounced by spelling it out as "I-O."

SoC is pronounced by spelling it out as "S-O-C."

---

[4] The term *glue logic* refers to the relatively small amounts of simple logic that are used to connect ("glue")—and interface between—larger logical blocks, functions, or devices.

[5] Although the term *system-on-chip (SoC)* would tend to imply an entire electronic system on a single device, the current reality is that you invariably require additional components. Thus, more accurate appellations might be *subsystem-on-chip (SSoC)* or *part of a system-on-chip (PoaSoC)*.

To be just a tad more specific, FPGAs are currently eating into four major market segments: ASIC and custom silicon, DSP, embedded microcontroller applications, and physical layer communication chips. Furthermore, FPGAs have created a new market in their own right: *reconfigurable computing (RC)*.

- **ASIC and custom silicon:** As was discussed in the previous section, today's FPGAs are increasingly being used to implement a variety of designs that could previously have been realized using only ASICs and custom silicon.

- **Digital signal processing:** High-speed DSP has traditionally been implemented using specially tailored microprocessors called *digital signal processors (DSPs)*. However, today's FPGAs can contain embedded multipliers, dedicated arithmetic routing, and large amounts of on-chip RAM, all of which facilitate DSP operations. When these features are coupled with the massive parallelism provided by FPGAs, the result is to outperform the fastest DSP chips by a factor of 500 or more.

- **Embedded microcontrollers:** Small control functions have traditionally been handled by special-purpose embedded processors called *microcontrollers*. These low-cost devices contain on-chip program and instruction memories, timers, and I/O peripherals wrapped around a processor core. FPGA prices are falling, however, and even the smallest devices now have more than enough capability to implement a soft processor core combined with a selection of custom I/O functions. The end result is that FPGAs are becoming increasingly attractive for embedded control applications.

- **Physical layer communications:** FPGAs have long been used to implement the glue logic that interfaces between physical layer communication chips and high-level networking protocol layers. The fact that today's high-end FPGAs can contain multiple high-speed transceivers means that communications and network-

RC is pronounced by spelling it out as "R-C."

DSP is pronounced by spelling it out as "D-S-P."

RAM is pronounced to rhyme with "ham."

ing functions can be consolidated into a single device.

- **Reconfigurable computing:** This refers to exploiting the inherent parallelism and reconfigurability provided by FPGAs to "hardware accelerate" software algorithms. Various companies are currently building huge FPGA-based reconfigurable computing engines for tasks ranging from hardware simulation to cryptography analysis to discovering new drugs.

What's in this book?

Anyone involved in the electronics design or *electronic design automation (EDA)* arenas knows that things are becoming evermore complex as the years go by, and FPGAs are no exception to this rule.

EDA is pronounced by spelling it out as "E-D-A."

Life was relatively uncomplicated in the early days—circa the mid-1980s—when FPGAs had only recently leaped onto the stage. The first devices contained only a few thousand simple logic gates (or the equivalent thereof), and the flows used to design these components—predominantly based on the use of schematic capture—were easy to understand and use. By comparison, today's FPGAs are incredibly complex, and there are more design tools, flows, and techniques than you can swing a stick at.

This book commences by introducing fundamental concepts and the various flavors of FPGA architectures and devices that are available. It then explores the myriad of design tools and flows that may be employed depending on what the design engineers are hoping to achieve. Furthermore, in addition to looking "inside the FPGA," this book also considers the implications associated with integrating the device into the rest of the system in the form of a circuit board, including discussions on the gigabit interfaces that have only recently become available.

Last but not least, electronic conversations are jam-packed with TLAs, which is a tongue-in-cheek joke that stands for "three-letter acronyms." If you say things the wrong way when talking to someone in the industry, you immediately brand yourself as an outsider (one of "them" as opposed to one of "us"). For this reason, whenever we introduce new TLAs—or their larger cousins—we also include a note on how to pronounce them.[6]

TLA is pronounced by spelling it out as "T-L-A."

What's not in this book?

This tome does not focus on particular FPGA vendors or specific FPGA devices, because new features and chip types appear so rapidly that anything written here would be out of date before the book hit the streets (sometimes before the author had completed the relevant sentence).

Similarly, as far as possible (and insofar as it makes sense to do so), this book does not mention individual EDA vendors or reference their tools by name because these vendors are constantly acquiring each other, changing the names of—or otherwise transmogrifying—their companies, or varying the names of their design and analysis tools. Similarly, things evolve so quickly in this industry that there is little point in saying "Tool A has this feature, but Tool B doesn't," because in just a few months' time Tool B will probably have been enhanced, while Tool A may well have been put out to pasture.

For all of these reasons, this book primarily introduces different flavors of FPGA devices and a variety of design tool concepts and flows, but it leaves it up to the reader to research which FPGA vendors support specific architectural constructs and which EDA vendors and tools support specific features (useful Web addresses are presented in chapter 6).

---

[6] In certain cases, the pronunciation for a particular TLA may appear in multiple chapters to help readers who are "cherry-picking" specific topics, rather than slogging their way through the book from cover to cover.

2,400,000 BC:
Hominids in Africa

Who's this book for?

This book is intended for a wide-ranging audience, which includes

- Small FPGA design consultants
- Hardware and software design engineers in larger system houses
- ASIC designers who are migrating into the FPGA arena
- DSP designers who are starting to use FPGAs
- Students in colleges and universities
- Sales, marketing, and other guys and gals working for EDA and FPGA companies
- Analysts and magazine editors

Fundamental Concepts

The key thing about FPGAs

The thing that really distinguishes an FPGA from an ASIC is … the crucial aspect that resides at the core of their reason for being is … embodied in their name:

Field **Programmable** Gate Array

All joking aside, the point is that in order to be programmable, we need some mechanism that allows us to configure (program) a prebuilt silicon chip.

A simple programmable function

As a basis for these discussions, let's start by considering a very simple programmable function with two inputs called *a* and *b* and a single output *y* (Figure 2-1).

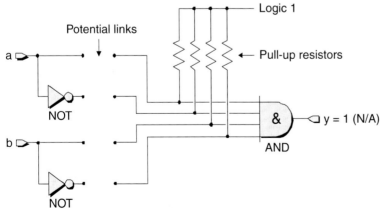

Figure 2-1. A simple programmable function.

25,000 BC:
The first boomerang is used by people in what is now Poland, 13,000 years before the Australians.

The inverting (NOT) gates associated with the inputs mean that each input is available in both its *true* (unmodified) and *complemented* (inverted) form. Observe the locations of the potential links. In the absence of any of these links, all of the inputs to the AND gate are connected via pull-up resistors to a logic 1 value. In turn, this means that the output y will always be driving a logic 1, which makes this circuit a very boring one in its current state. In order to make our function more interesting, we need some mechanism that allows us to establish one or more of the potential links.

Fusible link technologies

One of the first techniques that allowed users to program their own devices was—and still is—known as *fusible-link technology*. In this case, the device is manufactured with all of the links in place, where each link is referred to as a *fuse* (Figure 2-2).

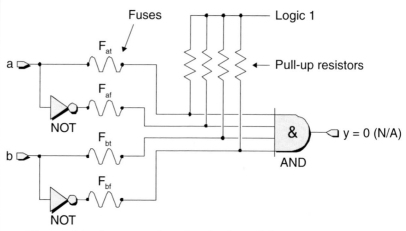

Figure 2-2. Augmenting the device with unprogrammed fusible links.

These fuses are similar in concept to the fuses you find in household products like a television. If anything untoward occurs such that the television starts consuming too much power, its fuse will burn out. This results in an open circuit (a break in the wire), which protects the rest of the unit from harm. Of course, the fuses in a silicon chip are formed using the same processes that are employed to create the transistors and wires on the chip, so they are microscopically small.

When an engineer purchases a programmable device based on fusible links, all of the fuses are initially intact. This means that, in its unprogrammed state, the output from our example function will always be logic 0. (Any 0 presented to the input of an AND gate will cause its output to be 0, so if input *a* is 0, the output from the AND will be 0. Alternatively, if input *a* is 1, then the output from its NOT gate—which we shall call *!a*—will be 0, and once again the output from the AND will be 0. A similar situation occurs in the case of input *b*.)

The point is that design engineers can selectively remove undesired fuses by applying pulses of relatively high voltage and current to the device's inputs. For example, consider what happens if we remove fuses F_{af} and F_{bt} (Figure 2-3).

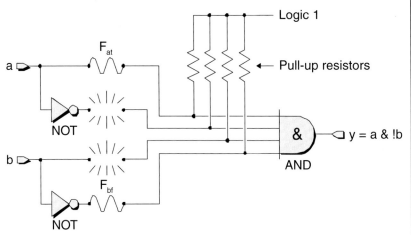

Figure 2-3. Programmed fusible links.

Removing these fuses disconnects the complementary version of input *a* and the true version of input *b* from the AND gate (the pull-up resistors associated with these signals cause their associated inputs to the AND to be presented with logic 1 values). This leaves the device to perform its new function, which is *y = a & !b*. (The "&" character in this equation is used to represent the AND, while the "!" character is used to represent the NOT. This syntax is discussed in a little more detail in chapter 3). This process of removing fuses is typically referred to as *programming* the device, but it may also be referred to as *blowing* the fuses or *burning* the device.

Devices based on fusible-link technologies are said to be *one-time programmable*, or OTP, because once a fuse has been blown, it cannot be replaced and there's no going back.

As fate would have it, although modern FPGAs are based on a wide variety of programming technologies, the fusible-link approach isn't one of them. The reasons for mentioning it here are that it sets the scene for what is to come, and it's relevant in the context of the precursor device technologies referenced in chapter 3.

OTP is pronounced by spelling it out as "O-T-P."

Antifuse technologies

As a diametric alternative to fusible-link technologies, we have their antifuse counterparts, in which each configurable path has an associated link called an *antifuse*. In its unprogrammed state, an antifuse has such a high resistance that it may be considered an open circuit (a break in the wire), as illustrated in Figure 2-4.

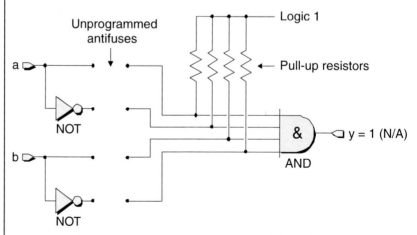

Figure 2-4. Unprogrammed antifuse links.

This is the way the device appears when it is first purchased. However, antifuses can be selectively "grown" (programmed) by applying pulses of relatively high voltage and current to the device's inputs. For example, if we add the antifuses associated with the complementary version of input *a* and the true version of input *b*, our device will now perform the function $y = !a \& b$ (Figure 2-5).

260 BC:
Archimedes works out the principle of the lever.

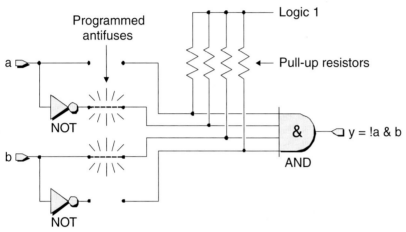

Figure 2-5. Programmed antifuse links.

An antifuse commences life as a microscopic column of amorphous (noncrystalline) silicon linking two metal tracks. In its unprogrammed state, the amorphous silicon acts as an insulator with a very high resistance in excess of one billion ohms (Figure 2-6a).

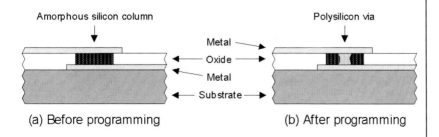

(a) Before programming   (b) After programming

Figure 2-6. Growing an antifuse.

The act of programming this particular element effectively "grows" a link—known as a *via*—by converting the insulating amorphous silicon into conducting polysilicon (Figure 2-6b).

Not surprisingly, devices based on antifuse technologies are OTP, because once an antifuse has been grown, it cannot be removed, and there's no changing your mind.

Mask-programmed devices

Before we proceed further, a little background may be advantageous in order to understand the basis for some of the nomenclature we're about to run into. Electronic systems in general—and computers in particular—make use of two major classes of memory devices: *read-only memory (ROM)* and *random-access memory (RAM)*.

ROMs are said to be *nonvolatile* because their data remains when power is removed from the system. Other components in the system can read data from ROM devices, but they cannot write new data into them. By comparison, data can be both written into and read out of RAM devices, which are said to be *volatile* because any data they contain is lost when the system is powered down.

Basic ROMs are also said to be *mask-programmed* because any data they contain is hard-coded into them during their construction by means of the photo-masks that are used to create the transistors and the metal tracks (referred to as the *metallization layers*) connecting them together on the silicon chip. For example, consider a transistor-based ROM cell that can hold a single *bit* of data (Figure 2-7).

The entire ROM consists of a number of *row* (word) and *column* (data) lines forming an array. Each column has a single pull-up resistor attempting to hold that column to a weak logic 1 value, and every row-column intersection has an associated transistor and, potentially, a mask-programmed connection.

The majority of the ROM can be preconstructed, and the same underlying architecture can be used for multiple customers. When it comes to customizing the device for use by a

ROM is pronounced to rhyme with "bomb."

RAM is pronounced to rhyme with "ham."

The concept of photo-masks and the way in which silicon chips are created are described in more detail in *Bebop to the Boolean Boogie (An Unconventional Guide to Electronics)*, ISBN 0-7506-7543-8

The term *bit* (meaning "binary digit") was coined by John Wilder Tukey, the American chemist, turned topologist, turned statistician in the 1940s.

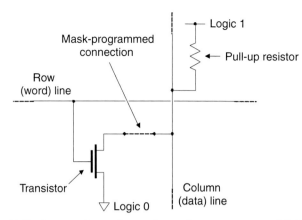

Figure 2-7. A transistor-based mask-programmed ROM cell.

Tukey had initially considered using "binit" or "bigit," but thankfully he settled on "bit," which is much easier to say and use.

The term *software* is also attributed to Tukey.

particular customer, a single photo-mask is used to define which cells are to include a mask-programmed connection and which cells are to be constructed without such a connection.

Now consider what happens when a row line is placed in its active state, thereby attempting to activate all of the transistors connected to that row. In the case of a cell that includes a mask-programmed connection, activating that cell's transistor will connect the column line through the transistor to logic 0, so the value appearing on that column as seen from the outside world will be a 0. By comparison, in the case of a cell that doesn't have a mask-programmed connection, that cell's transistor will have no effect, so the pull-up resistor associated with that column will hold the column line at logic 1, which is the value that will be presented to the outside world.

PROMs

The problem with mask-programmed devices is that creating them is a very expensive pastime unless you intend to produce them in extremely large quantities. Furthermore, such components are of little use in a development environment in which you often need to modify their contents.

For this reason, the first *programmable read-only memory (PROM)* devices were developed at Harris Semiconductor in 1970. These devices were created using a nichrome-based

PROM is pronounced just like the high school dance of the same name.

15 BC:
The Chinese invent the belt drive.

fusible-link technology. As a generic example, consider a somewhat simplified representation of a transistor-and-fusible-link–based PROM cell (Figure 2-8).

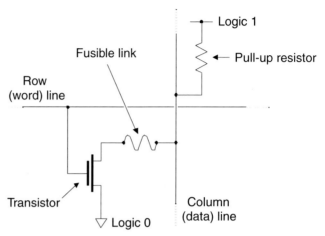

Figure 2-8. A transistor-and-fusible-link–based PROM cell.

   In its unprogrammed state as provided by the manufacturer, all of the fusible links in the device are present. In this case, placing a row line in its active state will turn on all of the transistors connected to that row, thereby causing all of the column lines to be pulled down to logic 0 via their respective transistors. As we previously discussed, however, design engineers can selectively remove undesired fuses by applying pulses of relatively high voltage and current to the device's inputs. Wherever a fuse is removed, that cell will appear to contain a logic 1.

   It's important to note that these devices were initially intended for use as memories to store computer programs and constant data values (hence the "ROM" portion of their appellation). However, design engineers also found them useful for implementing simple logical functions such as lookup tables and state machines. The fact that PROMs were relatively cheap meant that these devices could be used to fix bugs or test new implementations by simply burning a new device and plugging it into the system.

Over time, a variety of more general-purpose PLDs based on fusible-link and antifuse technologies became available (these devices are introduced in more detail in chapter 3).

EPROM-based technologies

As was previously noted, devices based on fusible-link or antifuse technologies can only be programmed a single time—once you've blown (or grown) a fuse, it's too late to change your mind. (In some cases, it's possible to incrementally modify devices by blowing, or growing, additional fuses, but the fates have to be smiling in your direction.) For this reason, people started to think that it would be nice if there were some way to create devices that could be programmed, erased, and reprogrammed with new data.

One alternative is a technology known as *erasable programmable read-only memory (EPROM)*, with the first such device—the 1702—being introduced by Intel in 1971. An EPROM transistor has the same basic structure as a standard MOS transistor, but with the addition of a second polysilicon *floating gate* isolated by layers of oxide (Figure 2-9).

EPROM is pronounced by spelling out the "E" to rhyme with "bee," followed by "PROM."

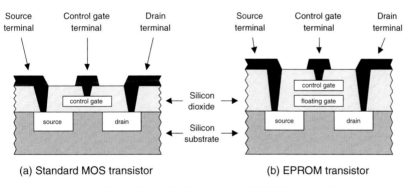

(a) Standard MOS transistor   (b) EPROM transistor

Figure 2-9. Standard MOS versus EPROM transistors.

In its unprogrammed state, the floating gate is uncharged and doesn't affect the normal operation of the control gate. In order to program the transistor, a relatively high voltage (the order of 12V) is applied between the control gate and drain

60 AD:
Hero, an Alexandrian
Greek, builds a toy
powered by stream.

terminals. This causes the transistor to be turned hard on, and energetic electrons force their way through the oxide into the floating gate in a process known as *hot (high energy) electron injection*. When the programming signal is removed, a negative charge remains on the floating gate. This charge is very stable and will not dissipate for more than a decade under normal operating conditions. The stored charge on the floating gate inhibits the normal operation of the control gate and, thus, distinguishes those cells that have been programmed from those that have not. This means we can use such a transistor to form a memory cell (Figure 2-10).

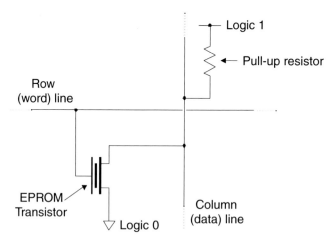

Figure 2-10. An EPROM transistor-based memory cell.

Observe that this cell no longer requires a fusible-link, antifuse, or mask-programmed connection. In its unprogrammed state, as provided by the manufacturer, all of the floating gates in the EPROM transistors are uncharged. In this case, placing a row line in its active state will turn on all of the transistors connected to that row, thereby causing all of the column lines to be pulled down to logic 0 via their respective transistors. In order to program the device, engineers can use the inputs to the device to charge the floating gates associated with selected transistors, thereby disabling those transistors. In these cases, the cells will appear to contain logic 1 values.

As they are an order of magnitude smaller than fusible links, EPROM cells are efficient in terms of silicon real estate. Their main claim to fame, however, is that they can be erased and reprogrammed. An EPROM cell is erased by discharging the electrons on that cell's floating gate. The energy required to discharge the electrons is provided by a source of *ultraviolet (UV)* radiation. An EPROM device is delivered in a ceramic or plastic package with a small quartz window in the top, where this window is usually covered with a piece of opaque sticky tape. In order for the device to be erased, it is first removed from its host circuit board, its quartz window is uncovered, and it is placed in an enclosed container with an intense UV source.

UV is pronounced by spelling it out as "U-V."

The main problems with EPROM devices are their expensive packages with quartz windows and the time it takes to erase them, which is in the order of 20 minutes. A foreseeable problem with future devices is paradoxically related to improvements in the process technologies that allow transistors to be made increasingly smaller. As the structures on the device become smaller and the density (number of transistors and interconnects) increases, a larger percentage of the surface of the die is covered by metal. This makes it difficult for the EPROM cells to absorb the UV light and increases the required exposure time.

Once again, these devices were initially intended for use as programmable memories (hence the "PROM" portion of their name). However, the same technology was later applied to more general-purpose PLDs, which therefore became known as *erasable PLDs (EPLDs)*.

EPLD is pronounced by spelling it out as "E-P-L-D."

EEPROM-based technologies

The next rung up the technology ladder appeared in the form of *electrically erasable programmable read-only memories (EEPROMs or E^{2}PROMs)*. An E^{2}PROM cell is approximately 2.5 times larger than an equivalent EPROM cell because it

EEPROM is pronounced by spelling out the "E-E" to rhyme with "bee-bee," followed by "PROM."

In the case of the alternative E²PROM designation, the "E²" stands for "E to the power of two," or "E-squared." Thus, E²PROM is pronounced "E-squared-PROM."

comprises two transistors and the space between them (Figure 2-11).

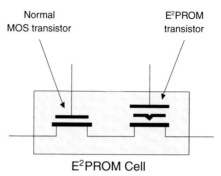

Figure 2-11. An E²PROM--cell.

The E²PROM transistor is similar to that of an EPROM transistor in that it contains a floating gate, but the insulating oxide layers surrounding this gate are very much thinner. The second transistor can be used to erase the cell electrically.

E²PROMs first saw the light of day as computer memories, but the same technology was subsequently applied to PLDs, which therefore became known as *electrically erasable PLDs* (*EEPLDs* or *E²PLDs*).

EEPLD is pronounced by spelling it out as "E-E-P-L-D."

E²PLD is pronounced "E-squared-P-L-D."

FLASH-based technologies

A development known as FLASH can trace its ancestry to both the EPROM and E²PROM technologies. The name "FLASH" was originally coined to reflect this technology's rapid erasure times compared to EPROM. Components based on FLASH can employ a variety of architectures. Some have a single floating gate transistor cell with the same area as an EPROM cell, but with the thinner oxide layers characteristic of an E²PROM component. These devices can be electrically erased, but only by clearing the whole device or large portions thereof. Other architectures feature a two-transistor cell similar to that of an E²PROM cell, thereby allowing them to be erased and reprogrammed on a word-by-word basis.

Initial versions of FLASH could only store a single bit of data per cell. By 2002, however, technologists were experimenting with a number of different ways of increasing this capacity. One technique involves storing distinct levels of charge in the FLASH transistor's floating gate to represent two bits per cell. An alternative approach involves creating two discrete storage nodes in a layer below the gate, thereby supporting two bits per cell.

SRAM-based technologies

There are two main versions of semiconductor RAM devices: *dynamic RAM (DRAM)* and *static RAM (SRAM)*. In the case of DRAMs, each cell is formed from a transistor-capacitor pair that consumes very little silicon real estate. The "dynamic" qualifier is used because the capacitor loses its charge over time, so each cell must be periodically recharged if it is to retain its data. This operation—known as *refreshing*—is a tad complex and requires a substantial amount of additional circuitry. When the "cost" of this refresh circuitry is amortized over tens of millions of bits in a DRAM memory device, this approach becomes very cost effective. However, DRAM technology is of little interest with regard to programmable logic.

By comparison, the "static" qualifier associated with SRAM is employed because—once a value has been loaded into an SRAM cell—it will remain unchanged unless it is specifically altered or until power is removed from the system. Consider the symbol for an SRAM-based programmable cell (Figure 2-12).

DRAM is pronounced by spelling out the "D" to rhyme with "knee," followed by "RAM" to rhyme with "spam."

SRAM is pronounced by spelling out the "S" to rhyme with "less," followed by "RAM" to rhyme with "Pam."

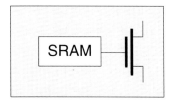

Figure 2-12. An SRAM-based programmable cell.

The entire cell comprises a multitransistor SRAM storage element whose output drives an additional control transistor. Depending on the contents of the storage element (logic 0 or logic 1), the control transistor will either be OFF (disabled) or ON (enabled).

One disadvantage of having a programmable device based on SRAM cells is that each cell consumes a significant amount of silicon real estate because these cells are formed from four or six transistors configured as a latch. Another disadvantage is that the device's configuration data (programmed state) will be lost when power is removed from the system. In turn, this means that these devices always have to be reprogrammed when the system is powered on. However, such devices have the corresponding advantage that they can be reprogrammed quickly and repeatedly as required.

The way in which these cells are used in SRAM-based FPGAs is discussed in more detail in the following chapters. For our purposes here, we need only note that such cells could conceptually be used to replace the fusible links in our example circuit shown in Figure 2-2, the antifuse links in Figure 2-4, or the transistor (and associated mask-programmed connection) associated with the ROM cell in Figure 2-7 (of course, this latter case, having an SRAM-based ROM, would be meaningless).

Summary

Table 2-1 shows the devices with which the various programming technologies are predominantly associated.

Additionally, we shouldn't forget that new technologies are constantly bobbing to the surface. Some float around for a bit, and then sink without a trace while you aren't looking; others thrust themselves onto center stage so rapidly that you aren't quite sure where they came from.

For example, one technology that is currently attracting a great deal of interest for the near-term future is *magnetic RAM (MRAM)*. The seeds of this technology were sown back in 1974, when IBM developed a component called a *magnetic*

MRAM is pronounced by spelling out the "M" to rhyme with "hem," followed by "RAM" to rhyme with "clam."

Technology	Symbol	Predominantly associated with ...
Fusible-link	—√√—	SPLDs
Antifuse	—❚—	FPGAs
EPROM	╢┠	SPLDs and CPLDs
E²PROM/ FLASH	╢┠	SPLDs and CPLDs (some FPGAs)
SRAM	[SRAM] ╢┠	FPGAs (some CPLDs)

Table 2-1. Summary of Programming Technologies

tunnel junction (MJT). This comprises a sandwich of two ferro-magnetic layers separated by a thin insulating layer. An MRAM memory cell can be created at the intersection of two tracks—say a row (word) line and a column (data) line—with an MJT sandwiched between them.

MRAM cells have the potential to combine the high speed of SRAM, the storage capacity of DRAM, and the nonvolatility of FLASH, all while consuming a miniscule amount of power. MRAM-based memory chips are predicted to become available circa 2005. Once these memory chips do reach the market, other devices—such as MRAM-based FPGAs—will probably start to appear shortly thereafter.

MJT is pronounced by spelling it out as "M-J-T."

The Origin of FPGAs

Related technologies

In order to get a good feel for the way in which FPGAs developed and the reasons why they appeared on the scene in the first place, it's advantageous to consider them in the context of other related technologies (Figure 3-1).

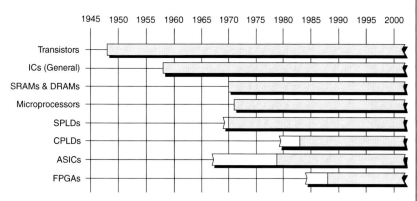

Figure 3-1. Technology timeline (dates are approximate).

The white portions of the timeline bars in this illustration indicate that although early incarnations of these technologies may have been available, for one reason or another they weren't enthusiastically received by the engineers working in the trenches during this period. For example, although Xilinx introduced the world's first FPGA as early as 1984, design engineers didn't really start using these little scamps with gusto and abandon until the early 1990s.

Transistors

On December 23, 1947, physicists William Shockley, Walter Brattain, and John Bardeen, working at Bell Laboratories in the United States, succeeded in creating the first transistor: a point-contact device formed from germanium (chemical symbol Ge).

BJT is pronounced by spelling it out as "B-J-T."

The year 1950 saw the introduction of a more sophisticated component called a *bipolar junction transistor (BJT)*, which was easier and cheaper to build and had the added advantage of being more reliable. By the late 1950s, transistors were being manufactured out of silicon (chemical symbol Si) rather than germanium. Even though germanium offered certain electrical advantages, silicon was cheaper and more amenable to work with.

TTL is pronounced by spelling it out as "T-T-L."

ECL is pronounced by spelling it out as "E-C-L."

If BJTs are connected together in a certain way, the resulting digital logic gates are classed as *transistor-transistor logic (TTL)*. An alternative method of connecting the same transistors results in *emitter-coupled logic (ECL)*. Logic gates constructed in TTL are fast and have strong drive capability, but they also consume a relatively large amount of power. Logic gates built in ECL are substantially faster than their TTL counterparts, but they consume correspondingly more power.

FET is pronounced to rhyme with "bet."

In 1962, Steven Hofstein and Fredric Heiman at the RCA research laboratory in Princeton, New Jersey, invented a new family of devices called *metal-oxide semiconductor field-effect transistors (MOSFETs)*. These are often just called FETs for short. Although the original FETs were somewhat slower than their bipolar cousins, they were cheaper, smaller, and used substantially less power.

NMOS, PMOS, and CMOS are pronounced by spelling out the "N," "P," "or "C" to rhyme with "hen," "pea," or "sea," respectively, followed by "MOS" to rhyme with "boss."

There are two main types of FETs, called NMOS and PMOS. Logic gates formed from NMOS and PMOS transistors connected together in a complementary manner are known as a *complementary metal-oxide semiconductor (CMOS)*. Logic gates implemented in CMOS used to be a tad slower than their TTL cousins, but both technologies are pretty much equivalent in this respect these days. However, CMOS logic gates have the advantage that their static (nonswitching) power consumption is extremely low.

Integrated circuits

The first transistors were provided as discrete components that were individually packaged in small metal cans. Over time, people started to think that it would be a good idea to fabricate entire circuits on a single piece of semiconductor. The first public discussion of this idea is credited to a British radar expert, G. W. A. Dummer, in a paper presented in 1952. But it was not until the summer of 1958 that Jack Kilby, working for *Texas Instruments (TI)*, succeeded in fabricating a phase-shift oscillator comprising five components on a single piece of semiconductor.

Around the same time that Kilby was working on his prototype, two of the founders of Fairchild Semiconductor—the Swiss physicist Jean Hoerni and the American physicist Robert Noyce—invented the underlying optical lithographic techniques that are now used to create transistors, insulating layers, and interconnections on modern ICs.

During the mid-1960s, TI introduced a large selection of basic building block ICs called the 54*xx* ("fifty-four hundred") series and the 74*xx* ("seventy-four hundred") series, which were specified for military and commercial use, respectively. These "jelly bean" devices, which were typically around 3/4" long, 3/8" wide, and had 14 or 16 pins, each contained small amounts of simple logic (for those readers of a pedantic disposition, some were longer, wider, and had more pins). For example, a 7400 device contained four 2-input NAND gates, a 7402 contained four 2-input NOR gates, and a 7404 contained six NOT (inverter) gates.

TI's 54*xx* and 74*xx* series were implemented in TTL. By comparison, in 1968, RCA introduced a somewhat equivalent CMOS-based library of parts called the 4000 ("four thousand") series.

IC is pronounced by spelling it out as "I-C."

SRAMs, DRAMs, and microprocessors

SRAM and DRAM are pronounced by spelling out the "S" or "D" to rhyme with "mess" or "bee," respectively, followed by "RAM" to rhyme with "spam."

The late 1960s and early 1970s were rampant with new developments in the digital IC arena. In 1970, for example, Intel announced the first 1024-bit DRAM (the 1103) and Fairchild introduced the first 256-bit SRAM (the 4100).

One year later, in 1971, Intel introduced the world's first *microprocessor (µP)*—the 4004—which was conceived and created by Marcian "Ted" Hoff, Stan Mazor, and Federico Faggin. Also referred to as a "computer-on-a-chip," the 4004 contained only around 2,300 transistors and could execute 60,000 operations per second.

Actually, although the 4004 is widely documented as being the first microprocessor, there were other contenders. In February 1968, for example, International Research Corporation developed an architecture for what they referred to as a *"computer-on-a-chip."* And in December 1970, a year before the 4004 saw the light of day, one Gilbert Hyatt filed an application for a patent entitled *"Single Chip Integrated Circuit Computer Architecture"* (wrangling about this patent continues to this day). What typically isn't disputed, however, is the fact that the 4004 was the first microprocessor to be physically constructed, to be commercially available, and to actually perform some useful tasks.

The reason SRAM and microprocessor technologies are of interest to us here is that the majority of today's FPGAs are SRAM-based, and some of today's high-end devices incorporate embedded microprocessor cores (both of these topics are discussed in more detail in chapter 4).

SPLDs and CPLDs

PLD and SPLD are pronounced by spelling them out as "P-L-D" and "S-P-L-D," respectively.

The first programmable ICs were generically referred to as *programmable logic devices (PLDs)*. The original components, which started arriving on the scene in 1970 in the form of PROMs, were rather simple, but everyone was too polite to mention it. It was only toward the end of the 1970s that significantly more complex versions became available. In order to distinguish them from their less-sophisticated ancestors, which still find use to this day, these new devices were referred to as *complex PLDs (CPLDs)*. Perhaps not surprisingly, it subsequently became common practice to refer to the original, less-pretentious versions as *simple PLDs (SPLDs)*.

Just to make life more confusing, some people understand the terms PLD and SPLD to be synonymous, while others regard PLD as being a superset that encompasses both SPLDs and CPLDs (unless otherwise noted, we shall embrace this latter interpretation).

And life just keeps on getting better and better because engineers love to use the same acronym to mean different things or different acronyms to mean the same thing (listening to a gaggle of engineers regaling each other in conversation can make even the strongest mind start to "throw a wobbly"). In the case of SPLDs, for example, there is a multiplicity of underlying architectures, many of which have acronyms formed from different combinations of the same three or four letters (Figure 3-2).

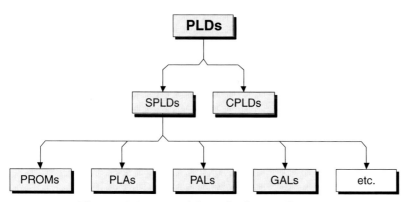

Figure 3-2. A positive plethora of PLDs.

Of course there are also EPLD, E^2PLD, and FLASH versions of many of these devices—for example, EPROMs and E^2PROMs—but these are omitted from figure 3-2 for purposes of simplicity (these concepts were introduced in chapter 2).

1500: Italy. Leonard da Vinci sketches details of a rudimentary mechanical calculator.

PROMs

PROM is pronounced like the high school dance of the same name.

The first of the simple PLDs were PROMs, which appeared on the scene in 1970. One way to visualize how these devices perform their magic is to consider them as consisting of a fixed array of AND functions driving a programmable array of OR functions. For example, consider a 3-input, 3-output PROM (Figure 3-3).

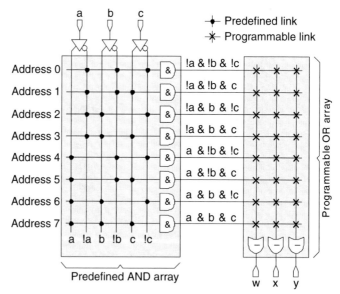

**Figure 3-3. Unprogrammed PROM
(predefined AND array, programmable OR array).**

The programmable links in the OR array can be implemented as fusible links, or as EPROM transistors and E²PROM cells in the case of EPROM and E²PROM devices, respectively. It is important to realize that this illustration is intended only to provide a high-level view of the way in which our example device works—it does not represent an actual circuit diagram. In reality, each AND function in the AND array has three inputs provided by the appropriate true or complemented versions of the a, b, and c device inputs. Similarly, each OR function in the OR array has eight inputs provided by the outputs from the AND array.

As was previously noted, PROMs were originally intended for use as computer memories in which to store program instructions and constant data values. However, design engineers also used them to implement simple logical functions such as lookup tables and state machines. In fact, a PROM can be used to implement any block of combinational (or combinational) logic so long as it doesn't have too many inputs or outputs. The simple 3-input, 3-output PROM shown in Figure 3-3, for example, can be used to implement any combinatorial function with up to 3 inputs and 3 outputs. In order to understand how this works, consider the small block of logic shown in Figure 3-4 (this circuit has no significance beyond the purposes of this example).

Some folks prefer to say "combinational logic," while others favor "combinatorial logic."

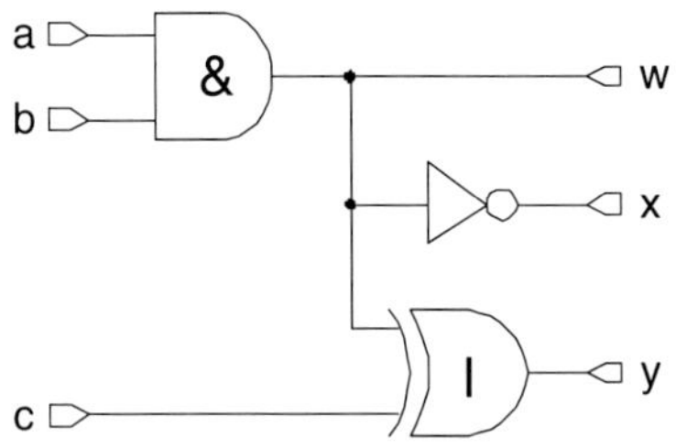

a	b	c	w	x	y
0	0	0	0	1	0
0	0	1	0	1	1
0	1	0	0	1	0
0	1	1	0	1	1
1	0	0	0	1	0
1	0	1	0	1	1
1	1	0	1	0	1
1	1	1	1	0	0

Figure 3-4. A small block of combinational logic.

We could replace this block of logic with our 3-input, 3-output PROM. We would only need to program the appropriate links in the OR array (Figure 3-5).

With regard to the equations shown in this figure, "&" represents AND, "|" represents OR, "^" represents XOR, and "!" represents NOT. This syntax (or numerous variations thereof) was very common in the early days of PLDs because it allowed logical equations to be easily and concisely represented in text files using standard computer keyboard characters.

The above example is, of course, very simple. Real PROMs can have significantly more inputs and outputs and can, therefore, be used to implement larger blocks of combinational logic. From the mid-1960s until the mid-1980s (or later),

The '&' (ampersand) character is commonly referred to as an "amp" or "amper."

The '|' (vertical line) character is commonly referred to as a "bar," "or," or "pipe."

The '^' (circumflex) character is commonly referred to as a "hat," "control," "up-arrow," or "caret." More rarely it may be referred to as a "chevron," "power of" (as in "to the power of"), or "shark-fin."

The '!' (exclamation mark) character is commonly referred to as a "bang," "ping," or "shriek".

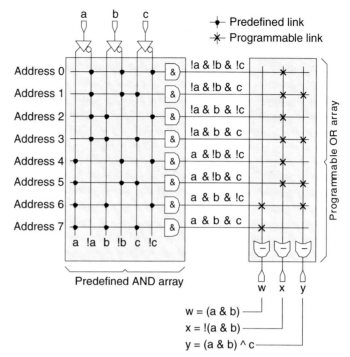

Figure 3-5. Programmed PROM.

combinational logic was commonly implemented by means of jelly bean ICs such as the TI 74xx series devices.

The fact that quite a large number of these jelly bean chips could be replaced with a single PROM resulted in circuit boards that were smaller, lighter, cheaper, and less prone to error (each solder joint on a circuit board provides a potential failure mechanism). Furthermore, if any logic errors were subsequently discovered in this portion of the design (if the design engineer had inadvertently used an AND function instead of a NAND, for example), then these slipups could easily be fixed by blowing a new PROM (or erasing and reprogramming an EPROM or E²PROM). This was preferable to the ways in which errors had to be addressed on boards based on jelly bean ICs. These included adding new devices to the board, cutting existing tracks with a scalpel, and adding wires by hand to connect the new devices into the rest of the circuit.

In logical terms, the AND ("&") operator is known as a *logical multiplication* or *product*, while the OR ("|") operator is known as a *logical addition* or *sum*. Furthermore, when we have a logical equation in the form

$$y = (a \,\&\, !b \,\&\, c) \mid (!a \,\&\, b \,\&\, c) \mid (a \,\&\, !b \,\&\, !c) \mid (a \,\&\, !b \,\&\, c)$$

then the term *literal* refers to any true or inverted variable (a, $!a$, b, $!b$, etc.), and a group of literals linked by "&" operators is referred to as a *product term*. Thus, the product term ($a \,\&\, !b \,\&\, c$) contains three literals—a, $!b$, and c—and the above equation is said to be in *sum-of-products* form.

The point is that, when they are employed to implement combinational logic as illustrated in figures 3-4 and 3-5, PROMs are useful for equations requiring a large number of product terms, but they can support relatively few inputs because every input combination is always decoded and used.

PLAs

In order to address the limitations imposed by the PROM architecture, the next step up the PLD evolutionary ladder was that of *programmable logic arrays (PLAs)*, which first became available circa 1975. These were the most user configurable of the simple PLDs because both the AND and OR arrays were programmable. First, consider a simple 3-input, 3-output PLA in its unprogrammed state (Figure 3.6).

PLA is pronounced by spelling it out as "P-L-A."

Unlike a PROM, the number of AND functions in the AND array is independent of the number of inputs to the device. Additional ANDs can be formed by simply introducing more rows into the array.

Similarly, the number of OR functions in the OR array is independent of both the number of inputs to the device and the number of AND functions in the AND array. Additional ORs can be formed by simply introducing more columns into the array.

1600:
John Napier invents a simple multiplication table called Napier's Bones.

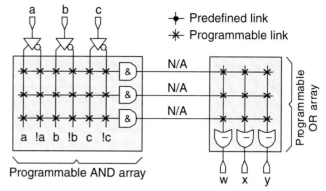

Figure 3-6. Unprogrammed PLA (programmable AND and OR arrays).

Now assume that we wish our example PLA to implement the three equations shown below. We can achieve this by programming the appropriate links as illustrated in Figure 3-7.

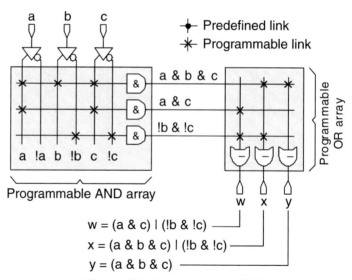

$$w = (a \,\&\, c) \mid (!b \,\&\, !c)$$
$$x = (a \,\&\, b \,\&\, c) \mid (!b \,\&\, !c)$$
$$y = (a \,\&\, b \,\&\, c)$$

Figure 3-7. Programmed PLA.

$$w = (a \,\&\, c) \mid (!b \,\&\, !c)$$
$$x = (a \,\&\, b \,\&\, c) \mathbin{!} (!b \,\&\, !c)$$
$$y = (a \,\&\, b \,\&\, c)$$

As fate would have it, PLAs never achieved any significant level of market presence, but several vendors experimented with different flavors of these devices for a while. For example, PLAs were not obliged to have AND arrays feeding OR arrays, and some alternative architectures such as AND arrays feeding NOR arrays were occasionally seen strutting their stuff. However, while it would be theoretically possible to field architectures such as OR-AND, NAND-OR, and NAND-NOR, these variations were relatively rare or nonexistent. One reason these devices tended to stick to AND-OR[1] (and AND-NOR) architectures was that the sum-of-products representations most often used to specify logical equations could be directly mapped onto these structures. Other equation formats—like product-of-sums—could be accommodated using standard algebraic techniques (this was typically performed by means of software programs that could perform these techniques with their metaphorical hands tied behind their backs).

PLAs were touted as being particularly useful for large designs whose logical equations featured a lot of common product terms that could be used by multiple outputs; for example, the product term ($!b$ & $!c$) is used by both the w and x outputs in Figure 3-7. This feature may be referred to as *product-term sharing.*

On the downside, signals take a relatively long time to pass through programmable links as opposed to their predefined counterparts. Thus, the fact that both their AND and OR arrays were programmable meant that PLAs were significantly slower than PROMs.

1614:
John Napier invents logarithms.

---

[1] Actually, one designer I talked to a few moments before penning these words told me that his team created a NOT-NOR-NOR-NOT architecture (this apparently offered a slight speed advantage), but they told their customers it was an AND-OR architecture (which is how it appeared to the outside world) because "that was what they were expecting." Even today, what device vendors say they build and what they actually build are not necessarily the same thing.

PAL, which is a registered trademark of Monolithic Memories, Inc., is pronounced the same way you'd greet a buddy ("Hiya pal").

Created by Lattice Semiconductor Corporation in 1983, *generic array logic (GAL)* devices offered sophisticated CMOS electrically erasable (E²) variations on the PAL concept.

GAL is pronounced the same way a guy thinks of his wife or girlfriend ("What a gal!").

PALs and GALs

In order to address the speed problems posed by PLAs, a new class of device called *programmable array logic (PAL)* was introduced in the late 1970s. Conceptually, a PAL is almost the exact opposite of a PROM because it has a programmable AND array and a predefined OR array. As an example, consider a simple 3-input, 3-output PAL in its unprogrammed state (Ffigure 3-8).

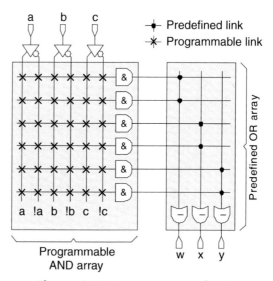

**Figure 3-8. Unprogrammed PAL
(programmable AND array, predefined OR array).**

The advantage of PALs (as compared to PLAs) is that they are faster because only one of their arrays is programmable. On the downside, PALs are more limited because they only allow a restricted number of product terms to be ORed together (but engineers are cunning people, and we have lots of tricks up our sleeves that, to a large extent, allow us to get around this sort of thing).

Additional programmable options

The PLA and PAL examples shown above were small and rudimentary for the purposes of simplicity. In addition to being a lot larger (having more inputs, outputs, and internal signals), real devices can offer a variety of additional programmable options, such as the ability to invert the outputs or to have tristatable outputs, or both.

Furthermore, some devices support registered or latched outputs (with associated programmable multiplexers that allow the user to specify whether to use the registered or nonregistered version of the output on a pin-by-pin basis). And some devices provide the ability to configure certain pins to act as either outputs or additional inputs, and the list of options goes on.

The problem here is that different devices may provide different subsets of the various options, which makes selecting the optimum device for a particular application something of a challenge. Engineers typically work around this by (a) restricting themselves to a limited selection of devices and then tailoring their designs to these devices, or (b) using a software program to help them decide which devices best fit their requirements on an application-by-application basis.

CPLDs

The one truism in electronics is that everyone is always looking for things to get bigger (in terms of functional capability), smaller (in terms of physical size), faster, more powerful, and cheaper—surely that's not too much to ask, is it? Thus, the tail end of the 1970s and the early 1980s began to see the emergence of more sophisticated PLD devices that became known as *complex PLDs (CPLDs).*

CPLD is pronounced by spelling it out as "C-P-L-D."

Leading the fray were the inventors of the original PAL devices—the guys and gals at *Monolithic Memories Inc. (MMI)*—who introduced a component they called a Mega-PAL. This was an 84-pin device that essentially comprised four standard PALs with some interconnect linking them together. Unfortunately, the MegaPAL consumed a disproportionate amount of power, and it was generally perceived to offer little advantage compared to using four individual devices.

The big leap forward occurred in 1984, when newly formed Altera introduced a CPLD based on a combination of CMOS

1621:
William Oughtred
invents the slide rule
(based on John Napier's
Logarithms).

and EPROM technologies. Using CMOS allowed Altera to achieve tremendous functional density and complexity while consuming relatively little power. And basing the programmability of these devices on EPROM cells made them ideal for use in development and prototyping environments.

Having said this, Altera's claim to fame wasn't due only to the combination of CMOS and EPROM. When engineers started to grow SPLD architectures into larger devices like the MegaPAL, it was originally assumed that the central interconnect array (also known as the *programmable interconnect matrix*) linking the individual SPLD blocks required 100 percent connectivity to the inputs and outputs associated with each block. The problem was that a twofold increase in the size of the SPLD blocks (equating to twice the inputs and twice the outputs) resulted in a fourfold increase in the size of the interconnect array. In turn, this resulted in a huge decrease in speed coupled with higher power dissipation and increased component costs.

Altera made the conceptual leap to using a central interconnect array with less than 100 percent connectivity (see the discussions associated with figure 3-10 for a tad more information on this concept). This increased the complexity of the software design tools, but it kept the speed, power, and cost of these devices scalable.

Although every CPLD manufacturer fields its own unique architecture, a generic device consists of a number of SPLD blocks (typically PALs) sharing a common programmable interconnection matrix (Figure 3-9).

In addition to programming the individual SPLD blocks, the connections between the blocks can be programmed by means of the programmable interconnect matrix.

Of course, figure 3-9 is a high-level representation. In reality, all of these structures are formed on the same piece of silicon, and various additional features are not shown here. For example, the programmable interconnect matrix may contain a lot of wires (say 100), but this is more than can be

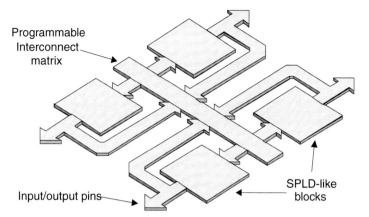

Programmable
Interconnect
matrix

Input/output pins

SPLD-like
blocks

Figure 3-9. A generic CPLD structure.

1623:
Wilhelm Schickard invents the first mechanical calculator.

handled by the individual SPLD blocks, which might only be able to accommodate a limited number of signals (say 30). Thus, the SPLD blocks are interfaced to the interconnect matrix using some form of programmable multiplexer (Figure 3-10).

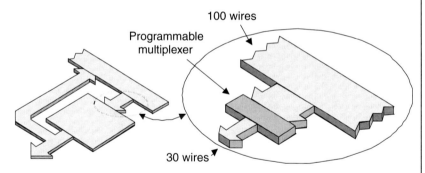

100 wires

Programmable
multiplexer

30 wires

Figure 3-10. Using programmable multiplexers.

Depending on the manufacturer and the device family, the CPLD's programmable switches may be based on EPROM, E²PROM, FLASH, or SRAM cells. In the case of SRAM-based devices, some variants increase their versatility by allowing the SRAM cells associated with each SPLD block to be used either as programmable switches or as an actual chunk of memory.

The Dark Ages refers to the period of history between classical antiquity and the Italian Renaissance. (Depending on the source, the starting point for the Dark Ages can vary by several hundred years.)

ABEL, CUPL, PALASM, JEDEC, etc.

In many respects, the early days of PLDs were the design engineers' equivalent of the Dark Ages. The specification for a new device typically commenced life in the form of a schematic (or state machine) diagram. These diagrams were created using pencil and paper because computer-aided electroni design capture tools, in the form we know them today, really didn't exist at that time.

Once a design had been captured in diagrammatic form, it was converted by hand into a tabular equivalent and subsequently typed into a text file. Among other things, this text file defined which fuses were to be blown or which antifuses were to be grown. In those days of yore, the text file was typed directly into a special box called a *device programmer*, which was subsequently used to program the chip. As time progressed, however, it became common to create the file on a host computer, which downloaded it into—and controlled—the device programmer as required (Figure 3-11).

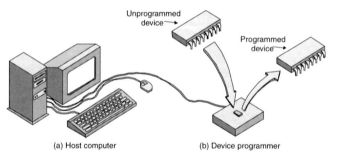

Figure 3-11. Programming a physical PLD.

Creating this programming file required the engineer to have an intimate knowledge of the device's internal links and the file format used by the device programmer. Just to increase the fun, every PLD vendor developed its own file format that typically worked only with its own devices. It was obvious to everyone concerned that this design flow was time-consuming and prone to error, and it certainly didn't facilitate locating and fixing any mistakes.

In 1980, a committee of the *Joint Electron Device Engineering Council (JEDEC)*—part of the Electronics Industry Association—proposed a standard format for PLD programming text files. It wasn't long before all of the device programmers were modified to accept this format.

Around the same time, John Birkner, the man who conceived the first PALs and managed their development, created *PAL Assembler (PALASM)*. PALASM referred to both a rudimentary *hardware description language (HDL)* and a software application. In its role as an HDL, PALASM allowed design engineers to specify the function of a circuit in the form of a text source file containing Boolean equations in sum-of-products form. In its role as a software application (what we would now refer to as an EDA tool), PALASM—which was written in only six pages of FORTRAN code—read in the text source file and automatically generated a text-based programming file for use with the device programmer.

In the context of its time, PALASM was a huge leap forward, but the original version only supported PAL devices made by MMI, and it didn't perform any minimization or optimization. In order to address these issues, Data I/O released its *Advanced Boolean Expression Language (ABEL)* in 1983. Around the same time, Assisted Technology released its *Common Universal tool for Programmable Logic (CUPL)*. ABEL and CUPL were both HDLs and software applications. In addition to supporting state machine constructs and automatic logic minimization algorithms, they both worked with multiple PLD types and manufacturers.

Although PALASM, ABEL, and CUPL are the best known of the early HDLs, there were many others, such as *Automated Map and Zap of Equations (AMAZE)* from Signetics. These simple languages and associated tools paved the way for the higher-level HDLs (such as Verilog and VHDL) and tools (such as logic synthesis) that are used for today's ASIC and FPGA designs.

JEDEC is pronounced "jed-eck"; that is, "jed" to rhyme with "bed" and "eck" to rhyme with "deck."

PALASM is pronounced "pal-as-em."

HDL is pronounced by spelling it out as "H-D-L."

Developed at IBM in the mid 1950s, FORTRAN, which stands for FORmula TRANslation language, was the first computer programming language higher than the assembly level.

ABEL is pronounced to rhyme with "fable."

CUPL is pronounced "koo-pel"; that is, "koo" to rhyme with "loo" and "pel" to rhyme with "bell."

ASIC is pronounced
by spelling out the "A" to
rhyme with "hay," fol-
lowed by "SIC" to rhyme
with "tick."

ASICs (gate arrays, etc.)

At the time of this writing, four main classes of *application-specific integrated circuit (ASIC)* deserve mention. In increasing order of complexity, these are *gate arrays*, *structured ASICs*, *standard cell devices*, and *full-custom chips* (Figure 3-12).

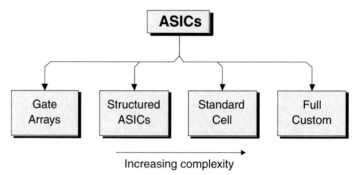

Figure 3-12. Different types of ASIC.

Although it would be possible to introduce these ASIC types in the order of increasing complexity reflected in this figure, it actually makes more sense to describe them in the sequence in which they appeared on the scene, which was full-custom chips, followed by gate arrays, then standard cell devices, and finally structured ASICs. (Note that it's arguable whether structured ASICs are more or less complex than traditional gate arrays.)

Full custom

In the early days of digital ICs, there were really only two classes of devices (excluding memory chips). The first were relatively simple building block–type components that were created by companies like TI and Fairchild and sold as standard off-the-shelf parts to anyone who wanted to use them. The second were full-custom ASICs like microprocessors, which were designed and built to order for use by a specific company.

In the case of full-custom devices, design engineers have complete control over every mask layer used to fabricate the silicon chip. The ASIC vendor does not prefabricate any components on the silicon and does not provide any libraries of predefined logic gates and functions.

By means of appropriate tools, the engineers can handcraft the dimensions of individual transistors and then create higher-level functions based on these elements. For example, if the engineers require a slightly faster logic gate, they can alter the dimensions of the transistors used to build that gate. The design tools used for full-custom devices are often created in-house by teh engineers themselves.

The design of full-custom devices is highly complex and time-consuming, but the resulting chips contain the maximum amount of logic with minimal waste of silicon real estate.

The Micromatrix and Micromosaic

Some time in the mid-1960s, Fairchild Semiconductor introduced a device called the *Micromatrix*, which comprised a limited number (around 100) of noninterconnected bare-bones transistors. In order to make this device perform a useful function, design engineers hand-drew the metallization layers used to connect the transistors on two plastic sheets.

The first sheet—drawn using a green pen—represented the Y-axis (north-south) tracks to be implemented on metal layer 1, while the second sheet—drawn using a red pen—represented the X-axis (east-west) tracks to be implemented on metal layer two. (Additional sheets were used to draw the vias (conducting columns) linking metal layer 1 to the transistors and the vias linking metal layers 1 and 2 together.)

Capturing a design in this way was painfully time-consuming and prone to error, but at least the hard, expensive, and really time-consuming work—creating the transistors—had already been performed. This meant that the Micromatrix allowed design engineers to create a custom device for a reasonable (though still expensive) cost in a reasonable (though still long) time frame.

1642:
Blaise Pascal invents a mechanical calculator called the Arithmetic Machine.

A few years later, in 1967, Fairchild introduced a device called the *Micromosaic*, which contained a few hundred noninterconnected transistors. These transistors could subsequently be connected together to implement around 150 AND, OR, and NOT gates. The key feature of the Micromosaic was that design engineers could specify the function the device was required to perform by means of a text file containing Boolean (logic) equations, and a computer program then determined the necessary transistor interconnections and constructed the photo-masks required to complete the device. This was revolutionary at the time, and the Micromosaic is now credited as being the forerunner of the modern *gate array* form of ASIC and also the first real application of *computer-aided design CAD*.

CAD is pronounced to rhyme with "bad."

Early gate arrays were sometimes known as *uncommitted logic arrays (ULAs)*, but this term has largely fallen into disuse.

Gate arrays

The gate array concept originated in companies like IBM and Fujitsu as far back as the late 1960s. However, these early devices were only available for internal consumption, and it wasn't until the mid-1970s that access to CMOS-based gate array technology became available to anyone willing to pay for it.

Gate arrays are based on the idea of a *basic cell* consisting of a collection of unconnected transistors and resistors. Each ASIC vendor determines what it considers to be the optimum mix of components provided in its particular basic cell (Figure 3-13).

The ASIC vendor commences by prefabricating silicon chips containing arrays of these basic cells. In the case of *channeled* gate arrays, the basic cells are typically presented as either single-column or dual-column arrays; the free areas between the arrays are known as the *channels* (Figure 3-14).

By comparison, in the case of *channel-less* or *channel-free* devices, the basic cells are presented as a single large array. The surface of the device is covered in a "sea" of basic cells, and there are no dedicated channels for the interconnections.

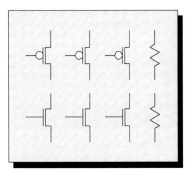 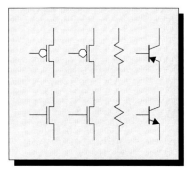

(a) Pure CMOS basic cell   (b) BiCMOS basic cell

Figure 3-13. Examples of simple gate array basic cells.

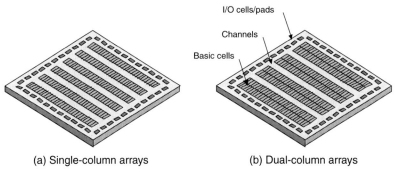

(a) Single-column arrays   (b) Dual-column arrays

Figure 3-14. Channeled gate array architectures.

1671:
Baron Gottfried von Leibniz invents a mechanical calculator called the Step Reckoner.

Thus, these devices are popularly referred to as *sea-of-gates* or *sea-of-cells*.

The ASIC vendor defines a set of logic functions such as primitive gates, multiplexers, and registers that can be used by the design engineers. Each of these building block functions is referred to as a *cell*—not to be confused with a basic cell—and the set of functions supported by the ASIC vendor is known as the *cell library*.

The means by which ASICs are actually designed is beyond the scope of this book. Suffice it to say that the design engineers eventually end up with a gate-level netlist, which describes the logic gates they wish to use and the connections between them. Special mapping, placement, and routing software tools are used to assign the logic gates to specific basic cells and define how the cells will be connected together. The results are used to generate the photo-masks that are in turn used to create the metallization layers that will link the components inside the basic cells and also link the basic cells to each other and to the device's inputs and outputs.

Gate arrays offer considerable cost advantages in that the transistors and other components are prefabricated, so only the metallization layers need to be customized. The disadvantage is that most designs leave significant amounts of internal resources unutilized, the placement of gates is constrained, and the routing of internal tracks is less than optimal. All of these factors negatively impact the performance and power consumption of the design.

Standard cell devices

In order to address the problems associated with gate arrays, standard cell devices became available in the early 1980s. These components bear many similarities to gate arrays. Once again, the ASIC vendor defines the cell library that can be used by the design engineers. The vendor also supplies hard-macro and soft-macro libraries, which include elements such as processors, communication functions, and a selection of RAM and ROM functions. Last but not least, the design engineers may decide to reuse previously designed functions or to purchase blocks of *intellectual property (IP)*.

Once again, by one means or another (which today involves incredibly sophisticated software tools), the design engineers end up with a gate-level netlist, which describes the logic gates they wish to use and the connections between them.

Unlike gate arrays, standard cell devices do not use the concept of a basic cell, and no components are prefabricated on the chip. Special tools are used to place each logic gate individually in the netlist and to determine the optimum way in which the gates are to be routed (connected together). The results are then used to create custom photo-masks for every layer in the device's fabrication.

When a team of electronics engineers is tasked with designing a complex integrated circuit, rather than reinventing the wheel, they may decide to purchase the plans for one or more functional blocks that have already been created by someone else. The plans for these functional blocks are known as intellectual property, or IP.

IP is pronounced by spelling it out as "I-P."

IP blocks can range all the way up to sophisticated communications functions and microprocessors. The more complex functions, like microprocessors, may be referred to as "cores."

The standard cell concept allows each logic function to be created using the minimum number of transistors with no redundant components, and the functions can be positioned so as to facilitate any connections between them. Standard cell devices, therefore, provide a closer-to-optimal utilization of the silicon than do gate arrays.

Structured ASICs

It's often said that there's nothing new under the sun. Ever since the introduction of standard cell devices, industry observers have been predicting the demise of gate arrays, but these little rascals continue to hold on to their market niche and, indeed, have seen something of a resurgence in recent years.

Structured ASICs (although they weren't called that at the time) spluttered into life around the beginning of the 1990s, slouched around for a while, and then returned to the nether regions from whence they came. A decade later—circa 2001 to 2002—a number of ASIC manufacturers started to investigate innovative ways of reducing ASIC design costs and development times. Not wishing to be associated with traditional gate arrays, everyone was happy when someone came up with the *structured ASIC* moniker somewhere around the middle of 2003.

As usual, of course, every vendor has its own proprietary architecture, so our discussions here will provide only a generic view of these components. Each device commences with a fundamental element called a *module* by some and a *tile* by others. This element may contain a mixture of prefabricated generic logic (implemented either as gates, multiplexers, or a lookup table), one or more registers, and possibly a little local RAM (Figure 3-15).

An array (sea) of these elements is then prefabricated across the face of the chip. Alternatively, some architectures commence with a *base cell* (or *base tile* or *base module*, or …) containing only generic logic in the form of prefabricated

1746: Holland. The Leyden jar is invented at University of Leyden.

1752: America.
Benjamin Franklin
performs his notorious
kite experiment.

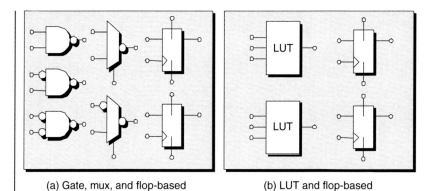

(a) Gate, mux, and flop-based   (b) LUT and flop-based

Figure 3-15. Examples of structured ASIC tiles.

gates, multiplexers, or lookup tables. An array of these base
units (say 4×4, 8×8, or 16×16)—in conjunction with some
special units containing registers, small memory elements, and
other logic—then make up a *master cell* (or *master tile* or *master
module* or …). Once again, an array (sea) of these master units
is then prefabricated across the face of the chip.

Also prefabricated (typically around the edge of the
device) are functions like RAM blocks, clock generators,
boundary scan logic, and so forth (Figure 3-16).

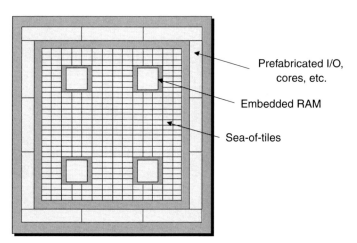

Prefabricated I/O,
cores, etc.

Embedded RAM

Sea-of-tiles

Figure 3-16. Generic structured ASIC.

The idea is that the device can be customized using only the metallization layers (just like a standard gate array). The difference is that, due to the greater sophistication of the structured ASIC tile, most of the metallization layers are also predefined.

Thus, many structured ASIC architectures require the customization of only two or three metallization layers (in one case, it is necessary to customize only a single via layer). This dramatically reduces the time and costs associated with creating the remaining photo-masks used to complete the device.

Although it's difficult to assign an exact value, the predefined and prefabricated logic associated with structured ASICs results in an overhead compared to standard cell devices in terms of power consumption, performance, and silicon real estate. Early indications are that structured ASICs require three times the real estate and consume two to three times the power of a standard cell device to perform the same function. In reality, these results will vary architecture-by-architecture, and also different types of designs may well favor different architectures. Unfortunately, no evaluations based on industry-standard reference designs have been performed across all of the emerging structured ASIC architectures at the time of this writing.

FPGAs

Around the beginning of the 1980s, it became apparent that there was a gap in the digital IC continuum. At one end, there were programmable devices like SPLDs and CPLDs, which were highly configurable and had fast design and modification times, but which couldn't support large or complex functions.

At the other end of the spectrum were ASICs. These could support extremely large and complex functions, but they were painfully expensive and time-consuming to design. Furthermore, once a design had been implemented as an ASIC it was effectively frozen in silicon (Figure 3-17).

1775: Italy. Count Alessandro Giuseppe Antonio Anastasio Volta invents a static electricity generator called the Electrophorus.

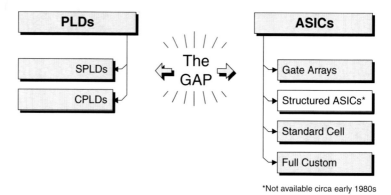

Figure 3-17. The gap between PLDs and ASICs.

FPGA is pronounced by spelling it out as "F-P-G-A."

In order to address this gap, Xilinx developed a new class of IC called a *field-programmable gate array*, or FPGA, which they made available to the market in 1984.

The various FPGAs available today are discussed in detail in chapter 4. For the nonce, we need only note that the first FPGAs were based on CMOS and used SRAM cells for configuration purposes. Although these early devices were comparatively simple and contained relatively few gates (or the equivalent thereof) by today's standards, many aspects of their underlying architecture are still employed to this day.

The early devices were based on the concept of a programmable logic block, which comprised a 3-input *lookup table (LUT)*, a register that could act as a flip-flop or a latch, and a multiplexer, along with a few other elements that are of little interest here. Figure 3-18 shows a very simple programmable logic block (the logic blocks in modern FPGAs can be significantly more complex—see chapter 4 for more details).

Each FPGA contained a large number of these programmable logic blocks, as discussed below. By means of appropriate SRAM programming cells, every logic block in the device could be configured to perform a different function. Each register could be configured to initialize containing a logic 0 or a logic 1 and to act as a flip-flop (as shown in Figure 3-18) or a latch. If the flip-flop option were selected, the register could be configured to be triggered by a positive- or

LUT is pronounced to rhyme with "nut."

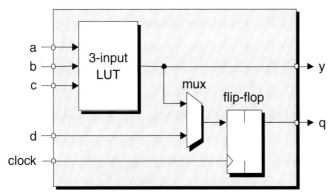

Figure 3-18. The key elements forming a simple programmable logic block.

1777:
Charles Stanhope invents a mechanical calculating machine.

negative-going clock (the clock signal was common to all of the logic blocks). The multiplexer feeding the flip-flop could be configured to accept the output from the LUT or a separate input to the logic block, and the LUT could be configured to represent any 3-input logical function.

For example, assume that a LUT was required to perform the function

$$y = (a \ \& \ b) \ | \ !c$$

This could be achieved by loading the LUT with the appropriate output values (figure 3-19).

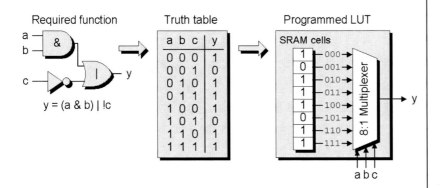

Figure 3-19. Configuring a LUT.

Note that the 8:1-multiplexer-based LUT illustrated in Figure 3-19 is used for purposes of simplicity; a more realistic implementation is shown in chapter 4. Furthermore, Chapter 5 presents in detail the ways in which FPGAs are actually programmed.

The complete FPGA comprised a large number of programmable logic block "islands" surrounded by a "sea" of programmable interconnects (Figure 3-20).

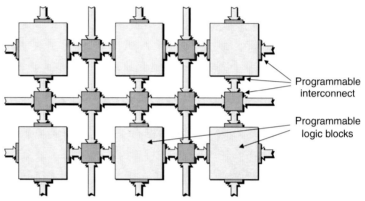

Figure 3-20. Top-down view of simple, generic FPGA architecture.

As usual, this high-level illustration is merely an abstract representation. In reality, all of the transistors and interconnects would be implemented on the same piece of silicon using standard IC creation techniques.

In addition to the local interconnect reflected in figure 3-20, there would also be global (high-speed) interconnection paths that could transport signals across the chip without having to go through multiple local switching elements.

The device would also include primary I/O pins and pads (not shown here). By means of its own SRAM cells, the interconnect could be programmed such that the primary inputs to the device were connected to the inputs of one or more programmable logic blocks, and the outputs from any logic block could be used to drive the inputs to any other logic block, the primary outputs from the device, or both.

The end result was that FPGAs successfully bridged the gap between PLDs and ASICs. On the one hand, they were highly configurable and had the fast design and modification times associated with PLDs. On the other hand, they could be used to implement large and complex functions that had previously been the domain only of ASICs. (ASICs were still required for the really large, complex, high-performance designs, but as FPGAs increased in sophistication, they started to encroach further and further into ASIC design space.)

Platform FPGAs

The concept of a *reference design* or *platform design* has long been used at the circuit board level. This refers to creating a base design configuration from which multiple products can be derived.

In addition to tremendous amounts of programmable logic, today's high-end FPGAs feature embedded (block) RAMs, embedded processor cores, high-speed I/O blocks, and so forth. Furthermore, designers have access to a wide range of IP. The end result is the concept of the *platform FPGA*. A company may use a platform FPGA design as a basis for multiple products inside that company, or it may supply an initial design to multiple other companies for them to customize and differentiate.

FPGA-ASIC hybrids

It would not make any sense to embed ASIC material inside an FPGA because designs created using such a device would face all of the classic problems (high NREs, long lead times, etc.) associated with ASIC design flows. However, there are a number of cases in which one or more FPGA cores have been used as part of a standard cell ASIC design.

One reason for embedding FPGA material inside an ASIC is that it facilitates the concept of platform design. The platform in this case would be the ASIC, and the embedded FPGA material could form one of the mechanisms used to customize and differentiate subdesigns.

Late 1700s:
Charles Stanhope invents a logic machine called the Stanhope Demonstrator.

1800: Italy.
Count Alessandro
Giuseppe Antonio
Anastasio Volta invents
the first battery.

Another reason is that the last few years have seen an increasing incidence of FPGAs being used to augment ASIC designs. In this scenario, a large, complex ASIC has an associated FPGA located in close proximity on the board (Figure 3-21).

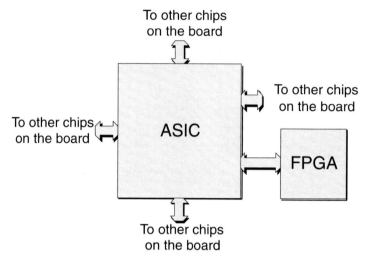

Figure 3-21. Using an FPGA to augment an ASIC design.

The reason for this scenario is that it's incredibly time-consuming and expensive to fix a bug in the ASIC or to modify its functionality to accommodate any changes to its original design specification. If the ASIC is designed in the right way, however, its associated FPGA can be used to implement any downstream modifications and enhancements. One problem with this approach is the time taken for signals to pass back and forth between the ASIC and the FPGA. The solution is to embed the FPGA core inside the ASIC, thereby resulting in an FPGA-ASIC hybrid.

One concern that has faced these hybrids, however, is that ASIC and FPGA design tools and flows have significant underlying differences. For example, ASICs are said to be *fine-grained* because (ultimately) they are implemented at the level of primitive logic gates. This means that traditional design technologies like logic synthesis and implementation technologies like place-and-route are also geared toward fine-grained architectures.

By comparison, FPGAs are said to be *medium-grained* (or *coarse-grained* depending on whom you are talking to) because they are physically realized using higher-level blocks like the programmable logic blocks introduced earlier in this chapter. In this case, the best design results are realized when using FPGA-specific synthesis and place-and-route technologies that view their world in terms of these higher-level blocks.

One area of interest for FPGA-ASIC hybrids is that of structured ASICs because they too may be considered to be block based. This means that, when looking around for design tools, structured ASIC vendors are talking to purveyors of FPGA-centric synthesis and place-and-route technology rather than their traditional ASIC counterparts. In turn, this means that FPGA-ASIC hybrids based on structured ASICs would automatically tend toward a unified tool and design flow because the same block-based synthesis and place-and-route engines could be used for both the ASIC and FPGA portions of the design.

How FPGA vendors design their chips

Last but not least, one question that is commonly asked—but is rarely (if ever) addressed in books on FPGAs—is, how do FPGA vendors actually go about designing a new generation of devices?

To put this another way, do they handcraft each transistor and track using a design flow similar to that of a full-custom ASIC, or do they create an RTL description, synthesize it into a gate-level netlist, and then use place-and-route software along the lines of a classic ASIC (gate array or standard cell) design flow (the concepts behind these tools are discussed in more detail in Section 2).

The short answer is yes! The slightly longer answer is that there are some portions of the device, like the programmable logic blocks and the basic routing structure, where the FPGA vendors fight tooth and nail for every square micron and every

1801: France. Joseph-Marie Jacquard invents a loom controlled by punch cards.

1820: France.
Andre Ampere
investigates the force
of an electric current
in a magnetic field.

fraction of a nanosecond. These sections of the design are handcrafted at the transistor and track level using full-custom ASIC techniques. On the bright side, these portions of the design are both relatively small and highly repetitive, so once created they are replicated thousands of times across the face of the chip.

Then there are housekeeping portions of the device, such as the configuration control circuitry, that only occur once per device and are not particularly size or performance critical. These sections of the design are created using standard cell ASIC-style techniques.

Alternative FPGA Architectures

A word of warning

In this chapter we introduce a plethora of architectural features. Certain options—such as using antifuse versus SRAM configuration cells—are mutually exclusive. Some FPGA vendors specialize in one or the other; others may offer multiple device families based on these different technologies. (Unless otherwise noted, the majority of these discussions assume SRAM-based devices.)

In the case of embedded blocks such as multipliers, adders, memory, and microprocessor cores, different vendors offer alternative "flavors" of these blocks with different "recipes" of ingredients. (Much like different brands of chocolate chip cookies featuring larger or smaller chocolate chips, for example, some FPGA families will have bigger/better/badder embedded RAM blocks, while others might feature more multipliers, or support more I/O standards, or ...)

The problem is that the features supported by each vendor and each family change on an almost daily basis. This means that once you've decided what features you need, you then need to do a little research to see which vendor's offerings currently come closest to satisfying your requirements.

A little background information

Before hurling ourselves into the body of this chapter, we need to define a couple of concepts to ensure that we're all marching to the same drumbeat. For example, you're going to see the term *fabric* used throughout this book. In the context of

The word "fabric" comes from the Middle English *fabryke*, meaning "something constructed."

a silicon chip, this refers to the underlying structure of the device (sort of like the phrase "the fabric of civilized society").

When you first hear someone using "fabric" in this way, it might sound a little snooty or pretentious (in fact, some engineers regard it as nothing more than yet another marketing term promoted by ASIC and FPGA vendors to make their devices sound more sophisticated than they really are). Truth to tell, however, once you get used to it, this is really quite a useful word.

When we talk about the *geometry* of an IC, we are referring to the size of the individual structures constructed on the chip—such as the portion of a *field-effect transistor* (FET) known as its *channel*. These structures are incredibly small. In the early to mid-1980s, devices were based on 3 μm geometries, which means that their smallest structures were 3 millionths of a meter in size. (In conversation, we would say, "This IC is based on a three-micron technology.")

The "μ" symbol stands for "micro" from the Greek *micros*, meaning "small" (hence the use of "μP" as an abbreviation for microprocessor.")

In the metric system, "μ" stands for "one millionth part of," so 1 μm represents "one millionth of a meter."

Each new geometry is referred to as a *technology node*. By the 1990s, devices based on 1 μm geometries were starting to appear, and feature sizes continued to plummet throughout the course of the decade. As we moved into the twenty-first century, high-performance ICs had geometries as small as 0.18 μm. By 2002, this had shrunk to 0.13 μm, and by 2003, devices at 0.09 μm were starting to appear.

Any geometry smaller than around 0.5 μm is referred to as *deep submicron* (DSM). At some point that is not well defined (or that has multiple definitions depending on whom one is talking to), we move into the *ultradeep submicron* (UDSM) realm.

DSM is pronounced by spelling it out as "D-S-M."

UDSM is pronounced by spelling it out as "U-D-S-M."

Things started to become a little awkward once geometries dropped below 1 μm, not the least because it's a pain to keep having to say things like "zero point one three microns." For this reason, when conversing it's becoming common to talk in terms of *nano*, where one nano (short for nanometer) equates to a thousandth of a micron—that is, one thousandth of a millionth of a meter. Thus, instead of mumbling, "point zero nine microns" (0.09 μm), one can simply proclaim, "ninety nano" (90 nano) and have done with it. Of course, these both mean exactly the same thing, but if you feel moved to regale your friends on these topics, it's best to use the vernacular of the day and present yourself as hip and trendy rather than as an old fuddy-duddy from the last millennium.

Antifuse versus SRAM versus …

SRAM-based devices

The majority of FPGAs are based on the use of SRAM configuration cells, which means that they can be configured over and over again. The main advantages of this technique are that new design ideas can be quickly implemented and tested, while evolving standards and protocols can be accommodated relatively easily. Furthermore, when the system is first powered up, the FPGA can initially be programmed to perform one function such as a self-test or board/system test, and it can then be reprogrammed to perform its main task.

Another big advantage of the SRAM-based approach is that these devices are at the forefront of technology. FPGA vendors can leverage the fact that many other companies specializing in memory devices expend tremendous resources on *research and development (R&D)* in this area. Furthermore, the SRAM cells are created using exactly the same CMOS technologies as the rest of the device, so no special processing steps are required in order to create these components.

R&D is pronounced by spelling it out as "R-and-D."

In the past, memory devices were often used to qualify the manufacturing processes associated with a new technology node. More recently, the mixture of size, complexity, and regularity associated with the latest FPGA generations has resulted in these devices being used for this task. One advantage of using FPGAs over memory devices to qualify the manufacturing process is that, if there's a defect, the structure of FPGAs is such that it's easier to identify and locate the problem (that is, figure out what and where it is). For example, when IBM and UMC were rolling out their 0.09 μm (90 nano) processes,

FPGAs from Xilinx were the first devices to race out of the starting gate.

Unfortunately, there's no such thing as a free lunch. One downside of SRAM-based devices is that they have to be reconfigured every time the system is powered up. This either requires the use of a special external memory device (which has an associated cost and consumes real estate on the board) or of an on-board microprocessor (or some variation of these techniques—see also chapter 5).

Security issues and solutions with SRAM-based devices

Another consideration with regard to SRAM-based devices is that it can be difficult to protect your intellectual property, or IP, in the form of your design. This is because the configuration file used to program the device is stored in some form of external memory.

Currently, there are no commercially available tools that will read the contents of a configuration file and generate a corresponding schematic or netlist representation. Having said this, understanding and extracting the logic from the configuration file, while not a trivial task, would not be beyond the bounds of possibility given the combination of clever folks and computing horsepower available today.

Let's not forget that there are reverse-engineering companies all over the world specializing in the recovery of "design IP." And there are also a number of countries whose governments turn a blind eye to IP theft so long as the money keeps rolling in (you know who you are). So if a design is a high-profit item, you can bet that there are folks out there who are ready and eager to replicate it while you're not looking.

In reality, the real issue here is not related to someone stealing your IP by reverse-engineering the contents of the configuration file, but rather their ability to clone your design, irrespective of whether they understand how it performs its magic. Using readily available technology, it is relatively easy

IP is pronounced by spelling it out as "I-P."

for someone to take a circuit board, put it on a "bed of nails" tester, and quickly extract a complete netlist for the board. This netlist can subsequently be used to reproduce the board. Now the only task remaining for the nefarious scoundrels is to copy your FPGA configuration file from its boot PROM (or EPROM, E²PROM, or whatever), and they have a duplicate of the entire design.

On the bright side, some of today's SRAM-based FPGAs support the concept of *bitstream encryption*. In this case, the final configuration data is encrypted before being stored in the external memory device. The encryption key itself is loaded into a special SRAM-based register in the FPGA via its JTAG port (see also Chapter 5). In conjunction with some associated logic, this key allows the incoming encrypted configuration bitstream to be decrypted as it's being loaded into the device.

JTAG is pronounced "J-TAG"; that is, by spelling out the 'J' followed by "tag" to rhyme with "bag."

The command/process of loading an encrypted bitstream automatically disables the FPGA's read-back capability. This means that you will typically use unencrypted configuration data during development (where you need to use read-back) and then start to use encrypted data when you move into production. (You can load an unencrypted bitstream at any time, so you can easily load a test configuration and then reload the encrypted version.)

The main downside to this scheme is that you require a battery backup on the circuit board to maintain the contents of the encryption key register in the FPGA when power is removed from the system. This battery will have a lifetime of years or decades because it need only maintain a single register in the device, but it does add to the size, weight, complexity, and cost of the board.

Antifuse-based devices

Unlike SRAM-based devices, which are programmed while resident in the system, antifuse-based devices are programmed off-line using a special device programmer.

The proponents of antifuse-based FPGAs are proud to point to an assortment of (not-insignificant) advantages. First of all, these devices are nonvolatile (their configuration data remains when the system is powered down), which means that they are immediately available as soon as power is applied to the system. Following from their nonvolatility, these devices don't require an external memory chip to store their configuration data, which saves the cost of an additional component and also saves real estate on the board.

One noteworthy advantage of antifuse-based FPGAs is the fact that their interconnect structure is naturally "rad hard," which means they are relatively immune to the effects of radiation. This is of particular interest in the case of military and aerospace applications because the state of a configuration cell in an SRAM-based component can be "flipped" if that cell is hit by radiation (of which there is a lot in space). By comparison, once an antifuse has been programmed, it cannot be altered in this way. Having said this, it should also be noted that any flip-flops in these devices remain sensitive to radiation, so chips intended for radiation-intensive environments must have their flip-flops protected by *triple redundancy design*. This refers to having three copies of each register and taking a majority vote (ideally all three registers will contain identical values, but if one has been "flipped" such that two registers say 0 and the third says 1, then the 0s have it, or vice versa if two registers say 1 and the third says 0).

But perhaps the most significant advantage of antifuse-based FPGAs is that their configuration data is buried deep inside them. By default, it is possible for the device programmer to read this data out because this is actually how the programmer works. As each antifuse is being processed, the device programmer keeps on testing it to determine when that element has been fully programmed; then it moves onto the next antifuse. Furthermore, the device programmer can be used to automatically verify that the configuration was performed successfully (this is well worth doing when you're talking about devices containing 50 million plus programmable elements). In order to do this, the device programmer

Radiation can come in the form of gamma rays (very high-energy photons), beta particles (high-energy electrons), and alpha particles.

It should be noted that rad-hard devices are not limited to antifuse technologies. Other components, such as those based on SRAM architectures, are available with special rad-hard packaging and triple redundancy design.

requires the ability to read the actual states of the antifuses and compare them to the required states defined in the configuration file.

Once the device has been programmed, however, it is possible to set (grow) a special security antifuse that subsequently prevents any programming data (in the form of the presence or absence of antifuses) from being read out of the device. Even if the device is decapped (its top is removed), programmed and unprogrammed antifuses appear to be identical, and the fact that all of the antifuses are buried in the internal metallization layers makes it almost impossible to reverse-engineer the design.

Vendors of antifuse-based FPGAs may also tout a couple of other advantages relating to power consumption and speed, but if you aren't careful this can be a case of the quickness of the hand deceiving the eye. For example, they might tease you with the fact that an antifuse-based device consumes only 20 percent (approximately) of the standby power of an equivalent SRAM-based component, that their operational power consumption is also significantly lower, and that their interconnect-related delays are smaller. Also, they might casually mention that an antifuse is much smaller and thus occupies much less real estate on the chip than an equivalent SRAM cell (although they may neglect to mention that antifuse devices also require extra programming circuitry, including a large, hairy programming transistor for each antifuse). They will follow this by noting that when you have a device containing tens of millions of configuration elements, using antifuses means that the rest of the logic can be much closer together. This serves to reduce the interconnect delays, thereby making these devices faster than their SRAM cousins.

And both of the above points would be true … if one were comparing two devices implemented at the same technology node. But therein lies the rub, because antifuse technology requires the use of around three additional process steps after the main manufacturing process has been qualified. For this (and related) reasons, antifuse devices are always at least

It's worth noting that when the MRAM technologies introduced in Chapter 2 come to fruition, these may well change the FPGA landscape.

This is because MRAM fuses would be much smaller than SRAM cells (thereby increasing component density and reducing track delays), and they would also consume much less power.

Furthermore, MRAM-based devices could be preprogrammed like antifuse-based devices (great for security) and reprogrammed like SRAM-based components (good for prototyping).

1821: England.
Michael Faraday invents
the first electric motor.

one—and usually several—generations (technology nodes) behind SRAM-based components, which effectively wipes out any speed or power consumption advantages that might otherwise be of interest.

Of course, the main disadvantage associated with antifuse-based devices is that they are OTP, so once you've programmed one of these little scallywags, its function is set in stone. This makes these components a poor choice for use in a development or prototyping environment.

EPROM-based devices

This section is short and sweet because no one currently makes—or has plans to make—EPROM-based FPGAs.

E²PROM/FLASH-based devices

E²PROM- or FLASH-based FPGAs are similar to their SRAM counterparts in that their configuration cells are connected together in a long shift-register-style chain. These devices can be configured off-line using a device programmer. Alternatively, some versions are in-system programmable, or ISP, but their programming time is about three times that of an SRAM-based component.

Once programmed, the data they contain is nonvolatile, so these devices would be "instant on" when power is first applied to the system. With regard to protection, some of these devices use the concept of a multibit key, which can range from around 50 bits to several hundred bits in size. Once you've programmed the device, you can load your user-defined key (bit-pattern) to secure its configuration data. After the key has been loaded, the only way to read data out of the device, or to write new data into it, is to load a copy of your key via the JTAG port (this port is discussed later in this chapter and also in chapter 5). The fact that the JTAG port in today's devices runs at around 20 MHz means that it would take billions of years to crack the key by exhaustively trying every possible value.

Two-transistor E²PROM and FLASH cells are approximately 2.5 times the size of their one-transistor EPROM cousins, but they are still way smaller than their SRAM counterparts. This means that the rest of the logic can be much closer together, thereby reducing interconnect delays.

On the downside, these devices require around five additional process steps on top of standard CMOS technology, which results in their lagging behind SRAM-based devices by one or more generations (technology nodes). Last but not least, these devices tend to have relatively high static power consumption due to their containing vast numbers of internal pull-up resistors.

Hybrid FLASH-SRAM devices

Last but not least, there's always someone who wants to add yet one more ingredient into the cooking pot. In the case of FPGAs, some vendors offer esoteric combinations of programming technologies. For example, consider a device where each configuration element is formed from the combination of a FLASH (or E²PROM) cell and an associated SRAM cell.

In this case, the FLASH elements can be preprogrammed. Then, when the system is powered up, the contents of the FLASH cells are copied in a massively parallel fashion into their corresponding SRAM cells. This technique gives you the nonvolatility associated with antifuse devices, which means the device is immediately available when power is first applied to the system. But unlike an antifuse-based component, you can subsequently use the SRAM cells to reconfigure the device while it remains resident in the system. Alternatively, you can reconfigure the device using its FLASH cells either while it remains in the system or off-line by means of a device programmer.

Summary

Table 4.1 briefly summarizes the key points associated with the various programming technologies described above:

1821: England. Michael Faraday plots the magnetic field around a conductor.

Feature	SRAM	Antifuse	E2PROM / FLASH
Technology node	State-of-the-art	One or more generations behind	One or more generations behind
Reprogrammable	Yes (in system)	No	Yes (in-system or offline)
Reprogramming speed (inc. erasing)	Fast	----	3x slower than SRAM
Volatile (must be programmed on power-up)	Yes	No	No (but can be if required)
Requires external configuration file	Yes	No	No
Good for prototyping	Yes (very good)	No	Yes (reasonable)
Instant-on	No	Yes	Yes
IP Security	Acceptable (especially when using bitstream encryption)	Very Good	Very Good
Size of configuration cell	Large (six transistors)	Very small	Medium-small (two transistors)
Power consumption	Medium	Low	Medium
Rad Hard	No	Yes	Not really

Table 4-1. Summary of programming technologies

Fine-, medium-, and coarse-grained architectures

It is common to categorize FPGA offerings as being either *fine grained* or *coarse grained*. In order to understand what this means, we first need to remind ourselves that the main feature that distinguishes FPGAs from other devices is that their underlying fabric predominantly consists of large numbers of relatively simple programmable logic block "islands" embedded in a "sea" of programmable interconnect. (Figure 4-1).

In the case of a fine-grained architecture, each logic block can be used to implement only a very simple function. For example, it might be possible to configure the block to act as any 3-input function, such as a primitive logic gate (AND, OR, NAND, etc.) or a storage element (D-type flip-flop, D-type latch, etc.).

In reality, the vast majority of the configuration cells in an FPGA are associated with its interconnect (as opposed to its configurable logic blocks). For this reason, engineers joke that FPGA vendors actually sell only the interconnect, and they throw in the rest of the logic for free!

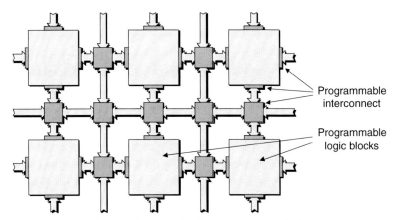

Programmable
interconnect

Programmable
logic blocks

Figure 4-1. Underlying FPGA fabric.

1821: England.
Sir Charles Wheatstone
reproduces sound.

In addition to implementing glue logic and irregular structures like state machines, fine-grained architectures are said to be particularly efficient when executing *systolic algorithms* (functions that benefit from massively parallel implementations). These architectures are also said to offer some advantages with regard to traditional logic synthesis technology, which is geared toward fine-grained ASIC architectures.

The mid-1990s saw a lot of interest in fine-grained FPGA architectures, but over time the vast majority faded away into the sunset, leaving only their coarse-grained cousins. In the case of a coarse-grained architecture, each logic block contains a relatively large amount of logic compared to their fine-grained counterparts. For example, a logic block might contain four 4-input LUTs, four multiplexers, four D-type flip-flops, and some fast carry logic (see the following topics in this chapter for more details).

An important consideration with regard to architectural granularity is that fine-grained implementations require a relatively large number of connections into and out of each block compared to the amount of functionality that can be supported by those blocks. As the granularity of the blocks increases to medium-grained and higher, the amount of connections into the blocks decreases compared to the amount of functionality they can support. This is important because the programmable interblock interconnect accounts for the vast majority of the delays associated with signals as they propagate through an FPGA.

One slight fly in the soup is that a number of companies have recently started developing really coarse-grained device architectures comprising arrays of nodes, where each node is a highly complex processing element ranging from an algorithmic function such as a *fast Fourier transform (FFT)* all the way up to a complete general-purpose microprocessor core (see also Chapters 6 and 23). Although these devices aren't classed as FPGAs, they do serve to muddy the waters. For this reason, LUT-based FPGA architectures are now often classed as medium-grained, thereby leaving the coarse-grained appellation free to be applied to these new node-based devices.

MUX- versus LUT-based logic blocks

There are two fundamental incarnations of the programmable logic blocks used to form the medium-grained architectures referenced in the previous section: MUX (multiplexer) based and LUT (lookup table) based.

MUX is pronounced to rhyme with "flux."

LUT is pronounced to rhyme with "nut."

MUX-based

As an example of a MUX-based approach, consider one way in which the 3-input function $y = (a \ \& \ b) \ | \ c$ could be implemented using a block containing only multiplexers (Figure 4-2).

The device can be programmed such that each input to the block is presented with a logic 0, a logic 1, or the true or inverse version of a signal (a, b, or c in this case) coming from another block or from a primary input to the device. This allows each block to be configured in myriad ways to implement a plethora of possible functions. (The x shown on the input to the central multiplexer in figure 4-2 indicates that we don't care whether this input is connected to a 0 or a 1.)

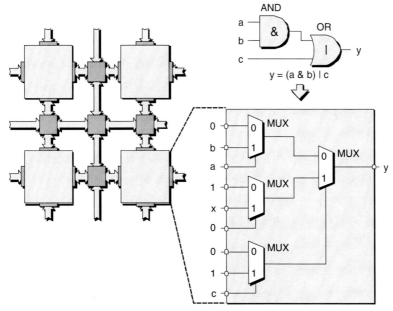

Figure 4-2. MUX-based logic block.

LUT-based

The underlying concept behind a LUT is relatively simple. A group of input signals is used as an index (pointer) to a lookup table. The contents of this table are arranged such that the cell pointed to by each input combination contains the desired value. For example, let's assume that we wish to implement the function:

$$y = (a \, \& \, b) \mid c$$

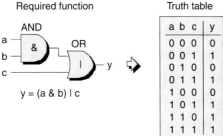

Figure 4-3. Required function and associated truth table.

If you take a group of logic gates several layers deep, then a LUT approach can be very efficient in terms of resource utilization and input-to-output delays. (In this context, "deep" refers to the number of logic gates between the inputs and the outputs. Thus, the function illustrated in figure 4-3 would be said to be two layers deep.)

However, one downside to a LUT-based architecture is that if you only want to implement a small function—such as a 2-input AND gate—somewhere in your design, you'll end up using an entire LUT to do so. In addition to being wasteful in terms of resources, the resulting delays are high for such a simple function.

By comparison, in the case of mux-based architectures containing a mixture of muxes and logic gates, it's often possible to gain access to intermediate values from the signals linking the logic gates and the muxes. In this case, each logic block can be broken down into smaller fragments, each of which can be used to implement a simple function. Thus, these architectures may offer advantages in terms of performance and silicon utilization for designs containing large numbers of independent simple logic functions.

This can be achieved by loading a 3-input LUT with the appropriate values. For the purposes of the following examples, we shall assume that the LUT is formed from SRAM cells (but it could be formed using antifuses, E²PROM, or FLASH cells, as discussed earlier in this chapter). A commonly used technique is to use the inputs to select the desired SRAM cell using a cascade of transmission gates as shown in Figure 4-4. (Note that the SRAM cells will also be connected together in a chain for configuration purposes—that is, to load them with the required values—but these connections have been omitted from this illustration to keep things simple.)

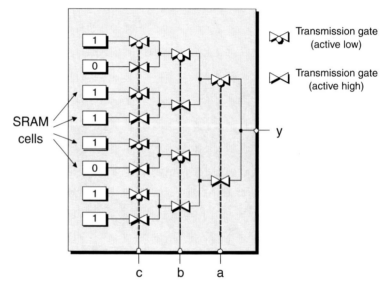

Figure 4-4. A transmission gate-based LUT (programming chain omitted for purposes of clarity).

If a transmission gate is enabled (active), it passes the signal seen on its input through to its output. If the gate is disabled, its output is electrically disconnected from the wire it is driving.

The transmission gate symbols shown with a small circle (called a "bobble" or a "bubble") indicate that these gates will be activated by a logic 0 on their control input. By comparison, symbols without bobbles indicate that these gates will be activated by a logic 1. Based on this understanding, it's easy to see how different input combinations can be used to select the contents of the various SRAM cells.

MUX-based versus LUT-based?

Once upon a time—when engineers handcrafted their circuits prior to the advent of today's sophisticated CAD tools—some folks say that it was possible to achieve the best results using MUX-based architectures. (Sad to relate, they usually don't explain exactly how these results were better, so this is largely left to our imaginations.) It is also said that MUX-based architectures have an advantage when it comes to implementing control logic along the lines of "if this input is *true* and this input is *false*, then make that output *true* …"[1] However, some of these architectures don't provide high-speed carry logic chains, in which case their LUT-based counterparts are left as the leaders in anything to do with arithmetic processing.

Throughout much of the 1990s, FPGAs were widely used in the telecommunications and networking markets. Both of these application areas involve pushing lots of data around, in which case LUT-based architectures hold the high ground. Furthermore, as designs (and device capacities) grew larger and synthesis technology increased in sophistication, handcrafting circuits largely became a thing of the past. The end result is that the majority of today's FPGA architectures are LUT-based, as discussed below.

3-, 4-, 5-, or 6-input LUTs?

The great thing about an *n*-input LUT is that it can implement any possible *n*-input combinational (or combinatorial)

As was noted in Chapter 3, some folks prefer to say "combinational logic," while others favor "combinatorial logic."

[1] Some MUX-based architectures—such as those fielded by QuickLogic (www.quicklogic.com)—feature logic blocks containing multiple layers of MUXes preceded by primitive logic gates like ANDs. This provides them with a large fan-in capability, which gives them an advantage for address decoding and state machine decoding applications.

1822: England.
Charles Babbage starts
to build a mechanic
calculating machine
called the Difference
Engine.

logic function. Adding more inputs allows you to represent more complex functions, but every time you add an input, you double the number of SRAM cells.

The first FPGAs were based on 3-input LUTs. FPGA vendors and university students subsequently researched the relative merits of 3-, 4-, 5-, and even 6-input LUTs into the ground (whatever you do, don't get trapped in conversation with a bunch of FPGA architects at a party). The current consensus is that 4-input LUTs offer the optimal balance of pros and cons.

In the past, some devices were created using a mixture of different LUT sizes, such as 3-input and 4-input LUTs, because this offered the promise of optimal device utilization. However, one of the main tools in the design engineer's treasure chest is logic synthesis, and uniformity and regularity are what a synthesis tool likes best. Thus, all of the really successful architectures are currently based only on the use of 4-input LUTs. (This is not to say that mixed-size LUT architectures won't reemerge in the future as design software continues to increase in sophistication.)

LUT versus distributed RAM versus SR

The fact that the core of a LUT in an SRAM-based device comprises a number of SRAM cells offers a number of interesting possibilities. In addition to its primary role as a lookup table, some vendors allow the cells forming the LUT to be used as a small block of RAM (the 16 cells forming a 4-input LUT, for example, could be cast in the role of a 16 × 1 RAM). This is referred to as *distributed RAM* because (a) the LUTs are strewn (distributed) across the surface of the chip, and (b) this differentiates it from the larger chunks of *block RAM* (introduced later in this chapter).

Yet another possibility devolves from the fact that all of the FPGA's configuration cells—including those forming the LUT—are effectively strung together in a long chain (Figure 4-5).

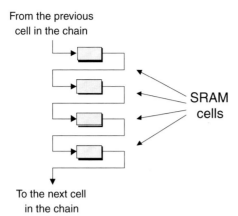

Figure 4-5. Configuration cells linked in a chain.

1822: France. Andre Ampere discovers that two wires carrying electric currents attract each other.

This aspect of the architecture is discussed in more detail in chapter 5. The point here is that, once the device has been programmed, some vendors allow the SRAM cells forming a LUT to be treated independently of the main body of the chain and to be used in the form of a shift register. Thus, each LUT may be considered to be multifaceted (figure 4-6).

CLBs versus LABs versus slices

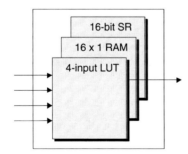

Figure 4-6. A multifaceted LUT.

"Man can not live by LUTs alone," as the Bard would surely say if he were to be reincarnated accidentally as an FPGA designer. For this reason, in addition to one or more LUTs, a programmable logic block will contain other elements, such as multiplexers and registers. But before we delve

1827: England.
Sir Charles Wheatstone
constructs a
microphone.

into this topic, we first need to wrap our brains around some terminology.

A Xilinx logic cell

One niggle when it comes to FPGAs is that each vendor has its own names for things. But we have to start somewhere, so let's kick off by saying that the core building block in a modern FPGA from Xilinx is called a *logic cell (LC)*. Among other things, an LC comprises a 4-input LUT (which can also act as a 16×1 RAM or a 16-bit shift register), a multiplexer, and a register (Figure 4-7).

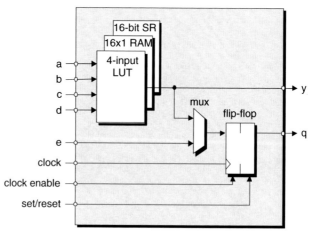

Figure 4-7. A simplified view of a Xilinx LC.

It must be noted that the illustration presented in Figure 4-7 is a gross simplification, but it serves our purposes here. The register can be configured to act as a flip-flop, as shown in Figure 4-7 or as a latch. The polarity of the *clock* (rising-edge triggered or falling-edge triggered) can be configured, as can the polarity of the *clock enable* and *set/reset* signals (active-high or active-low).

In addition to the LUT, MUX, and register, the LC also contains a smattering of other elements, including some special fast carry logic for use in arithmetic operations (this is discussed in more detail a little later).

An Altera logic element

Just for reference, the equivalent core building block in an FPGA from Altera is called a *logic element (LE)*. There are a number of differences between a Xilinx LC and an Altera LE, but the overall concepts are very similar.

Slicing and dicing

The next step up the hierarchy is what Xilinx calls a *slice* (Altera and the other vendors doubtless have their own equivalent names). Why "slice"? Well, they had to call it something, and—whichever way you look at it—the term *slice* is "something." At the time of this writing, a slice contains two logic cells (Figure 4-8).

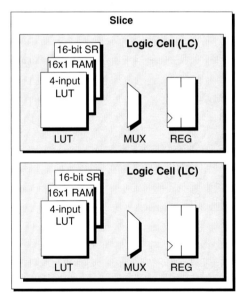

Figure 4-8. A slice containing two logic cells.

The reason for the "at the time of this writing" qualifier is that these definitions can—and do—change with the seasons.

1827: Germany. Georg Ohm investigates electrical resistance and defines Ohm's Law.

The definition of what forms a CLB varies from year to year. In the early days, a CLB consisted of two 3-input LUTs and one register. Later versions sported two 4-input LUTs and two registers.

Now, each CLB can contain two or four slices, where each slice contains two 4-input LUTS and two registers. And as for the morrow … well, it would take a braver man than I even to dream of speculating.

The internal wires have been omitted from this illustration to keep things simple; it should be noted, however, that although each logic cell's LUT, MUX, and register have their own data inputs and outputs, the slice has one set of *clock*, *clock enable*, and *set/reset* signals common to both logic cells.

CLBs and LABs

And moving one more level up the hierarchy, we come to what Xilinx calls a *configurable logic block (CLB)* and what Altera refers to as a *logic array block (LAB)*. (Other FPGA vendors doubtless have their own equivalent names for each of these entities, but these are of interest only if you are actually working with their devices.)

Using CLBs as an example, some Xilinx FPGAs have two slices in each CLB, while others have four. At the time of this writing, a CLB equates to a single logic block in our original visualization of "islands" of programmable logic in a "sea" of programmable interconnect (Figure 4-9).

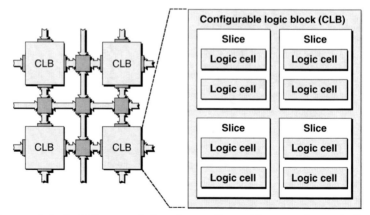

Figure 4-9. A CLB containing four slices (the number of slices depends on the FPGA family).

The point where a set of data or control signals enters or exits a logic function is commonly referred to as a "port.

"In the case of a single-port RAM, data is written in and read out of the function using a common data bus.

There is also some fast programmable interconnect within the CLB. This interconnect (not shown in Figure 4-9 for reasons of clarity) is used to connect neighboring slices.

The reason for having this type of logic-block hierarchy—LC→ Slice (with two LCs)→ CLB (with four slices)—is that it is complemented by an equivalent hierarchy in the interconnect. Thus, there is fast interconnect between the LCs in a slice, then slightly slower interconnect between slices in a CLB, followed by the interconnect between CLBs. The idea is to achieve the optimum trade-off between making it easy to connect things together without incurring excessive interconnect-related delays.

Distributed RAMs and shift registers

We previously noted that each 4-bit LUT can be used as a 16×1 RAM. And things just keep on getting better and better because, assuming the four-slices-per-CLB configuration illustrated in figure 4-9, all of the LUTs within a CLB can be configured together to implement the following:

- Single-port 16×8 bit RAM
- Single-port 32×4 bit RAM
- Single-port 64×2 bit RAM
- Single-port 128×1 bit RAM
- Dual-port 16×4 bit RAM
- Dual-port 32×2 bit RAM
- Dual-port 64×1 bit RAM

Alternatively, each 4-bit LUT can be used as a 16-bit shift register. In this case, there are special dedicated connections between the logic cells within a slice and between the slices themselves that allow the last bit of one shift register to be connected to the first bit of another without using the ordinary LUT output (which can be used to view the contents of a selected bit within that 16-bit register). This allows the LUTs within a single CLB to be configured together to implement a shift register containing up to 128 bits as required.

Fast carry chains

A key feature of modern FPGAs is that they include the special logic and interconnect required to implement fast carry

In the case of a dual-port RAM, data is written into the function using one data bus (port) and read out using a second data bus (port). In fact, the read and write operations each have an associated address bus (used to point to a word of interest inside the RAM). This means that the read and write operations can be performed simultaneously.

chains. In the context of the CLBs introduced in the previous section, each LC contains special carry logic. This is complemented by dedicated interconnect between the two LCs in each slice, between the slices in each CLB, and between the CLBs themselves.

This special carry logic and dedicated routing boosts the performance of logical functions such as counters and arithmetic functions such as adders. The availability of these fast carry chains—in conjunction with features like the shift register incarnations of LUTs (discussed above) and embedded multipliers and the like (introduced below)—provided the wherewithal for FPGAs to be used for applications like DSP.

DSP is pronounced by spelling it out as "D-S-P."

Embedded RAMs

A lot of applications require the use of memory, so FPGAs now include relatively large chunks of embedded RAM called *e-RAM* or *block RAM*. Depending on the architecture of the component, these blocks might be positioned around the periphery of the device, scattered across the face of the chip in relative isolation, or organized in columns, as shown in Figure 4-10.

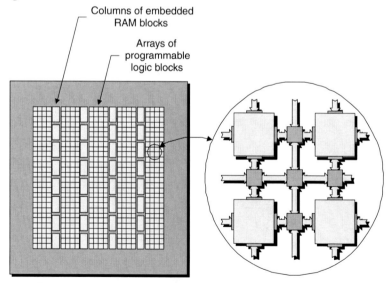

Columns of embedded
RAM blocks

Arrays of
programmable
logic blocks

**Figure 4-10. Bird's-eye view of chip with columns of
embedded RAM blocks.**

Depending on the device, such a RAM might be able to hold anywhere from a few thousand to tens of thousands of bits. Furthermore, a device might contain anywhere from tens to hundreds of these RAM blocks, thereby providing a total storage capacity of a few hundred thousand bits all the way up to several million bits.

Each block of RAM can be used independently, or multiple blocks can be combined together to implement larger blocks. These blocks can be used for a variety of purposes, such as implementing standard single- or dual-port RAMs, *first-in first-out (FIFO)* functions, state machines, and so forth.

FIFO is pronounced "fi" to rhyme with "hi," followed by "fo" to rhyme with "no" (like the "Hi-Ho" song in "Snow White and the Seven Dwarfs").

Embedded multipliers, adders, MACs, etc.

Some functions, like multipliers, are inherently slow if they are implemented by connecting a large number of programmable logic blocks together. Since these functions are required by a lot of applications, many FPGAs incorporate special hardwired multiplier blocks. These are typically located in close proximity to the embedded RAM blocks introduced in the previous point because these functions are often used in conjunction with each other (Figure 4-11).

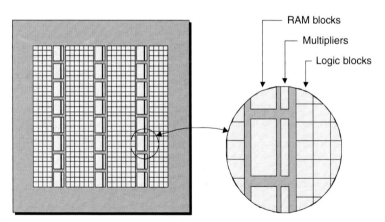

RAM blocks

Multipliers

Logic blocks

Figure 4-11. Bird's-eye view of chip with columns of embedded multipliers and RAM blocks.

Similarly, some FPGAs offer dedicated adder blocks. One operation that is very common in DSP-type applications is

1829: England.
Sir Charles Wheatstone
invents the concertina.

called a multiply-and-accumulate (MAC) (Figure 4-12). As
its name would suggest, this function multiplies two numbers
together and adds the result to a running total stored in an
accumulator.

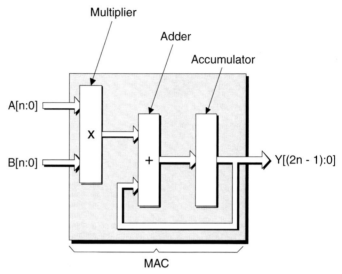

Figure 4-12. The functions forming a MAC.

If the FPGA you are working with supplies only embedded
multipliers, you will have to implement this function by combining the multiplier with an adder formed from a number of
programmable logic blocks, while the result is stored in some
associated flip-flops, in a block RAM, or in a number of distributed RAMs. Life becomes a little easier if the FPGA also
provides embedded adders, and some FPGAs provide entire
MACs as embedded functions.

Embedded processor cores (hard and soft)

Almost any portion of an electronic design can be realized
in hardware (using logic gates and registers, etc.) or software
(as instructions to be executed on a microprocessor). One of
the main partitioning criteria is how fast you wish the various
functions to perform their tasks:

- *Picosecond and nanosecond logic:* This has to run insanely fast, which mandates that it be implemented in hardware (in the FPGA fabric).
- *Microsecond logic:* This is reasonably fast and can be implemented either in hardware or software (this type of logic is where you spend the bulk of your time deciding which way to go).
- *Millisecond logic:* This is the logic used to implement interfaces such as reading switch positions and flashing *light-emitting diodes (LEDs)*. It's a pain slowing the hardware down to implement this sort of function (using huge counters to generate delays, for example). Thus, it's often better to implement these tasks as microprocessor code (because processors give you lousy speed—compared to dedicated hardware—but fantastic complexity).

1831: England. Michael Faraday creates the first electric dynamo.

The fact is that the majority of designs make use of microprocessors in one form or another. Until recently, these appeared as discrete devices on the circuit board. Of late, high-end FPGAs have become available that contain one or more embedded microprocessors, which are typically referred to as *microprocessor cores*. In this case, it often makes sense to move all of the tasks that used to be performed by the external microprocessor into the internal core. This provides a number of advantages, not the least being that it saves the cost of having two devices; it eliminates large numbers of tracks, pads, and pins on the circuit board; and it makes the board smaller and lighter.

Hard microprocessor cores

A hard microprocessor core is implemented as a dedicated, predefined block. There are two main approaches for integrating such a core into the FPGA. The first is to locate it in a strip (actually called "The Stripe") to the side of the main FPGA fabric (Figure 4-13).

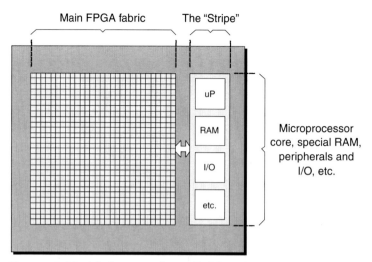

Figure 4-13. Birds-eye view of chip with embedded core outside of the main fabric.

In this scenario, all of the components are typically formed on the same silicon chip, although they could also be formed on two chips and packaged as a *multichip module* (MCM). The main FPGA fabric would also include the embedded RAM blocks, multipliers, and the like introduced earlier, but these have been omitted from this illustration to keep things simple.

One advantage of this implementation is that the main FPGA fabric is identical for devices with and without the embedded microprocessor core, which can help make things easier for the design tools used by the engineers. The other advantage is that the FPGA vendor can bundle a whole load of additional functions in the strip to complement the microprocessor core, such as memory, special peripherals, and so forth.

An alternative is to embed one or more microprocessor cores directly into the main FPGA fabric. One, two, and even four core implementations are currently available as I pen these words (Figure 4-14).

Once again, the main FPGA fabric would also include the embedded RAM blocks, multipliers, and the like introduced

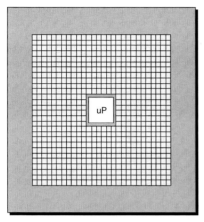

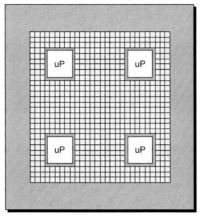

(a) One embedded core (b) Four embedded cores

Figure 4-14. Bird's-eye view of chips with embedded cores inside the main fabric.

1831: England. Michael Faraday creates the first electrical transformer.

earlier, but these have been omitted from this illustration to keep things simple.

In this case, the design tools have to be able to take account of the presence of these blocks in the fabric; any memory used by the core is formed from embedded RAM blocks, and any peripheral functions are formed from groups of general-purpose programmable logic blocks. Proponents of this scheme will argue that there are inherent speed advantages to be gained from having the microprocessor core in intimate proximity to the main FPGA fabric.

Soft microprocessor cores

As opposed to embedding a microprocessor physically into the fabric of the chip, it is possible to configure a group of programmable logic blocks to act as a microprocessor. These are typically called *soft cores*, but they may be more precisely categorized as either "soft" or "firm" depending on the way in which the microprocessor's functionality is mapped onto the logic blocks (see also the discussions associated with the hard IP, soft IP, and firm IP topics later in this chapter).

1831: England.
Michael Faraday
discovers magnetic
lines of force.

Soft cores are simpler (more primitive) and slower than their hard-core counterparts.[2] However, they have the advantage that you only need to implement a core if you need it and also that you can instantiate as many cores as you require until you run out of resources in the form of programmable logic blocks.

Clock trees and clock managers

All of the synchronous elements inside an FPGA—for example, the registers configured to act as flip-flops inside the programmable logic blocks—need to be driven by a clock signal. Such a clock signal typically originates in the outside world, comes into the FPGA via a special clock input pin, and is then routed through the device and connected to the appropriate registers.

Clock trees

Consider a simplified representation that omits the programmable logic blocks and shows only the clock tree and the registers to which it is connected (Figure 4-15).

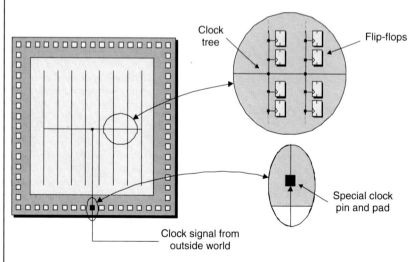

Figure 4-15. A simple clock tree.

---
[2] A soft core typically runs at 30 to 50 percent of the speed of a hard core.

This is called a "clock tree" because the main clock signal branches again and again (the flip-flops can be consider, to be the "leaves" on the end of the branches). This structure is used to ensure that all of the flip-flops see their versions of the clock signal as close together as possible. If the clock were distributed as a single long track driving all of the flip-flops one after another, then the flip-flop closest to the clock pin would see the clock signal much sooner than the one at the end of the chain. This is referred to as *skew*, and it can cause all sorts of problems (even when using a clock tree, there will be a certain amount of skew between the registers on a branch and also between branches).

The clock tree is implemented using special tracks and is separate from the general-purpose programmable interconnect. The scenario shown above is actually very simplistic. In reality, multiple clock pins are available (unused clock pins can be employed as general-purpose I/O pins), and there are multiple *clock domains* (clock trees) inside the device.

Clock managers

Instead of configuring a clock pin to connect directly into an internal clock tree, that pin can be used to drive a special hard-wired function (block) called a *clock manager* that generates a number of *daughter clocks* (Figure 4-16).

A clock manager as described here is referred to as a *digital clock manager (DCM)* in the Xilinx world. DCM is pronounced by spelling it out as "D-C-M."

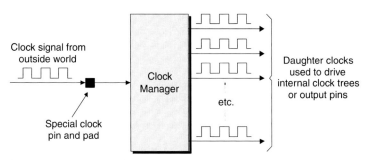

Clock signal from
outside world

Clock
Manager

etc.

Daughter clocks
used to drive
internal clock trees
or output pins

Special clock
pin and pad

Figure 4-16. A clock manager generates daughter clocks.

These daughter clocks may be used to drive internal clock trees or external output pins that can be used to provide clocking services to other devices on the host circuit board. Each family of FPGAs has its own type of clock manager (there may be multiple clock manager blocks in a device), where different clock managers may support only a subset of the following features:

Jitter removal: For the purposes of a simple example, assume that the clock signal has a frequency of 1 MHz (in reality, of course, this could be much, much higher). In an ideal environment each clock edge from the outside world would arrive exactly one millionth of a second after its predecessor. In the real world, however, clock edges may arrive a little early or a little late.

As one way to visualize this effect—known as *jitter*—imagine if we were to superimpose multiple edges on top of each other; the result would be a "fuzzy" clock (Figure 4-17).

The term hertz was taken from the name of Heinrich Rudolf Hertz, a professor of physics at Karlsruhe Polytechnic in Germany, who first transmitted and received radio waves in a laboratory environment in 1888.

One hertz (Hz) equates to "one cycle per second," so MHz stands for megahertz or "million Hertz."

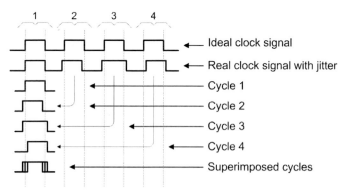

Figure 4-17. Jitter results in a fuzzy clock.

The FPGA's clock manager can be used to detect and correct for this jitter and to provide "clean" daughter clock signals for use inside the device (Figure 4-18).

Frequency synthesis: It may be that the frequency of the clock signal being presented to the FPGA from the outside world is not exactly what the design engineers wish for. In this case, the clock manager can be used to generate daughter clocks with frequencies that are derived by multiplying or dividing the original signal.

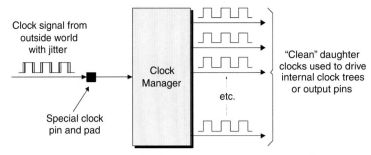

Figure 4-18. The clock manager can remove jitter.

1831: England. Michael Faraday discovers the principal of electro-magnetic induction.

As a really simple example, consider three daughter clock signals: the first with a frequency equal to that of the original clock, the second multiplied to be twice that of the original clock, and the third divided to be half that of the original clock (Figure 4-19).

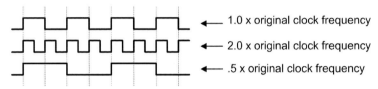

Figure 4-19. Using the clock manager to perform frequency synthesis.

Once again, Figure 4-19 reflects very simple examples. In the real world, one can synthesize all sorts of internal clocks, such as an output that is four-fifths the frequency of the original clock.

Phase shifting: Certain designs require the use of clocks that are phase shifted (delayed) with respect to each other. Some clock managers allow you to select from fixed phase shifts of common values such as 120° and 240° (for a three-phase clocking scheme) or 90°, 180°, and 270° (if a four-phase clocking scheme is required). Others allow you to configure the exact amount of phase shift you require for each daughter clock.

For example, let's assume that we are deriving four internal clocks from a master clock, where the first is in phase with the original clock, the second is phase shifted by 90°, the third by 180°, and so forth (Figure 4-20).

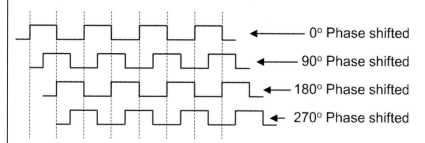

Figure 4-20. Using the clock manager to phase-shift the daughter clocks.

Auto-skew correction: For the sake of simplicity, let's assume that we're talking about a daughter clock that has been configured to have the same frequency and phase as the main clock signal coming into the FPGA. By default, however, the clock manager will add some element of delay to the signal as it performs its machinations. Also, more significant delays will be added by the driving gates and interconnect employed in the clock's distribution. The result is that—if nothing is done to correct it—the daughter clock will lag behind the input clock by some amount. Once again, the difference between the two signals is known as *skew*.

Depending on how the main clock and the daughter clock are used in the FPGA (and on the rest of the circuit board), this can cause a variety of problems. Thus, the clock manager may allow a special input to feed the daughter clock. In this case, the clock manager will compare the two signals and specifically add additional delay to the daughter clock sufficient to realign it with the main clock (Figure 4-21).

To be a tad more specific, only the *prime* (zero phase-shifted) daughter clock will be treated in this way, and all of the other daughter clocks will be phase aligned to this prime daughter clock.

Some FPGA clock managers are based on *phase-locked loops (PLLs)*, while others are based on *digital delay-locked loops*

PLL is pronounced by spelling it out as "P-L-L."

"DLL is pronounced by spelling it out as "D-L-L."

At this time, I do not know why digital delay-locked loop is not abbreviated to "DDLL."

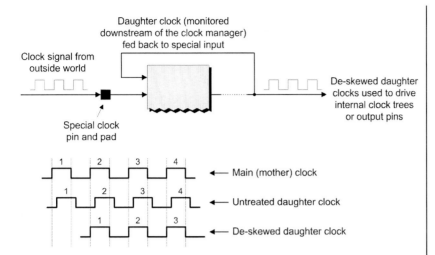

Figure 4-21. Deskewing with reference to the mother clock.

(DLLs). PLLs have been used since the 1940s in analog implementations, but recent emphasis on digital methods has made it desirable to match signal phases digitally. PLLs can be implemented using either analog or digital techniques, while DLLs are by definition digital in nature. The proponents of DLLs say that they offer advantages in terms of precision, stability, power management, noise insensitivity, and jitter performance.

General-purpose I/O

Today's FPGA packages can have 1,000 or more pins, which are arranged as an array across the base of the package. Similarly, when it comes to the silicon chip inside the package, flip-chip packaging strategies allow the power, ground, clock, and I/O pins to be presented across the surface of the chip. Purely for the purposes of these discussions (and illustrations), however, it makes things simpler if we assume that all of the connections to the chip are presented in a ring around the circumference of the device, as indeed they were for many years.

1831: England. Michael Faraday discovers that a moving magnet induces an electric current.

Configurable I/O standards

Let's consider for a moment an electronic product from the perspective of the architects and engineers designing the circuit board. Depending on what they are trying to do, the devices they are using, the environment the board will operate in, and so on, these guys and gals will select a particular standard to be used to transfer data signals. (In this context, "standard" refers to electrical aspects of the signals, such as their logic 0 and logic 1 voltage levels.)

The problem is that there is a wide variety of such standards, and it would be painful to have to create special FPGAs to accommodate each variation. For this reason, an FPGA's general-purpose I/O can be configured to accept and generate signals conforming to whichever standard is required. These general-purpose I/O signals will be split into a number of banks—we'll assume eight such banks numbered from 0 to 7 (Figure 4-22).

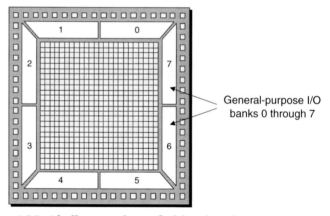

General-purpose I/O banks 0 through 7

Figure 4-22. Bird's-eye view of chip showing general-purpose I/O banks.

The interesting point is that each bank can be configured individually to support a particular I/O standard. Thus, in addition to allowing the FPGA to work with devices using multiple I/O standards, this allows the FPGA to actually be used to interface between different I/O standards (and also to translate between different protocols that may be based on particular electrical standards).

Configurable I/O impedances

The signals used to connect devices on today's circuit board often have fast *edge rates* (this refers to the time it takes the signal to switch between one logic value and another). In order to prevent signals reflecting back (bouncing around), it is necessary to apply appropriate terminating resistors to the FPGA's input or output pins.

In the past, these resistors were applied as discrete components that were attached to the circuit board outside the FPGA. However, this technique became increasingly problematic as the number of pins started to increase and their *pitch* (the distance between them) shrank. For this reason, today's FPGAs allow the use of internal terminating resistors whose values can be configured by the user to accommodate different circuit board environments and I/O standards.

Core versus I/O supply voltages

In the days of yore—circa 1965 to 1995—the majority of digital ICs used a ground voltage of 0V and a supply voltage of +5V. Furthermore, their I/O signals also switched between 0V (logic 0) and +5V (logic 1), which made life really simple.

Over time, the geometries of the structures on silicon chips became smaller because smaller transistors have lower costs, higher speed, and lower power consumption. However, these processes demanded lower supply voltages, which have continued to fall over the years (Table 4.2).

The point is that this supply (which is actually provided using large numbers of power and ground pins) is used to power the FPGA's internal logic. For this reason, this is known as the *core voltage*. However, different I/O standards may use signals with voltage levels significantly different from the core voltage, so each bank of general-purpose I/Os can have its own additional supply pins.

1832: England. Charles Babbage conceives the first mechanical computer, the Analytical Engine.

1832: England
Joseph Henry
discovers self-induction
or inductance.

Year	Supply (Core Voltage (V))	Technology Node (nm)
1998	3.3	350
1999	2.5	250
2000	1.8	180
2001	1.5	150
2003	1.2	130

Table 4.2. Supply voltages versus technology nodes.

It's interesting to note that—from the 350 nm node onwards—the core voltage has scaled fairly linearly with the process technology. However, there are physical reasons not to go much below 1V (these reasons are based on technology aspects such as transistor input switching thresholds and voltage drops), so this "voltage staircase" might start to tail off in the not-so-distant future.

Gigabit transceivers

The traditional way to move large amounts of data between devices is to use a *bus*, a collection of signals that carry similar data and perform a common function (Figure 4-23).

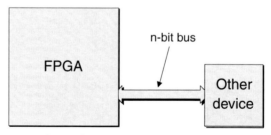

Figure 4-23: Using a bus to communicate between devices.

Early microprocessor-based systems circa 1975 used 8-bit buses to pass data around. As the need to push more data around and to move it faster grew, buses grew to 16 bits in width, then 32 bits, then 64 bits, and so forth. The problem is that this requires a lot of pins on the device and a lot of tracks connecting the devices together. Routing these tracks so that they all have the same length and impedance becomes increasingly painful as boards grow in complexity. Furthermore, it becomes increasingly difficult to manage signal integrity issues (such as susceptibility to noise) when you are dealing with large numbers of bus-based tracks.

For this reason, today's high-end FPGAs include special hard-wired gigabit transceiver blocks. These blocks use one pair of differential signals (which means a pair of signals that always carry opposite logical values) to *transmit (TX)* data and another pair to *receive (RX)* data (Figure 4-24).

1833: England. Michael Faraday defines the laws of electrolysis.

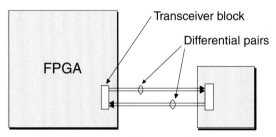

Figure 4-24: Using high-speed transceivers to communicate between devices.

These transceivers operate at incredibly high speeds, allowing them to transmit and receive billions of bits of data per second. Furthermore, each block actually supports a number (say four) of such transceivers, and an FPGA may contain a number of these transceiver blocks.

Hard IP, soft IP, and firm IP

Each FPGA vendor offers its own selection of *hard, firm,* and *soft IP.* Hard IP comes in the form of preimplemented blocks such as microprocessor cores, gigabit interfaces, multipliers, adders, MAC functions, and the like. These blocks are designed to be as efficient as possible in terms of power consumption, silicon real estate, and performance. Each FPGA family will feature different combinations of such blocks. together with various quantities of programmable logic blocks.

IP is pronounced by spelling it out as "I-P."

HDL is pronounced by spelling it out as "H-D-L."

VHDL is pronounced by spelling it out as "V-H-D-L."

RTL is pronounced by spelling it out as "R-T-L."

At the other end of the spectrum, soft IP refers to a source-level library of high-level functions that can be included to the users' designs. These functions are typically represented using a hardware description language, or HDL, such as Verilog or VHDL at the *register transfer level (RTL)* of abstraction. Any soft IP functions the design engineers decide to use are incorporated into the main body of the design—which is also specified in RTL—and subsequently synthesized down into a group of programmable logic blocks (possibly combined with some hard IP blocks like multipliers, etc.).

Holding somewhat of a middle ground is firm IP, which also comes in the form of a library of high-level functions. Unlike their soft IP equivalents, however, these functions have already been optimally mapped, placed, and routed into a group of programmable logic blocks (possibly combined with some hard IP blocks like multipliers, etc.). One or more copies of each predefined firm IP block can be instantiated (called up) into the design as required.

The problem is that it can be hard to draw the line between those functions that are best implemented as hard IP and those that should be implemented as soft or firm IP (using a number of general-purpose programmable logic blocks). In the case of functions like the multipliers, adders, and MACs discussed earlier in this chapter, these are generally useful for a wide range of applications. On the other hand, some FPGAs contain dedicated blocks to handle specific interface protocols like the PCI standard. It can, of course, make your life a lot easier if this happens to be the interface with which you wish to connect your device to the rest of the board. On the other hand, if you decide you need to use some other interface, a dedicated PCI block will serve only to waste space, block traffic, and burn power in your chip.

PCI is pronounced by spelling it out as "P-C-I."

Generally speaking, once FPGA vendors add a function like this into their device, they've essentially placed the component into a niche. Sometimes you have to do this to achieve the desired performance, but this is a classic problem because the next generation of the device is often fast enough to perform this function in its main (programmable) fabric.

System gates versus real gates

One common metric used to measure the size of a device in the ASIC world is that of *equivalent gates*. The idea is that different vendors provide different functions in their cell libraries, where each implementation of each function requires a different number of transistors. This makes it difficult to compare the relative capacity and complexity of two devices.

The answer is to assign each function an equivalent gate value along the lines of "Function A equates to five equivalent gates; function B equates to three equivalent gates ..." The next step is to count all of the instances of each function, convert them into their equivalent gate values, sum all of these values together, and proudly proclaim, "My ASIC contains 10 million equivalent gates, which makes it *much* bigger than your ASIC!"

Unfortunately, nothing is simple because the definition of what actually constitutes an equivalent gate can vary depending on whom one is talking to. One common convention is for a 2-input NAND function to represent one equivalent gate. Alternatively, some vendors define an equivalent gate as equaling an arbitrary number of transistors. And a more esoteric convention defines an ECL equivalent gate as being "one-eleventh the minimum logic required to implement a single-bit full adder" (who on earth came up with this one?). As usual, the best policy here is to make sure that everyone is talking about the same thing before releasing your grip on your hard-earned money.

And so we come to FPGAs. One of the problems FPGA vendors run into occurs when they are trying to establish a basis for comparison between their devices and ASICs. For example, if someone has an existing ASIC design that contains 500,000 equivalent gates and he wishes to migrate this design into an FPGA implementation, how can he tell if his design will fit into a particular FPGA. The fact that each

1837: America. Samual Finley Breese Morse exhibits an electric telegraph.

1837: England
Sir Charles Wheatstone
and Sir William
Fothergill Cooke patent
the five-needle electric
telegraph.

4-input LUT can be used to represent anywhere between one and more than twenty 2-input primitive logic gates makes such a comparison rather tricky.

In order to address this issue, FPGA vendors started talking about *system gates* in the early 1990s. Some folks say that this was a noble attempt to use terminology that ASIC designers could relate to, while others say that it was purely a marketing ploy that didn't do anyone any favors.

Sad to relate, there appears to be no clear definition as to exactly what a system gate is. The situation was difficult enough when FPGAs essentially contained only generic programmable logic in the form of LUTs and registers. Even then, it was hard to state whether or not a particular ASIC design containing x equivalent gates could fit into an FPGA containing y system gates. This is because some ASIC designs may be predominantly combinatorial, while others may make excessively heavy use of registers. Both cases may result in a suboptimal mapping onto the FPGA.

The problem became worse when FPGAs started containing embedded blocks of RAM, because some functions can be implemented much more efficiently in RAM than in general-purpose logic. And the fact that LUTs can act as distributed RAM only serves to muddy the waters; for example, one vendor's system gate count values now include the qualifier, "Assumes 20 percent to 30 percent of LUTs are used as RAM." And, of course, the problems are exacerbated when we come to consider FPGAs containing embedded processor cores and similar functions, to the extent that some vendors now say, "System gate values are not meaningful for these devices."

Is there a rule of thumb that allows you to convert system gates to equivalent gates and vice versa? Sure, there are lots of them! Some folks say that if you are feeling optimistic, then you should divide the system gate value by three (in which case three million FPGA system gates would equate to one million ASIC equivalent gates, for example). Or if you're feeling a tad more on the pessimistic side, you could divide the system gates by five (in which case three million system gates would equate to 600,000 equivalent gates).

However, other folks would say that the above is only true if you assume that the system gates value encompasses all of the functions that you can implement using both the general-purpose programmable logic and the block RAMs. These folks would go on to say that if you remove the block RAMs from the equation, then you should divide the system gates value by ten (in which case, three million system gates would equate to only 300,000 equivalent gates), but in this case you still have the block RAMs to play with … arrggghhhh!

Ultimately, this topic spirals down into such a quagmire that even the FPGA vendors are trying desperately not to talk about system gates any more. When FPGAs were new on the scene, people were comfortable with the thought of equivalent gates and not so at ease considering designs in terms of LUTs, slices, and the like; however, the vast number of FPGA designs that have been undertaken over the years means that engineers are now much happier thinking in FPGA terms. For this reason, speaking as someone living in the trenches, I would prefer to see FPGAs specified and compared using only simple counts of:

- Number of logic cells or logic elements or whatever (which equates to the number of 4-input LUTs and associated flip-flops/latches)
- Number (and size) of embedded RAM blocks
- Number (and size) of embedded multipliers
- Number (and size) of embedded adders
- Number (and size) of embedded MACs
- etc.

Why is this so hard? And it would be really useful to take a diverse suite of real-world ASIC design examples, giving their equivalent gate values, along with details as to their flops/latches, primitive gates, and other more complex functions, then to relate each of these examples to the number of

1842: England Joseph Henry discovers that an electrical spark between two conductors is able to induce magnetism in needles—this effect is detected at a distance of 30 meters.

1842: Scotland. Alexander Bail demonstrates first electromechanical means to capture, transmit, and reproduce an image.

LUTs and flip-flops/latches required in equivalent FPGA implementations, along with the amount of embedded RAM and the number of other embedded functions.

Even this would be less than ideal, of course, because one tends to design things differently for FPGA and ASIC targets, but it would be a start.

FPGA years

We've all heard it said that each year for a dog is equivalent to seven human years, the idea being that a 10-year-old pooch would be 70 years old in human terms. Thinking like this doesn't actually do anyone much good. On the other hand, it does provide a useful frame of reference so that when your hound can no longer keep up with you on a long walk, you can say, "Well, it's only to be expected because the poor old fellow is almost 100 years old" (or whatever).

Similarly, in the case of FPGAs, it may help to think that one of their years equates to approximately 15 human years. Thus, if you're working with an FPGA that was only introduced to the market within the last year, you should view it as a teenager. On the one hand, if you have high hopes for the future, he or she may end up with a Nobel Peace Prize or as the President of the United States. On the other hand, the object of your affections will typically have a few quirks that you have to get used to and learn to work around.

By the time an FPGA has been on the market for two years (equating to 30 years in human terms), you can start to think of it as reasonably mature and a good all-rounder at the peak of its abilities. After three years (45 years old), an FPGA is becoming somewhat staid and middle-aged, and by four years (60 years old). you should treat it with respect and make sure that you don't try to work it like a carthorse!

Programming (Configuring) an FPGA

Weasel words

Before plunging headfirst into this topic, it's probably appropriate to preface our discussions with a few "weasel words" (always remember the saying, "Eagles may soar, but weasels rarely get sucked into jet engines at 10,000 feet!").

The point is that each FPGA vendor has its own unique terminology and its own techniques and protocols for doing things. To make life even more exciting, the detailed mechanisms for programming FPGAs can vary on a family-by-family basis. For these reasons, the following discussions are intended to provide only a generic introduction to this subject.

Configuration files, etc.

Section 2 of this book describes a variety of tools and flows that may be used to capture and implement FPGA designs. The end result of all of these techniques is a *configuration file* (sometimes called a *bit file*), which contains the information that will be uploaded into the FPGA in order to program it to perform a specific function.

In the case of SRAM-based FPGAs, the configuration file contains a mixture of *configuration data* (bits that are used to define the state of programmable logic elements directly) and *configuration commands* (instructions that tell the device what to do with the configuration data). When the configuration file is in the process of being loaded into the device, the information being transferred is referred to as the *configuration bitstream*.

1843: England. Augusta Ada Lovelace publishes her notes explaining the concept of a computer.

E²-based and FLASH-based devices are programmed in a similar manner to their SRAM-based cousins. By comparison, in the case of antifuse-based FPGAs, the configuration file predominantly contains only a representation of the configuration data that will be used to grow the antifuses.

Configuration cells

The underlying concept associated with programming an FPGA is relatively simple (i.e., load the configuration file into the device). It can, however, be a little tricky to wrap one's brain around all of the different facets associated with this process, so we'll start with the basics and work our way up. Initially, let's assume we have a rudimentary device consisting only of an array of very simple programmable logic blocks surrounded by programmable interconnect (Figure 5-1).

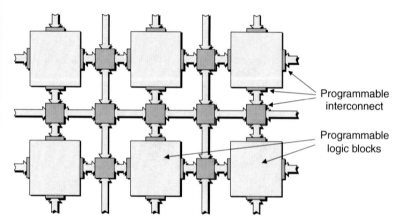

Figure 5-1. Top-down view of simple FPGA architecture.

Any facets of the device that may be programmed are done so by means of special configuration cells. The majority of FPGAs are based on the use of SRAM cells, but some employ FLASH (or E²) cells, while others use antifuses.

Irrespective of the underlying technology, the device's interconnect has a large number of associated cells that can be used to configure it so as to connect the device's primary inputs and outputs to the programmable logic blocks and these logic blocks to each other. (In the case of the device's primary I/Os, which are not shown in Figure 5-1, each has a number of associated cells that can be used to configure them to accommodate specific I/O interface standards and so forth.)

For the purpose of this portion of our discussions, we shall assume that each programmable logic block comprises only a 4-input LUT, a multiplexer, and a register (Figure 5-2). The multiplexer requires an associated configuration cell to specify which input is to be selected. The register requires associated cells to specify whether it is to act as an edge-triggered flip-flop (as shown in Figure 5-2) or a level-sensitive latch, whether it is to be triggered by a positive- or negative-going clock edge (in the case of the flip-flop option) or an active-low or active-high enable (if the register is instructed to act as a latch), and whether it is to be initialized with a logic 0 or a logic 1. Meanwhile, the 4-input LUT is itself based on 16 configuration cells.

LUT is pronounced to rhyme with "nut."

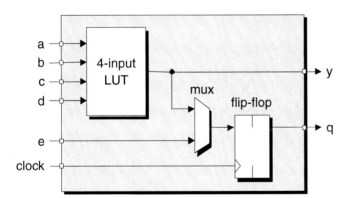

Figure 5-2. A very simple programmable logic block.

Antifuse-based FPGAs

In the case of antifuse-based FPGAs, the antifuse cells can be visualized as scattered across the face of the device at strategic locations. The device is placed in a special device programmer, the configuration (bit) file is uploaded into the device programmer from the host computer, and the device programmer uses this file to guide it in applying pulses of relatively high voltage and current to selected pins to grow each antifuse in turn.

A very simplified way of thinking about this is that each antifuse has a "virtual" x-y location on the surface of the chip, where these x-y values are specified as integers. Based on this scenario, we can visualize using one group of I/O pins to represent the x value associated with a particular antifuse and another group of pins to represent the y value. (Things are more complicated in the real world, but this is a nice way to think about things that doesn't tax our brains too much.)

Once all of the fuses have been grown, the FPGA is removed from the device programmer and attached to a circuit board. Antifuse-based devices are, of course, *one-time programmable (OTP)* because once you've started the programming process, you're committed and it's too late to change your mind.

SRAM-based FPGAs

For the remainder of this chapter we shall consider only SRAM-based FPGAs. Remember that these devices are volatile, which means that they have to be programmed in-system (on the circuit board), and they always need to be reprogrammed when power is first applied to the system.

From the outside world, we can visualize all of the SRAM configuration cells as comprising a single (long) shift register. Consider a simple bird's-eye view of the surface of the chip showing only the I/O pins/pads and the SRAM configuration cells (Figure 5-3).

As a starting point, we shall assume that the beginning and end of this register chain are directly accessible from the outside world. However, it's important to note that this is only the case when using the *configuration port* programming mechanism in conjunction with the *serial load with FPGA as master* or *serial load with FPGA as slave* programming modes, as discussed below.

Also note that the *configuration data out* pin/signal shown in Figure 5-3 is only used if multiple FPGAs are to be config-

FLASH (and E²)–based devices are typically programmed in a similar manner to their SRAM cousins.

Unlike SRAM-based FPGAs, FLASH-based devices are nonvolatile. They retain their configuration when power is removed from the system, and they don't need to be reprogrammed when power is reapplied to the system (although they can be if required).

Also, FLASH-based devices can be programmed in-system (on the circuit board) or outside the system by means of a device programmer.

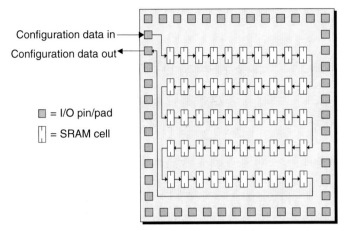

Figure 5-3. Visualizing the SRAM cells as a long shift register.

ured by cascading (daisy-chaining) them together or if it is required to be able to read the configuration data back out of the device for any reason.

The quickness of the hand deceives the eye

It isn't really necessary to know this bit, so if you're in a hurry, you can bounce over into the next section, but I found this interesting and thought you might find it to be so also. As figure 5-3 shows, the easiest way to visualize the internal organization of the SRAM programming cells is as a long shift register. If this were really the case, then each cell would be implemented as a flip-flop, and all of the flop-flops in the chain would be driven by a common clock.

The problem is that an FPGA can contain a humongous number of configuration cells. By 2003, for example, a reasonably high-end device could easily contain 25 million such cells! The core of a flip-flop requires eight transistors, while the core of a latch requires only four transistors. For this reason, the configuration cells in an SRAM-based FPGA are formed from latches. (In our example device with 25 million configuration cells, this results in a saving of 100 million transistors, which is nothing to sneeze at.)

Programming an FPGA can take a significant amount of time. Consider a reasonably high-end device containing 25 million SRAM-based configuration cells. Programming such a device using a serial mode and a 25 MHz clock would take one second. This isn't too bad when you are first powering up a system, but it means that you really don't want to keep on reconfiguring the FPGA when the system is in operation.

1843: England.
Sir Charles Wheatstone
and Sir William
Fothergill Cooke patent
the 2-needle electrical
telegraph.

The problem is that you can't create a shift register out of latches (well, actually you can, as is discussed a little later in this chapter, but not one that's 25 million cells long). The way the FPGA vendors get around this is to have a group of flip-flops—say 1,024—sharing a common clock and configured as a classic shift register. This group is called a *frame*.

The 25 million configuration cells/latches in our example device are also divided up into frames, each being the same length as the flip-flop frame. From the viewpoint of the outside world, you simply clock the 25 million bits of configuration data into the device. Inside the device, however, as soon as the first 1,024 bits have been serially loaded into the flop-flop frame, special-purpose internal circuitry automatically parallel copies/loads this data into the first latch frame. When the next 1,024 bits have been loaded into the flip-flop frame, they are automatically parallel copied/loaded into the second latch frame, and so on for the rest of the device. (The process is reversed when data is read out of the device.)

Programming embedded (block) RAMs, distributed RAMs, etc.

In the case of FPGAs containing large blocks of embedded (block) RAM, the cores of these blocks are implemented out of SRAM latches, and each of these latches is a configuration cell that forms a part of our "imaginary" register chain (as discussed in the previous section).

One interesting point is that each 4-input LUT (see Figure 5-2) can be configured to act as a LUT, as a small (16×1) chunk of distributed RAM, or as a 16-bit shift register. All of these manifestations employ the same group of 16 SRAM latches, where each of these latches is a configuration cell that forms a part of our imaginary register chain.

"But what about the 16-bit shift register incarnation," you cry. "Doesn't this need to be implemented using real flip-flops?" Well, that's a good question—I'm glad you asked. In fact, a trick circuit is employed using the concept of a capacitive latch that prevents classic race conditions (this is pretty much the same way designers built flip-flops out of discrete transistors, resistors, and capacitors in the early 1960s).

Multiple programming chains

Figure 5-3 shows the configuration cells presented as a single programming chain. As there can be tens of millions of configuration cells, this chain can be very long indeed. Some FPGAs are architected so that the configuration port actually drives a number of smaller chains. This allows individual portions of the device to be configured and facilitates a variety of concepts such as modular and incremental design (these concepts are discussed in greater detail in Section 2).

Quickly reinitializing the device

As was previously noted, the register in the programmable logic block has an associated configuration cell that specifies whether it is to be initialized with a logic 0 or a logic 1. Each FPGA family typically provides some mechanism such as an *initialization pin* that, when placed in its active state, causes all of these registers to be returned to their initialization values (this mechanism does not reinitialize any embedded [block] or distributed RAMs).

Using the configuration port

The early FPGAs made use of something called the *configuration port*. Even today, when more sophisticated techniques are available (like the JTAG interface discussed later in this chapter), the configuration port is still widely used because it's relatively simple and is well understood by stalwarts in the FPGA fraternity.

We start with a small group of dedicated *configuration mode* pins that are used to inform the device which configuration mode is going to be used. In the early days, only two pins were employed to provide four modes, as shown in Table 5-1.

Note that the names of the modes shown in this table—and also the relationship between the codes on the

1844: America. Morse Telegraph connects Washington and Baltimore.

1845: England.
Michael Faraday
discovers the rotation of
polarised light by
magnetism.

Mode Pins	Mode
0 0	Serial load with FPGA as master
0 1	Serial load with FPGA as slave
1 0	Parallel load with FPGA as master
1 1	Parallel load with FPGA as slave

Table 5-1. The four original configuration modes

mode pins and the modes themselves—are intended for use only as an example. The actual codes and mode names vary from vendor to vendor.

The mode pins are typically hardwired to the desired logic 0 and logic 1 values at the circuit board level. (These pins can be driven from some other logic that allows the programming mode to be modified, but this is rarely done in practice.)

In addition to the hard-wired mode pins, an additional pin is used to instruct the FPGA to actually commence the configuration, while yet another pin is used by the device to report back when it's finished (there are also ways to determine if an error occurred during the process). This means that in addition to configuring the FPGA when the system is first powered up, the device may also be reinitialized using the original configuration data, if such an occurrence is deemed necessary.

The configuration port also makes use of additional pins to control the loading of the data and to input the data itself. The number of these pins depends on the configuration mode selected, as discussed below. The important point here is that once the configuration has been completed, most of these pins can subsequently be used as general-purpose I/O pins (we will return to this point a little later).

Serial load with FPGA as master

This is perhaps the simplest programming mode. In the early days, it involved the use of an external PROM. This was subsequently superceded by an EPROM, then an E^{2}PROM, and now—most commonly—a FLASH-based device. This special-purpose memory component has a single data output pin that is connected to a *configuration data in* pin on the FPGA (Figure 5-4).

1845: England. First use of the electronic telegraph to help apprehend a criminal.

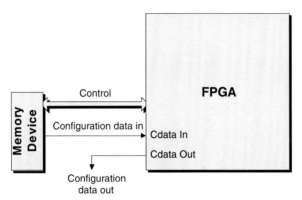

Figure 5-4. Serial load with FPGA as master.

The FPGA also uses several bits to control the external memory device, such as a reset signal to inform it when the FPGA is ready to start reading data and a clock signal to clock the data out.

The idea with this mode is that the FPGA doesn't need to supply the external memory device with a series of addresses. Instead, it simply pulses the reset signal to indicate that it wishes to start reading data from the beginning, and then it sends a series of clock pulses to clock the configuration data out of the memory device.

The *configuration data out* signal coming from the FPGA need only be connected if it is required to read the configuration data from the device for any reason. One such scenario occurs when there are multiple FPGAs on the circuit board. In this case, each could have its own dedicated external memory device and be configured in isolation, as shown in Figure 5-4. Alternatively, the FPGAs could be cascaded (daisy-chained) together and share a single external memory (Figure 5-5).

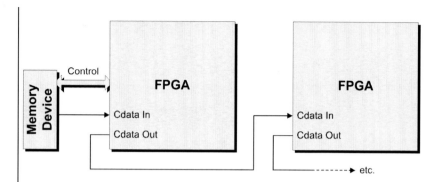

Figure 5-5. Daisy-chaining FPGAs.

In this scenario, the first FPGA in the chain (the one connected directly to the external memory) would be configured to use the *serial master* mode, while the others would be *serial slaves*, as discussed later in this chapter.

Parallel load with FPGA as master

In many respects, this is very similar to the previous mode, except that the data is read in 8-bit chunks from a memory device with eight output pins. Groups of eight bits are very common and are referred to as *bytes*. In addition to providing control signals, the original FPGAs also supplied the external memory device with an address that was used to point to whichever byte of configuration data was to be loaded next (Figure 5-6).

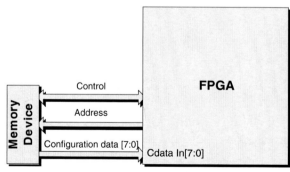

Figure 5-6. Parallel load with FPGA as master (original technique).

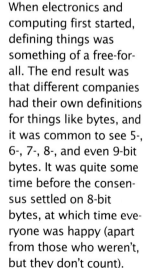

Groups of four bits are also common and are given the special name of nybble (sometimes nibble). The idea is that "two nybbles make a byte," which is a (little) joke. This goes to show that engineers do have a sense of humor; it's just not very sophisticated.

When electronics and computing first started, defining things was something of a free-for-all. The end result was that different companies had their own definitions for things like bytes, and it was common to see 5-, 6-, 7-, 8-, and even 9-bit bytes. It was quite some time before the consensus settled on 8-bit bytes, at which time everyone was happy (apart from those who weren't, but they don't count).

The way this worked was that the FPGA had an internal counter that was used to generate the address for the external memory. (The original FPGAs had 24-bit counters, which allowed them to address 16 million bytes of data.) At the beginning of the configuration sequence, this counter would be initialized with zero. After the byte of data pointed to by the counter had been read, the counter would be incremented to point to the next byte of data. This process would continue until all of the configuration data had been loaded.

It's easy to assume that this parallel-loading technique offers speed advantages, but it didn't for quite some time. This is because—in early devices—as soon as a byte of data had been read into the device, it was clocked into the internal configuration shift register in a serial manner. Happily, this situation has been rectified in more modern FPGA families. On the other hand, although the eight pins can be used as general-purpose I/O pins once the configuration data has been loaded, in reality this is less than ideal. This is because these pins still have the tracks connecting them to the external memory device, which can cause a variety of signal integrity problems.

The real reason why this technique was so popular in the days of yore is that the special-purpose memory devices used in the *serial load with FPGA as master* mode were quite expensive. By comparison, this parallel technique allowed design engineers to use off-the-shelf memory devices, which were much cheaper.

Having said this, special-purpose memory devices created for use with FPGAs are now relatively inexpensive (and being FLASH-based, they are also reusable). Thus, modern FPGAs now use a new variation on this parallel-loading technique. In this case, the external memory is a special-purpose device that doesn't require an external address, which means that the FPGAs no longer requires an internal counter for this purpose (Figure 5-7).

As for the serial mode discussed earlier, the FPGA simply pulses the external memory device's reset signal to indicate

1845: England/France. First telegraph cable is laid across the English Channel.

1846: Germany.
Gustav Kirchhoff
defines Kirchoff's laws
of electrical networks.

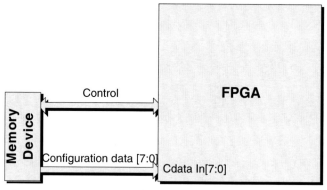

**Figure 5-7. Parallel load with FPGA as the master
(modern technique).**

that it wishes to start reading data from the beginning, and
then it sends a series of clock pulses to clock the configuration
data out of the memory device.

Parallel load with FPGA as slave

The modes discussed above, in which the FPGA is the
master, are attractive because of their inherent simplicity and
also because they only require the FPGA itself, along with a
single external memory device.

However, a large number of circuit boards also include a
microprocessor, which is typically already used to perform a
wide variety of housekeeping tasks. In this case, the design
engineers might decide to use the microprocessor to load the
FPGA (Figure 5-8).

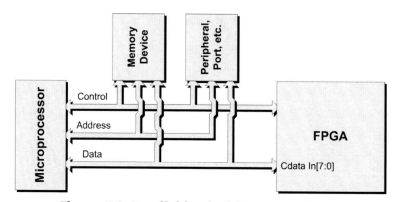

Figure 5-8. Parallel load with FPGA as slave.

The idea here is that the microprocessor is in control. The microprocessor informs the FPGA when it wishes to commence the configuration process. It then reads a byte of data from the appropriate memory device (or peripheral, or whatever), writes that data into the FPGA, reads the next byte of data from the memory device, writes that byte into the FPGA, and so on until the configuration is complete.

This scenario conveys a number of advantages, not the least being that the microprocessor might be used to query the environment in which its surrounding system resides and to then select the configuration data to be loaded into the FPGA accordingly.

Serial load with FPGA as slave

This mode is almost identical to its parallel counterpart, except that only a single bit is used to load data into the FPGA (the microprocessor still reads data out of the memory device one byte at a time, but it then converts this data into a series of bits to be written to the FPGA).

The main advantage of this approach is that it uses fewer I/O pins on the FPGA. This means that—following the configuration process—only a single I/O pin has the additional track required to connect it to the microprocessor's data bus.

Using the JTAG port

Like many other modern devices, today's FPGAs are equipped with a JTAG port. Standing for the *Joint Test Action Group* and officially known to engineers by its IEEE 1149.1 specification designator, JTAG was originally designed to implement the *boundary scan* technique for testing circuit boards and ICs.

A detailed description of JTAG and boundary scan is beyond the scope of this book. For our purposes here, it is sufficient to understand that the FPGA has a number of pins that are used as a JTAG port. One of these pins is used to input JTAG data, and another is used to output that data. Each of the FPGAs remaining I/O pins has an associated JTAG regis-

JTAG is pronounced by spelling out the "J," followed by "tag" to rhyme with "bag."

1847: England.
George Boole publishes
his first ideas on
symbolic logic.

ter (a flip-flop), where these registers are daisy-chained
together (Figure 5-9).

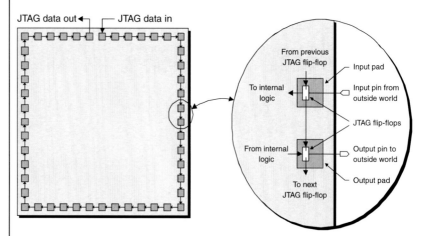

Figure 5-9. JTAG boundary scan registers.

The idea behind boundary scan is that, by means of the
JTAG port, it's possible to serially clock data into the JTAG
registers associated with the input pins, let the device (the
FPGA in this case) operate on that data, store the results from
this processing in the JTAG registers associated with the output pins, and, ultimately, to serially clock this result data back
out of the JTAG port.

However, JTAG devices also contain a variety of additional JTAG-related control logic, and, with regard to
FPGAs, JTAG can be used for much more than boundary
scans. For example, it's possible to issue special commands
that are loaded into a special JTAG command register (not
shown in Figure 5-9) by means of the JTAG port's data-in
pin. One such command instructs the FPGA to connect its
internal SRAM configuration shift register to the JTAG scan
chain. In this case, the JTAG port can be used to program the
FPGA. Thus, today's FPGAs now support five different programming modes and, therefore, require the use of three mode
pins, as shown in Table 5-2 (additional modes may be added
in the future).

Mode Pins	Mode
0 0 0	Serial load with FPGA as master
0 0 1	Serial load with FPGA as slave
0 1 0	Parallel load with FPGA as master
0 1 1	Parallel load with FPGA as slave
1 x x	Use only the JTAG port

Table 5-2. Today's five configuration modes

1850: England. Francis Galton invents the Teletype printer.

Note that the JTAG port is always available, so the device can initially be configured via the traditional configuration port using one of the standard configuration modes, and it can subsequently be reconfigured using the JTAG port as required. Alternately, the device can be configured using only the JTAG port.

Using an embedded processor

But wait, there's more! In chapter 4, we discussed the fact that some FPGAs sport embedded processor cores, and each of these cores will have its own dedicated JTAG boundary scan chain. Consider an FPGA containing just one embedded processor (Figure 5-10).

The FPGA itself would only have a single external JTAG port. If required, a JTAG command can be loaded via this port that instructs the device to link the processor's local JTAG chain into the device's main JTAG chain. (Depending on the vendor, the two chains could be linked by default, in which case a complementary command could be used to disengage the internal chain.)

The idea here is that the JTAG port can be used to initialize the internal microprocessor core (and associated peripherals) to the extent that the main body of the configuration process can then be handed over to the core. In

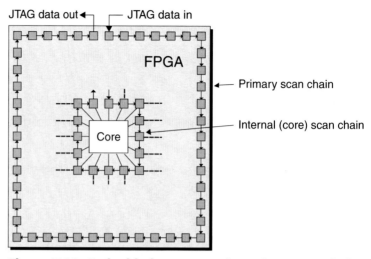

Figure 5-10. Embedded processor boundary scan chain.

some cases, the core might be used to query the environment in which the FPGA resides and to then select the configuration data to be loaded into the FPGA accordingly.

Who Are All the Players?

Introduction

As was noted in chapter 1, this tome does not focus on particular FPGA vendors or specific FPGA devices because new features and chip types are constantly becoming available. Insofar as is possible, the book also tries not to mention individual EDA vendors or reference their tools by name because this arena is so volatile that tool names and feature sets can change from one day to the next.

Having said this, this chapter offers pointers to some of the key FPGA and EDA vendors associated with FPGAs or related areas.

FPGA and FPAA vendors

The bulk of this book focuses on digital FPGAs. It is interesting to note, however, that *field-programmable analog arrays (FPAAs)* are also available. Furthermore, as opposed to supplying FPGA devices, some companies specialize in providing FPGA IP cores to be employed as part of standard cell ASIC or structured ASIC designs.

Company	Web site	Comment
Actel Corp.	www.actel.com	FPGAs
Altera Corp.	www.altera.com	FPGAs
Anadigm Inc.	www.anadigm.com	FPAAs
Atmel Corp.	www.atmel.com	FPGAs
Lattice Semiconductor Corp.	www.latticesemi.com	FPGAs
Leopard Logic Inc.	www.leopardlogic.com	Embedded FPGA cores
QuickLogic Corp.	www.quicklogic.com	FPGAs
Xilinx Inc.	www.xilinx.com	FPGAs

FPGA is pronounced by spelling it out as "F-P-G-A."

FPAA is pronounced by spelling it out as "F-P-A-A."

FPNA is pronounced
by spelling it out as
"F-P-N-A." (These are not
to be confused with field
programmable neural
arrays, which share the
FPNA acronym.)

FFT is pronounced
by spelling it out as
"F-F-T."

FPNA vendors

This is a bit of a tricky category, not the least because the name *field programmable nodal arrays (FPNAs)* was invented just a few seconds ago by the author as he penned these words (he's just that sort of a fellow). The idea here is that each of these devices features a mega-coarse-grained architecture comprising an array of nodes, where each node is a complex processing element ranging from an ALU-type operation, to an algorithmic function such as a FFT, all the way up to a complete general-purpose microprocessor core.

These devices aren't FPGAs in the classic sense. Yet, the definition of what is and what isn't an FPGA is a bit fluffy around the edges on a good day, to the extent that it would be fair to say that modern FPGAs with embedded RAMs, embedded processors, and gigabit transceivers aren't FPGAs in the "classic sense." In the case of FPNAs, these devices are both digital and field programmable, so they deserve at least some mention here.

At the time of this writing, 30 to 50 companies are seriously experimenting with different flavors of FPNAs; a representative sample of the more interesting ones is as follows (see also Chapter 23):

Company	Web site	Comment
Exilent Ltd.	www.elixent.com	ALU-based nodes
IPflex Inc	www.ipflex.com	Operation-based nodes
Motorola	www.motorola.com	Processor-based nodes
PACT XPP Technologies AG	www.pactxpp.com	ALU-based nodes
picoChip Designs Ltd.	www.picochip.com	Processor-based nodes
QuickSilver Technology Inc.	www.qstech.com	Algorithmic element nodes

EDA is pronounced
by spelling it out as
"E-D-A."

OEM, which stands for
"original equipment
manufacturer," is
pronounced by spelling
it out as "O-E-M."

Full-line EDA vendors

Each FPGA, FPAA, and FPNA vendor supplies a selection of design tools associated with its particular devices. In the case of FPGAs, these tools invariably include the place-and-route engines. The FPGA vendor may also OEM tools (often "lite" versions) from external EDA companies. (In this context, OEM means that the FPGA vendors license this software from a third party and then package it and provide it as part of their own environments.)

First of all, we have the big boys—the full-line EDA vendors who can supply complete solutions from soup to nuts (in certain cases, these solutions may include OEM'd point tools from the specialist EDA vendors discussed in the next section).

Company	Web site	Comment
Altium Ltd.	www.altium.com	System-on-FPGA hardware-software design environment
Cadence Design Systems	www.cadence.com	FPGA design entry and simulation (OEM synthesis)
Mentor Graphics Corp.	www.mentor.com	FPGA design entry, simulation, and synthesis
Synopsys Inc.	www.synopsys.com	FPGA design entry, simulation, and synthesis

Nothing is simple in this life. For example, it may seem strange to group a relatively small company like Altium with comparative giants like the "big three." In the context of FPGAs, however, Altium supplies a complete hardware and software codesign environment for system-on-FPGA development. This includes design entry, simulation, synthesis, compilation/assembly, and comprehensive debugging facilities, along with an associated multi-FPGA vendor-capable development board.

FPGA-specialist and independent EDA vendors

As opposed to purchasing an existing solution, some design teams prefer to create their own customized environments using point tools from a number of EDA vendors. In many cases, these tools are cheaper than their counterparts from the full-line vendors, but they may also be less sophisticated and less powerful. At the same time, smaller vendors sometimes come out with incredibly cool and compelling offerings, and they may be more accessible and responsive to their customers.

1850:
The paper bag is invented.

("You pay your money and you make your choice," as the old saying goes.)

Company	Web site	Comment
0-In Design Automation	www.0-In.com	Assertion-based verification
AccelChip Inc.	www.accelchip.com	FPGA-based DSP design
Aldec Inc.	www.aldec.com	Mixed-language simulation
Celoxica Ltd.	www.celoxica.com	FPGA-based system design and synthesis
Elanix Inc.	www.elanix.com	DSP design and algorithmic verification
Fintronic USA Inc.	www.fintronic.com	Simulation
First Silicon Solutions Inc.	www.fs2.com	On-chip instrumentation and debugging for FPGA logic and embedded processors
Green Hills Software Inc.	www.ghs.com	RTOS and embedded software specialists
Hier Design Inc.	www.hierdesign.com	FPGA-based silicon virtual prototyping (SVP)
Novas Software Inc.	www.novas.com	Verification results analysis
Simucad Inc.	www.simucad.com	Simulation
Synplicity Inc.	www.synplicity.com	FPGA-based synthesis
The MathWorks Inc.	www.mathworks.com	System design and algorithmic verification
TransEDA PLC	www.transeda.com	Verification IP
Verisity Design Inc.	www.verisity.com	Verification languages and environments
Wind River Systems Inc.	www.windriver.com	RTOS and embedded software specialists

RTOS, which stands for "real-time operating system," is pronounced by spelling out the "R," followed by "toss" to rhyme with "boss."

FPGA design consultants with special tools

There are a lot of small design houses specializing in FPGA designs. Some of these boast rather cunning internally developed design tools that are well worth taking a look at.

Company	Web site	Comment
Dillon Engineering Inc.	www.dilloneng.com	ParaCore Architect
Launchbird Inc.	www.launchbird.com	Confluence system design language and compiler

Open-source, free, and low-cost design tools

Last but not least, let's assume that you wish to establish a small FPGA design team or to set yourself up as a small FPGA design consultant, but you are a little short of funds just at the

You could probably get through the rest of your day without hearing this, but on the off chance you are interested, a groat was an English silver coin (worth four old pennies) that was used between the fourteenth and seventeenth centuries.

moment (trust me, I can relate to this). In this case, it is possible to use a variety of open-source, free, and low-cost technologies to get a new FPGA design house off the ground without paying more than a few groats for design tools.

Company	Website	Comment
Altera Corp.	www.altera.com	Synthesis and place-and-route
Gentoo	www.gentoo.com	Linux development platform
Icarus	http://icarus.com/eda/verilog	Verilog simulator
Xilinx Inc.	www.xilinx.com	Synthesis and place-and-route
——	www.cs.man.ac.uk/apt/tools/gtkwave/	GTKWave waveform viewer
——	www.opencores.org	Open-source hardware cores and EDA tools
——	www.opencollector.org	Database of open-source hardware cores and EDA tools
——	www.python.org	Python programming language (for custom tooling and DSP programming)
——	www.veripool.com/dinotrace	Dinotrace waveform viewer
——	www.veripool.com/verilator.html	Verilator (Verilog to cycle-accurate C translator)

With regard to using Linux as the development platform, the two main FPGA vendors—Xilinx and Altera—are now porting their tools to Linux. Xilinx and Altera also offer free versions of their ISE and Quartus-II FPGA design environments, respectively (and even the full-up versions of these environments are within the budgets of most startups).

1853:
Scotland/Ireland. Sir Charles Tilston Bright lays the first deepwater cable between Scotland and Ireland.

FPGA Versus ASIC Design Styles

Introduction

My mother is incredibly proud of the fact that "I R an electronics engineer." This comes equipped with an absolute and unshakable faith that I can understand—and fix—any piece of electronic equipment (from any era) on the planet. In reality, of course, the truth is far less glamorous because very few among us are experts at everything.[1]

In a similar vein, some design engineers have spent the best years of their young lives developing a seemingly endless series of ASICs, while others have languished in their cubicles learning the arcane secrets that are the province of the FPGA maestro.

The problem arises when an engineer steeped in one of these implementation technologies is suddenly thrust into the antipodal domain. For example, a common scenario these days is for engineers who bask in the knowledge that they know everything there is to know about ASICs to be suddenly tasked with creating a design targeted toward an FPGA implementation.

Meaning a direct or diametrical opposite, the word "antipodal" comes to us from the Greek, from the plural of *antipous*, meaning "with the feet opposite."

This is a tricky topic because there are so many facets to it; the best we can hope for here is to provide an overview as to some of the more significant differences between ASIC and FPGA design styles.

---

[1] Only the other day, for example, I ran into an old *Wortsel Grinder Mark 4* (with the filigreed flanges and reverberating notchet tattles). I didn't have a clue what to do with it, so you can only imagine how foolish I felt.

1854: Crimea. Telegraph used in the Crimea War.

Coding styles

When it comes to the language-driven design flows discussed in chapter 9, ASIC designers tend to write very portable code (in VHDL or Verilog) and to make the minimum use of instantiated (specifically named) cells.

By comparison, FPGA designers are more likely to instantiate specific low-level cells. For example, FPGA users may not be happy with the way the synthesis tool generates something like a multiplexer, so they may handcraft their own version and then instantiate it from within their code. Furthermore, pure FPGA users tend to use far more technology-specific attributes with regard to their synthesis engine than do their ASIC counterparts.

Pipelining and levels of logic

What is pipelining?

One tends to hear the word *pipelining* quite a lot, but this term is rarely explained. Of course, engineers know what this means, but as this book is intended for a wide audience, we'll take a few seconds to make sure that we're all tap-dancing to the same tune.[2] Let's suppose that we're building something like a car, and we have all of the parts lying around at hand. Let's further assume that the main steps in the process are as follows:

1. Attach the wheels to the chassis.
2. Attach the engine to the chassis.
3. Attach the seats to the chassis.
4. Attach the body to the chassis.
5. Paint everything.

---

[2] As a young man, my dad and his brothers used to be tap-dancers in the variety halls of England before WW II (but I bet they never expected to find this fact noted in an electronics book in the 21st Century).

Yes ... I know, I know. For all of you engineers out there whom I can hear moaning and groaning (you know who you are), I'm aware that we haven't got a steering wheel or lights, etc., but this is just an example for goodness' sake!

Now let's assume that we require a specialist to perform each of these tasks. One approach would be for everyone to be sitting around playing cards. The first guy (or gal, of course)[3] gets up and attaches the wheels to the chassis, and then returns to the game. On his return, the second guy gets up and adds the engine, then he returns to the game. Now the third guy wanders over to attach the seats. Upon the third guy's return, the fourth guy ambles over to connect the body, and so forth. Once the first car has been completed, they start all over again.

Obviously, this is a very inefficient scenario. If, for example, we assume that each step takes one hour, then the whole process will take five hours. Furthermore, for each of these hours, only one man is working, while the other four are hanging around amusing themselves. It would be much more efficient to have five cars on the assembly line at any one time. In this case, as soon as the first guy has attached the wheels to the first chassis, the second guy would start to add the engine to that chassis while the first guy would begin to add the wheels to the second chassis. Once the assembly line is in full flow, everyone will be working all of the time and a new car will be created every hour.

Pipelining in electronic systems

The point is that we can often replicate this scenario in electronic systems. For example, let's assume that we have a design (or a function forming part of a design) that can be implemented as a series of blocks of combinational (or combinatorial) logic (Figure 7-1).

---

[3] Except where such interpretation is inconsistent with the context, the singular shall be deemed to include the plural, the masculine shall be deemed to include the feminine, and the spelling (and the punctuation) shall be deemed to be correct!

1855: England. James Clerk Maxwell explains Faraday's lines of force using mathematics.

1858: America Cunard agents in New York send first commercial telegraph message to report a collision between two steam ships.

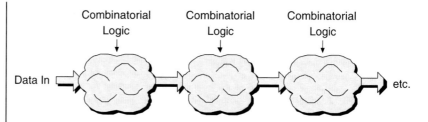

Figure 7-1. A function implemented using only combinatorial logic.

Let's say that each block takes Y nanoseconds to perform its task and that we have five such blocks (of which only three are shown in Figure 7-1, of course). In this case, it will take 5 × Y nanoseconds for a word of data to propagate through the function, starting with its arrival at the inputs to the first block and ending with its departure from the outputs of the last block.

Generally speaking, we wouldn't want to present a new word of data to the inputs until we have stored the output results associated with the first word of data.[4] This means that we end up with the same result as our inefficient car assembly scenario in that it takes a long time to process each word of data, and the majority of the workers (logic blocks) are sitting around twiddling their metaphorical thumbs for most of the time. The answer is to use a pipelined design technique in which "islands" of combinational logic are sandwiched between blocks of registers (Figure 7-2).

All of the register banks are driven by a common clock signal. On each active clock edge, the registers feeding a block of logic are loaded with the results from the previous stage. These values then propagate through that block of logic until they arrive at its outputs, at which point they are ready to be loaded into the next set of registers on the next clock.

---

[4] There is a technique called *wave-pipelining* in which we might have multiple "waves" of data passing through the logic at the same time. However, this is beyond the scope of this book (and it would not be applicable to an FPGA implementation in any case).

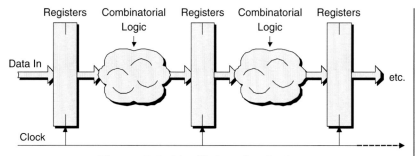

Figure 7-2. Pipelining the design.

In this case, as soon as "the pump has been primed" and the pipeline is fully loaded, a new word of data can be processed every Y nanoseconds.

Levels of logic

All of this boils down to the design engineer's having to perform a balancing act. Partitioning the combinational logic into smaller blocks and increasing the number of register stages will increase the performance of the design, but it will also consume more resources (and silicon real estate) on the chip and increase the *latency* of the design.

This is also the point where we start to run into the concept of *levels of logic*, which refers to the number of gates between the inputs and outputs of a logic block. For example, Figure 7-3 would be said to comprise three levels of logic because the worst-case path involves a signal having to pass through three gates before reaching the output.

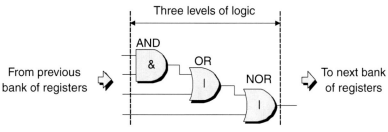

Figure 7-3. Levels of logic.

In the context of an electronic system, the term *latency* refers to the time (clock cycles) it takes for a specific block of data to work its way through a function, device, or system.

One way to think of latency is to return to the concept of an automobile assembly line. In this case, the throughput of the system might be one car rolling off the end of the line every minute. However, the latency of the system might be a full eight-hour shift since it takes hundreds of steps to finish a car (where each of these steps corresponds to a logic/register stage in a pipelined design).

In the case of an ASIC, a group of gates as shown in Figure 7-3 can be placed close to each other such that their track delays are very small. This means that, depending on the design, ASIC engineers can sometimes be a little sloppy about this sort of thing (it's not unheard of to have paths with, say, 15 or more levels of logic).

By comparison, if this sort of design were implemented on an FPGA with each of the gates implemented in a separate LUT, then it would "fly like a brick" (go incredibly slowly) because the track delays on FPGAs are much more significant, relatively speaking. In reality, of course, a LUT can actually represent several levels of logic (the function shown in Figure 7-3 could be implemented in a single 4-input LUT), so the position isn't quite as dire as it may seem at first.

Having said this, the bottom line is that in order to bring up (or maintain) performance, FPGA designs tend to be more highly pipelined than their ASIC counterparts. This is facilitated by the fact that every FPGA logic cell tends to comprise both a LUT and a register, which makes registering the output very easy.

Asynchronous design practices

Asynchronous structures

Depending on the task at hand, ASIC engineers may include asynchronous structures in their designs, where these constructs rely on the relative propagation delays of signals in order to function correctly. These techniques do not work in the FPGA world as the routing (and associated delays) can change dramatically with each new run of the place-and-route engines.

Of course, every latch is based on internal feedback—and every flip-flop is essentially formed from two latches—but this feedback is very tightly controlled by the device manufacturer.

Combinational loops

As a somewhat related topic, a combinational loop occurs when the generation of a signal depends on itself feeding back through one or more logic gates. These are a major source of critical race conditions where logic values depend on routing delays. Although the practice is frowned upon in some circles, ASIC engineers can be little rapscallions when it comes to using these structures because they can fix track routing (and therefore the associated propagation delays) very precisely. This is not the case in the FPGA domain, so all such feedback loops should include a register element.

Delay chains

Last but not least, ASIC engineers may use a series of buffer or inverter gates to create a delay chain. These delay chains may be used for a variety of purposes, such as addressing race conditions in asynchronous portions of the design. In addition to the delay from such a chain being hard to predict in the FPGA world, this type of structure increases the design's sensitivity to operating conditions, decreases its reliability, and can be a source of problems when migrating to another architecture or implementation technology.

Clock considerations

Clock domains

ASIC designs can feature a huge number of clocks (one hears of designs with more than 300 different clock domains). In the case of an FPGA, however, there are a limited number of dedicated global clock resources in any particular device. It is highly recommended that designers budget their clock systems to stay within the dedicated clock resources (as opposed to using general-purpose inputs as user-defined clocks).

Some FPGAs allow their clock trees to be fragmented into clock segments. If the target technology does support this feature, it should be identified and accounted for while mapping external or internal clocks.

Clock balancing

In the case of ASIC designs, special techniques must be used to balance clock delays throughout the device. By comparison, FPGAs feature device-wide, low-skew clock routing

1858: Atlantic. First transatlantic telegraph cable is laid (and later failed).

1858:
Queen Victoria
exchanges transatlantic
telegraph messages
with President Buchanan
in America.

resources. This makes clock balancing unnecessary by the design engineer because the FPGA vendor has already taken care of it.

Clock gating versus clock enabling

ASIC designs often use the technique of gated clocks to help reduce power dissipation, as shown in Figure 7-4a. However, these tend to give the design asynchronous characteristics and make it sensitive to glitches caused by inputs switching too closely together on the gating logic.

By comparison, FPGA designers tend to use the technique of enabling clocks. Originally this was performed by means of an external multiplexer as illustrated in Figure 7-4b; today, however, almost all FPGA architectures have a dedicated clock enable pin on the register itself, as shown in Figure 7-4c.

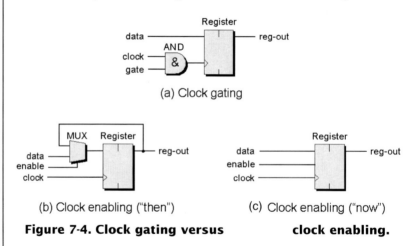

Figure 7-4. Clock gating versus clock enabling.

PLLs and clock conditioning circuitry

FPGAs typically include PLL or DLL functions—one for each dedicated global clock (see also the discussions in Chapter 4). If these resources are used for on-chip clock generation, then the design should also include some mechanism for disabling or bypassing them so as to facilitate chip testing and debugging.

Reliable data transfer across multiclock domains

In reality, this topic is true for both ASIC and FPGA designs, the point being that the exchange of data between two independent clock domains must be performed very carefully so as to avoid losing or corrupting data. Bad synchronization may lead to metastability issues and tricky timing analysis problems. In order to achieve reliable transfers across domains, it is recommended to employ handshaking, double flopping, or asynchronous FIFO techniques.

Register and latch considerations

Latches

ASIC engineers often make use of latches in their designs. As a general rule-of-thumb, if you are designing an FPGA, and you are tempted to use a latch, don't!

Flip-flops with both "set" and "reset" inputs

Many ASIC libraries offer a wide range of flip-flops, including a selection that offer both *set* and *reset* inputs (both synchronous and asynchronous versions are usually available).

By comparison, FPGA flip-flops can usually be configured with either a *set* input or a *reset* input. In this case, implementing both *set* and *reset* inputs requires the use of a LUT, so FPGA design engineers often try to work around this and come up with an alternative implementation.

Global resets and initial conditions

Every register in an FPGA is programmed with a default initial condition (that is, to contain a logic 0 or a logic 1). Furthermore, the FPGA typically has a global reset signal that will return all of the registers (but not the embedded RAMs) to their initial conditions. ASIC designers typically don't implement anything equivalent to this capability.

1859: Germany. Hittorf and Pucker invent the cathode ray tube (CRT).

Resource sharing (time-division multiplexing)

TDM is pronounced by spelling it out as "T-D-M."

In the context of communications, TDM refers to a method of taking multiple data streams and combining them into a single signal by dividing the streams into many segments (each having a very short duration) and multiplexing between them.

By comparison, in the context of resource sharing, TDM refers to sharing a resource like a multiplier by multiplexing its inputs and letting different data paths use the resource at different times.

Resource sharing is an optimization technique that uses a single functional block (such as an adder or comparator) to implement several operations. For example, a multiplier may first be used to process two values called A and B, and then the same multiplier may be used to process two other values called C and D. (A good example of resource sharing is provided in Chapter 12.)

Another name for resource sharing is *time-division multiplexing (TDM)*. Resources on an FPGA are more limited than on an ASIC. For this reason, FPGA designers tend to spend more effort on resource sharing than do their ASIC counterparts.

Use it or lose it!

Actually, things are a little subtler than the brief note above might suggest because there is a fundamental use-it-or-lose-it consideration with regard to FPGA hardware. This means that FPGAs only come in certain sizes, so if you can't drop down to the next lower size, then you might as well use everything that's available on the part you have.

For example, let's assume that you have a design that requires two embedded hard processor cores. In addition to these processors, you might decide that by means of resource sharing, you could squeeze by with say 10 multipliers and 2 megabytes of RAM. But if the only FPGA containing two processors also comes equipped with 50 multipliers and 10 megabytes of RAM, you can't get a refund, so you might as well make full use of the extra capabilities.

But wait, there's more

In the case of FPGAs, getting data from LUTs/CLBs to and from special components like multipliers and MACs is usually more expensive (in terms of connectivity) than connecting with other LUTs/CLBs. Since resource sharing increases the amount of connectivity, you need to keep a watchful eye on this situation.

In addition to the big components like multipliers and MACs, you can also share things like adders. Interestingly enough, in the carry-chain technologies (such as those fielded by Altera and Xilinx), as a first-order approximation, the cost of building an adder is pretty much equivalent to the cost of building a data bus's worth of sharing logic. For example, implementing two adders "as is" with completely independent inputs and outputs will cost you two adders and no resource-sharing multiplexers. But if you share, you will have one adder and two multiplexers (one for each set of inputs). In FPGA terms, this will be more expensive rather than less (in ASICs, the cost of a multiplexer is far less than the cost of an adder, so you would have a different trade-off point).

In the real world, the interactions between "using it or losing it" and connectivity costs are different for each technology and each situation; that is, Altea parts are different from Xilinx parts and so on.

State machine encoding

The encoding scheme used for state machines is a good example of an area where what's good for an ASIC design might not be well suited for an FPGA implementation.

As we know, every LUT in an FPGA has a companion flip-flop. This usually means that there are a reasonable number of flip-flops sitting around waiting for something to do. In turn, this means that in many cases, a "one-hot" encoding scheme will be the best option for an FPGA-based state machine, especially if the activities in the various states are inherently independent.

The "one-hot" encoding scheme refers to the fact that each state in a state machine has its own state variable in the form of a flip-flop, and only one state variable may be active ("hot") at any particular time.

Test methodologies

ASIC designers typically spend a lot of time working with tools that perform SCAN chain insertion and *automatic test pattern generation (ATPG)*. They may also include logic in their designs to perform *built-in self-test (BIST)*. A large proportion of these efforts are intended to test the device for manufacturing defects. By comparison, FPGA designers typically don't worry about this form of device testing because FPGAs are preverified by the vendor.

JTAG is pronounced "J-TAG"; that is, by spelling out the 'J' followed by "tag" to rhyme with "nag."

Similarly, ASIC engineers typically expend a lot of effort inserting and boundary scan (JTAG) into their designs and verifying them. By comparison, FPGAs already contain boundary scan capabilities in their fabric.

Schematic-Based Design Flows

In the days of yore

In order to set the stage, let's begin by considering the way in which digital ICs were designed in the days of old—circa the early 1960s. This information will interest nontechnical readers, as well as newbie engineers who are familiar with current design tools and flows, but who may not know how they evolved over time. Furthermore, these discussions establishe an underlying framework that will facilitate understanding the more advanced design flows introduced in subsequent chapters.

In those days, electronic circuits were crafted by hand. Circuit diagrams—also known as *schematic diagrams* or just *schematics*—were hand-drawn using pen, paper, and stencils (or the occasional tablecloth should someone happen to have a brilliant idea while in a restaurant). These diagrams showed the symbols for the logic gates and functions that were to be used to implement the design, along with the connections between them.

Each design team usually had at least one member who was really good at performing logic *minimization* and *optimization*, which ultimately boils down to replacing one group of gates with another that will perform the same task faster or using less real estate on the silicon.

Checking that the design would work as planned insofar as its logical implementation—*functional verification*—was typically performed by a group of engineers sitting around a table working their way through the schematics saying, "Well, that looks OK." Similarly, *timing verification*—checking that the

The wires connecting the logic gates on an integrated circuit may be referred to as *wires*, *tracks*, or *interconnect*, and all of these terms may be used interchangeably.

In certain cases, the term *metallization* may also be used to refer to these tracks because they are predominantly formed by means of the IC's metal (metallization) layers.

1865: England.
James Clerk Maxwell
predicts the existence of
electromagnetic waves
that travel in the same
way as light.

design met its required input-to-output and internal path
delays and that no violation times (such as *setup* and *hold*
parameters) associated with any of the internal registers were
violated—was performed using a pencil and paper (if you were
really lucky, you might also have access to a mechanical or
electromechanical calculator).

Finally, a set of drawings representing the structures used
to form the logic gates (or, more accurately, the transistors
forming the logic gates) and the interconnections between
them were drawn by hand. These drawings, which were
formed from groups of simple polygons such as squares and
rectangles, were subsequently used to create the photo-masks,
which were themselves used to create the actual silicon chip.

The early days of EDA

Front-end tools like logic simulation

Not surprisingly, the handcrafted way of designing discussed above was time-consuming and prone to error.
Something had to be done, and a number of companies and
universities leapt into the fray in a variety of different directions. In the case of functional verification, for example, the
late 1960s and early 1970s saw the advent of special programs
in the form of rudimentary *logic simulators*.

In order to understand how these work, let's assume that
we have a really simple *gate-level* design whose schematic diagram has been hand-drawn on paper (Figure 8-1).

By "gate-level" we mean that the design is represented as a
collection of primitive logic gates and functions and the connections between them. In order to use the logic simulator,
the engineers first need to create a textual representation of
the circuit called a *gate-level netlist*. In those far-off times, the
engineers would typically have been using a mainframe computer, and the netlist would have been captured as a set of
punched cards called a *deck* ("deck of cards" … get it?). As
computers (along with storage devices like hard disks) became

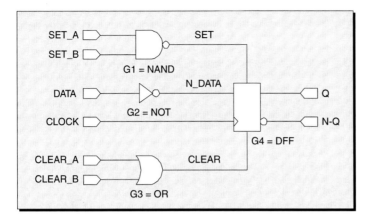

Figure 8-1. A simple schematic diagram (on paper).

```
BEGIN CIRCUIT=TEST
   INPUT  SET_A, SET-B, DATA, CLOCK, CLEAR_A, CLEAR_B;
   OUTPUT Q, N_Q;
   WIRE   SET, N_DATA, CLEAR;

   GATE G1=NAND (IN1=SET_A, IN2=SET_B, OUT1=SET);
   GATE G2=NOT  (IN1=DATA, OUT1=N_DATA);
   GATE G3=OR   (IN1=CLEAR_A, IN2=CLEAR_B, OUT1=CLEAR);
   GATE G4=DFF  (IN1=SET, IN2=N_DATA, IN3=CLOCK,
                 IN4=CLEAR, OUT1=Q, OUT2=N_Q);

END CIRCUIT=TEST;
```

Figure 8-2. A simple gate-level netlist (text file).

1865:
Atlantic cable links
Valencia (Ireland) and
Trinity Bay
(Newfoundland).

more accessible, netlists began to be stored as text files
(Figure 8-2).

It was also possible to associate delays with each logic gate.
These delays—which are omitted here in order to keep things
simple—were typically referenced as integer multiples of some
core simulation time unit (see also Chapter 19).

Note that the format shown in Figure 8-2 was made up
purely for the purposes of this example. This was in keeping
with the times because—just to keep everyone on their
toes—anyone who created a tool like a logic simulator also
tended to invent his or her own proprietary netlist language.

All of the early logic simulators had internal representations of primitive gates like AND, NAND, OR, NOR, etc. These were referred to as *simulation primitives*. Some simulators also had internal representations of more sophisticated functions like D-type flip-flops. In this case, the G4=DFF function in Figure 8-2 would map directly onto this internal representation.

Alternatively, one could create a subcircuit called DFF, whose functionality was captured as a netlist of primitive AND, NAND, etc. gates. In this case, the G4=DFF function in Figure 8-2 would actually be seen by the simulator as a call to instantiate a copy of this subcircuit.

Next, the user would create a set of *test vectors*—also known as *stimulus*—which were patterns of logic 0 and logic 1 values to be applied to the circuit's inputs. Once again, these test vectors were textual in nature, and they were typically presented in a tabular form looking something like that shown in Figure 8-3 (anything after a ";" character is considered to be a comment).

Instead of calling our test vectors "stimulus," we really should have said "stimuli," but we were engineers, not English majors!

```
                     C C
                     L L
          S S   C E E
          E E D L A A
          T T A O R R
          _ _ T C _ _
    TIME  A B A K A B
    ----- -----------
        0 1 1 1 0 0 0    ; Set up initial values
      500 1 1 1 1 0 0    ; Rising edge on clock (load 0)
     1000 1 1 1 0 0 0    ; Falling edge on clock
     1500 1 1 0 0 0 0    ; Set data to 0 (N_data = 1)
     2000 1 1 0 1 0 0    ; Rising edge on clock (load 1)
     2500 1 1 0 1 0 1    ; Clear_B goes active (load 0)
        :
     etc.
```

Figure 8-3. A simple set of test vectors (text file).

The times at which the stimulus values were to be applied were shown in the left-hand column. The names of the input signals are presented vertically to save space.

As we know from Figures 8-1 and 8-2, there is an inverting (NOT) gate between the DATA input and the D-type flip-flop. Thus, when the DATA input is presented with 1 at time zero, this value will be inverted to a 0, which is the value that will be loaded into the register when the clock undergoes a rising (0-to-1) edge at time 500. Similarly, when the DATA input is presented with 0 at time 1,500, this value will be inverted to a 1, which is the value that will be loaded into the register when the clock undergoes its next rising (0-to-1) transition at time 2,000.

In today's terminology, the file of test vectors shown in Figure 8-3 would be considered a rudimentary *testbench*. Once again, time values were typically specified as integer multiples of some core simulation time unit.

The engineer would then invoke the logic simulator, which would read in the gate-level netlist and construct a virtual representation of the circuit in the computer's memory. The simulator would then read in the first test vector (the first line from the stimulus file), apply those values to the appropriate virtual inputs, and propagate their effects through the circuit. This would be repeated for each of the subsequent test vectors forming the testbench (Figure 8-4).

The simulator would also use one or more control files (or online commands) to tell it which internal nodes (wires) and output pins to monitor, how long to simulate for, and so forth. The results, along with the original stimulus, would be stored in tabular form in a textual output file.

Let's assume that we've just travelled back in time and run one of the old simulators using the circuit represented in Figures 8-1 and 8-2 along with the stimulus shown in Figure 8-3. We will also assume that the NOT gate has a delay of five simulator time units associated with it, which means that a change on that gate's input will take five time units to propagate through the gate and appear on its output. Similarly,

1866: Ireland/USA. First permanent transatlantic telegraph cable is laid.

1869:
William Stanley Jevons
invents the Logic Piano.

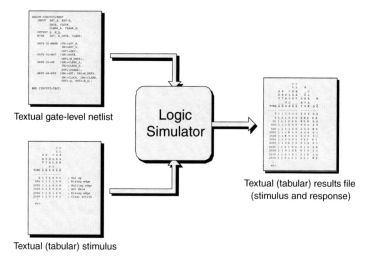

Textual gate-level netlist

Textual (tabular) stimulus

Textual (tabular) results file
(stimulus and response)

Figure 8-4. Running the logic simulator.

we'll assume that both the NAND and OR gates have associated delays of 10 time units, while the D-type flip-flop has associated delays of 20 time units.

In this case, if the simulator were instructed to monitor all of the internal nodes and output pins, the output file containing the simulation results would look something like that shown in figure 8-5.

For the purposes of our discussions, any changes to a signal's value are shown in bold font in this illustration, but this was not the case in the real world.

In this example, the initial values are applied to the input pins at time 0. At this time, all of the internal nodes and output pins show X values, which indicates unknown states. After five time units, the initial logic 1 that was applied to the DATA input propagates through the inverting NOT gate and appears as a logic 0 on the internal N_DATA node. Similarly, at time 10, the initial values that were applied to the SET_A and SET_B inputs propagate through the NAND gate to the internal SET node, while the values on the CLEAR_A and CLEAR_B inputs propagate through the OR gate to the internal CLEAR node.

```
          C C
          L L       N
   S S  C E E      _ C
   E E D L A A      D L
   T T A O R R  S A E    N
   _ _   T C _ _  E T A    _
TIME A B A K A B  T A R   Q Q
----- ----------- -----  ---
    0 1 1 1 0 0 0  X X X  X X  ; Set up initial values
    5 1 1 1 0 0 0  X 0 X  X X
   10 1 1 1 0 0 0  0 0 0  X X

  500 1 1 1 1 0 0  0 0 0  X X  ; Rising edge on clock
  520 1 1 1 1 0 0  0 0 0  0 1

 1000 1 1 1 0 0 0  0 0 0  0 1  ; Falling edge on clock

 1500 1 1 0 0 0 0  0 0 0  0 1  ; Set data to 0
 1505 1 1 0 0 0 0  0 1 0  0 1

 2000 1 1 0 1 0 0  0 1 0  0 1  ; Rising edge on clock
 2020 1 1 0 1 0 0  0 1 0  1 0

 2500 1 1 0 1 0 1  0 1 0  1 0  ; Clear_B goes active
 2510 1 1 0 1 0 1  0 1 1  1 0
 2530 1 1 0 1 0 1  0 1 1  0 1
    :
  etc.
```

Figure 8-5. Output results (text file).

1872:
First simultaneous transmission from both ends of a telegraph wire.

At time 500, a rising (0-to-1) edge on the CLOCK input causes the D-type flip-flop to load the value from the N_DATA node. The result appears on the Q and N_Q output pins 20 time units later. And so it goes.

Blank lines in the output file, such as the one shown between time 10 and time 500, were used to separate related groups of actions. For example, setting the initial values at time 0 caused signal changes at times 5 and 10. Then the transition on the CLOCK input at time 500 caused signal changes at time 520. As these two groups of actions were totally independent of each other, they were separated by a blank line.

It wasn't long before engineers were working with circuits that could contain thousands of gates and internal nodes along with simulation runs that could encompass thousands of time steps. (Oh, the hours I spent poring over files like this (a) trying to see if a circuit was working as expected, and (b) desperately attempting to track down the problem if it wasn't!)

Back-end tools like layout

As opposed to tools like logic simulators that were intended to aid the engineers who were defining the function of ICs (and circuit boards), some companies focused on creating tools that would help in the process of laying the ICs out. In this context, "layout" refers to determining where to place the logic gates (actually, the transistors forming the logic gates) on the surface of the chip and how to route the wires between them.

In the early 1970s, companies like Calma, ComputerVision, and Applicon created special computer programs that helped personnel in the drafting department capture digital representations of hand-drawn designs. In this case, a design was placed on a large-scale digitizing table, and then a mouse-like tool was used to digitize the boundaries of the shapes (polygons) used to define the transistors and the interconnect. These digital files were subsequently used to create the photo-masks, which were themselves used to create the actual silicon chip.

Over time, these early computer-aided drafting tools evolved into interactive programs called *polygon editors* that allowed users to draw the polygons directly onto the computer screen. Descendants of these tools eventually gained the ability to accept the same netlist used to drive the logic simulator and to perform the layout (place-and-route) tasks automatically.

The drafting department is referred to as the "drawing office" in the UK.

CAE + CAD = EDA

Tools like logic simulators that were used in the front-end (logical design capture and functional verification) portion of the design flow were originally gathered together under the umbrella name of *computer-aided engineering* (CAE). By comparison, tools like layout (place-and-route) that were used in

CAE is pronounced by spelling it out as "C-A-E."

CAD is pronounced to rhyme with "bad."

the back-end (physical) portion of the design flow were originally gathered together under the name of *computer-aided
design (CAD)*.

For historical reasons that are largely based on the origins
of the terms CAE and CAD, the term *design engineer*—or simply *engineer*—typically refers to someone who works in the
front-end of the design flow; that is, someone who performs
tasks like conceiving and describing (capturing) the functionality of an IC (what it does and how it does it). By comparison,
the term *layout designer*—or simply *designer*—commonly refers
to someone who is ensconced in the back-end of the design
flow; that is, someone who performs tasks such as laying out an
IC (determining the locations of the gates and the routes of
the tracks connecting them together).

Sometime during the 1980s, all of the CAE and CAD tools
used to design electronic components and systems were gathered under the name of *electronic design automation*, or EDA,
and everyone was happy (apart from the ones who weren't,
but no one listened to their moaning and groaning, so that
was alright).

A simple (early) schematic-driven ASIC flow

Toward the end of the 1970s and the beginning of the
1980s, companies like Daisy, Mentor, and Valid started providing graphical *schematic capture* programs that allowed
engineers to create circuit (schematic) diagrams interactively.
Using the mouse, an engineer could select symbols representing such entities as I/O pins and logic gates and functions from
a special *symbol library* and place them on the screen. The
engineer could then use the mouse to draw lines (wires) on the
screen connecting the symbols together.

Once the circuit had been entered, the schematic capture
package could be instructed to generate a corresponding gate-
level netlist. This netlist could first be used to drive a logic
simulator in order to verify the functionality of the design. The
same netlist could then be used to drive the place-and-route
software (Figure 8-6).

The term CAD is also
used to refer to
computer-aided design
tools used in a variety
of other engineering
disciplines, such as
mechanical and
architectural design.

EDA is pronounced
by spelling it out as
"E-D-A."

1873: England
James Clerk Maxwell
describes the
electromagnetic nature
of light and publishes
his theory of radio
waves.

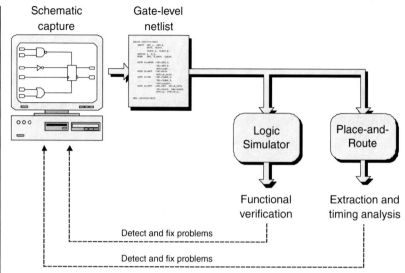

Figure 8-6. Simple (early) schematic-driven ASIC flow.

Any timing information that was initially used by the logic simulator would be estimated—particularly in the case of the tracks—and accurate timing analysis was only possible once all of the logic gates had been placed and the tracks connecting them had been routed. Thus, following place-and-route, an *extraction* program would be used to calculate the parasitic resistance and capacitance values associated with the structures (track segments, vias, transistors, etc.) forming the circuit. A timing analysis program would then use these values to generate a timing report for the device. In some flows, this timing information was also fed back to the logic simulator in order to perform a more accurate simulation.

The important thing to note here is that, when creating the original schematic, the user would access the symbols for the logic gates and functions from a special library that was associated with the targeted ASIC technology.[1] Similarly, the

___
[1] There are always different ways to do things. For example, some flows were based on the concept of using a generic symbol library containing a subset of logic functions common to all ASIC cell libraries. The netlist

simulator would be instructed to use a corresponding library of simulation models with the appropriate logical functionality[2] and timing for the targeted ASIC technology. The end result was that the gate-level netlist presented to the place-and-route software directly mapped onto the logic gates and functions being physically implemented on the silicon chip (this is a tad different from the FPGA flow, as is discussed in the following topic).

A simple (early) schematic-driven FPGA flow

When the first FPGAs arrived on the scene in 1984, it was natural that their design flows would be based on existing schematic-driven ASIC flows. Indeed, the early portions of the flows were very similar in that, once again, a schematic capture package was used to represent the circuit as a collection of primitive logic gates and functions and to generate a corresponding gate-level netlist. As before, this netlist was subsequently used to drive the logic simulator in order to perform the functional verification.

The differences began with the implementation portion of the flow because the FPGA fabric consisted of an array of *configurable logic blocks (CLBs)*, each of which was formed from a number of LUTs and registers. This required the introduction of some additional steps called *mapping* and *packing* into the flow (Figure 8-7).

---

generated from the schematic capture application could then be run through a translator that converted the generic cell names to their equivalents in the targeted ASIC library.

[2] With regard to functionality, one might expect a primitive logical entity like a 2-input AND gate to function identically across multiple libraries. This is certainly the case when "good" (logic 0 and 1) values are applied to the inputs, but things may vary when high-impedance 'Z' values or unknown 'X' values are applied to the inputs. And even with good 0 and 1 values applied to their inputs, more complex functions like D-type latches and flip-flops can behave very differently for "unusual" cases such as the set and clear inputs being driven active at the same time.

1874: America. Alexander Graham Bell conceives the idea of the telephone.

1875: America.
Edison invents the
Mimeograph.

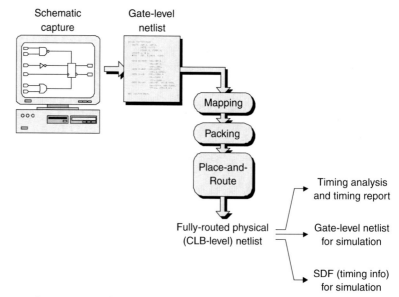

Figure 8-7. Simple (early) schematic-driven FPGA flow.

Mapping

In this context, *mapping* refers to the process of associating
entities such as the gate-level functions in the gate-level netlist with the LUT-level functions available on the FPGA. Of
course, this isn't a one-for-one mapping because each LUT
can be used to represent a number of logic gates (Figure 8-8).

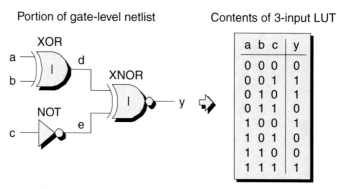

Figure 8-8. Mapping logic gates into LUTs.

Mapping (which is still performed today, but elsewhere in the flow, as will be discussed in later chapters) is a nontrivial problem because there are a large number of ways in which the logic gates forming a netlist can be partitioned into the smaller groups to be mapped into LUTs. As a simple example, the functionality of the NOT gate shown in Figure 8-8 might have been omitted from this LUT and instead incorporated into the upstream LUT driving wire c.

1875: England. James Clerk Maxwell states that atoms must have a structure.

Packing

Following the mapping phase, the next step was *packing,* in which the LUTs and registers were packed into the CLBs. Once again, packing (which is still performed today, but elsewhere in the flow, as will be discussed in later chapters) is a nontrivial problem because there are myriad potential combinations and permutations. For example, assume an incredibly simple design comprising only a couple of handfuls of logic gates that end up being mapped onto four 3-input LUTs that we'll call A, B, C, and D. Now assume that we're dealing with an FPGA whose CLBs can each contain two 3-input LUTs. In this case we'll need two CLBs (called 1 and 2) to contain our four LUTs. As a first pass, there are 4! (factorial four = 4 3 2 1 = 24) different ways in which our LUTs can be packed into the two CLBs (Figure 8-9).

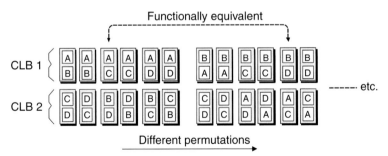

Figure 8-9. Packing LUTs into CLBs.

Only 12 of the 24 possible permutations are shown here (the remainder are left as an exercise for the reader). Further-

Prior to the advent of FPGAs, the equivalent functionality to place-and-route in "CPLD land" was performed by an application known as a "fitter."

When FPGAs first arrived on the scene, people used the same "fitter" appellation, but over time they migrated to using the term "place-and-route" because this more accurately reflected what was actually occurring.

As opposed to using a symbol library of primitive logic gates and registers, an interesting alternative circa the early 1990s was to use a symbol library corresponding to slightly more complex logical functions (say around 70 functions). The output from the schematic was a netlist of functional blocks that were already de facto mapped onto LUTs and packed into CLBs.

This had the advantage of giving a better idea of the number of levels of logic between register stages, but it limited such activities as optimization and swapping.

more, in reality there are actually only 12 permutations of significance because each has a "mirror image" that is functionally its equivalent, such as the AC-BD and BD-AC pairs shown in Figure 8-9. The reason for this is that when we come to place-and-route, the relative locations of the two CLBs can be exchanged.

Place-and-route

Following packing, we move to *place-and-route*. With regard to the previous point, let's assume that our two CLBs need to be connected together, but that—purely for the purposes of this portion of our discussions—they can only be placed horizontally or vertically adjacent to each other, in which case there are four possibilities (Figure 8-10).

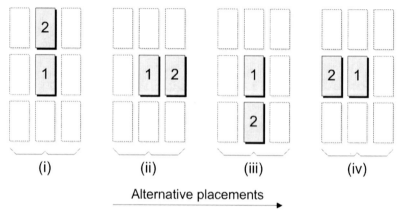

Alternative placements

Figure 8-10. Placing the CLBs.

In the case of placement (i) for example, if CLB 1 contained LUTs A-C and CLB 2 contained LUTs B-D, then this would be identical to swapping the positions of the two CLBs and exchanging their contents.

If we only had the two CLBs shown in figure 8-10, it would be easy to determine their optimal placement with respect to each other (which would have to be one of the four options shown above) and the absolute placement of this two-CLB group with respect to the entire chip.

The placement problem is much more complex in the real world because a real design can contain extremely large numbers of CLBs (hundreds *or* thousands in the early days, and hundreds of thousands by 2004). In addition to CLBs 1 and 2 being connected together, they will almost certainly need to be connected to other CLBs. For example, CLB 1 may also need to be connected to CLBs 3, 5 and 8, while CLB 2 may need to be connected to CLBs 4, 6, 7, and 8. And each of these new CLBs may need to be connected to each other or to yet more CLBs. Thus, although placing CLBs 1 and 2 next to each other would be best for them, it might be detrimental to their relationships with the other CLBs, and the most optimal solution overall might actually be to separate CLBs 1 and 2 by some amount.

Although placement is difficult, deciding on the optimal way to route the signals between the various CLBs poses an even more Byzantine problem. The complexity of these tasks is mind-boggling, so we'll leave it to those guys and gals who write the place-and-route algorithms (they are the ones sporting size-16 extra-wide brains with go-faster stripes) and quickly move onto other things.

Timing analysis and post-place-and-route simulation

Following place-and-route, we have a fully routed physical (CLB-level) netlist, as was illustrated in Figure 8-7. At this point, a *static timing analysis (STA)* utility will be run to calculate all of the input-to-output and internal path delays and also to check for any timing violations (setup, hold, etc.) associated with any of the internal registers.

STA is pronounced by spelling it out as "S-T-A" (see also Chapter 19).

One interesting point occurs if the design engineers wish to resimulate their design with accurate (post-place-and-route) timing information. In this case, they have to use the FPGA tool suite to generate a new gate-level netlist along with associated timing information in the form of an industry-standard file format called—perhaps not surprisingly—*standard delay format (SDF)*. The main reason for generating this new gate-

SDF is pronounced by spelling it out as "S-D-F" (see also Chapter 10).

1876: America. 10th March. Intelligible human speech heard over Alexander Graham Bell's telephone for the first time.

level netlist is that—once the original netlist has been coerced into its CLB-level equivalent—it simply isn't possible to relate the timings associated with this new representation back into the original gate-level incarnation.

Flat versus hierarchical schematics

Clunky flat schematics

The very first schematic packages essentially allowed a design to be captured as a humongous, flat circuit diagram split into a number of "pages." In order to visualize this, let's assume that you wish to draw a circuit diagram comprising 1,000 logic gates on a piece of paper. If you created a single large diagram, you would end up with a huge sheet of paper (say eight-feet square) with primary inputs to the circuit on the left, primary outputs from the circuit on the right, and the body of the circuit in the middle.

Carrying this circuit diagram around and showing it to your friends would obviously be a pain. Instead, you might want to cut it up into a number of pages and store them all together in a folder. In this case, you would make sure that your partitioning was logical such that each page contained all of the gates relating to a particular function in the design. Also, you would use interpage connectors (sort of like pseudo inputs and outputs) to link signals between the various pages.

This is the way the original schematic capture packages worked. You created a single flat schematic as a series of pages linked together by interpage connector symbols, where the names you gave these symbols told the system which ones were to be connected together. For example, consider a simple circuit sketched on a piece of paper (Figure 8-11).

Assume that the gates on the left represent some control logic, while the four registers on the right are implementing a 4-bit shift register. Obviously, this is a trivial example, and a real circuit would have many more logic gates. We're just trying to tie down some underlying concepts here, such as the fact that when you entered this circuit into your schematic

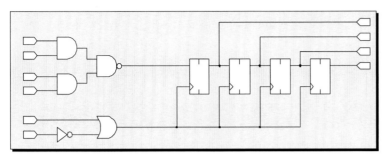

Figure 8-11. Simple schematic drawn on a piece of paper.

1876: America. Alexander Graham Bell patents the telephone.

capture system, you might split it into two pages (Figure 8-12).

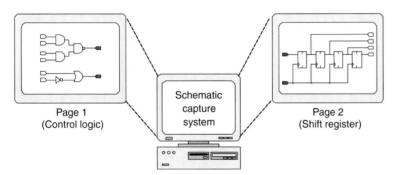

Figure 8-12. Simple two-page flat schematic.

Sleek hierarchical (block-based) schematics

There were a number of problems associated with the flat schematics discussed above, especially when dealing with real-world circuits requiring 50 or more pages:

- It was difficult to visualize a high-level, top-down view of the design.
- It was difficult to save and reuse portions of the design in future projects.
- In the case of designs in which some portion of the circuit was repeated multiple times (which is very common), that portion would have to be redrawn or copied onto multiple pages. This became really

1877: America.
First commercial
telephone service goes
into operation.

painful if you subsequently realized that you had to
make a change because you would have to make the
same change to all of the copies.

The answer was to enhance schematic capture packages to
support the concept of hierarchy. In the case of our shift register circuit, for example, you might start with a top-level page
in which you would create two blocks called control and shift,
each with the requisite number of input and output pins. You
would then connect these blocks to each other and also to
some primary inputs and outputs.

Next, you would instruct the system to "push down" into
the control block, which would open up a new schematic
page. If you were lucky, the system would automatically pre-
populate this page with input and output connector symbols
(and with associated names) corresponding to the pins on its
parent block. You would then create the schematic corresponding to that block as usual (Figure 8-13).

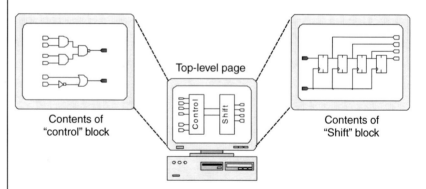

Figure 8-13. Simple hierarchical schematic.

In fact, each block could contain a further block-level
schematic, or a gate-level schematic, or (very commonly) a
mixture of both. These hierarchical block-based schematics
answered the problems associated with flat schematics:

- They made it easier to visualize a high-level, top-down view of the design and to work one's way through the design.
- They made it easier to save and reuse portions of the design in future projects.
- In the case of designs in which some portion of the circuit was repeated multiple times, it was only necessary to create that portion—as a discrete block—once and then to instantiate (call) that block multiple times. This made things easy if you subsequently realized that you had to make a change because you would only have to modify the contents of the initial block.

Schematic-driven FPGA design flows today

All of the original schematic, mapping, packing, and place-and-route applications were typically created and owned by the FPGA companies themselves. However, the general feeling is that a company can either be good at creating EDA tools or it can be good at creating silicon chips, but not both.

Another facet of the problem is that design tools were originally extremely expensive in the ASIC world (even tools like schematic capture, which today are commonly regarded as commodity products). By comparison, the FPGA vendors were focused on selling chips, so right from the get-go they offered their tools at a very low cost (in fact, if you were a big enough customer, they'd give you the entire design tool suite for free). While this had its obvious attractions to the end user, the downside was that the FPGA vendors weren't too keen spending vast amounts of money enhancing tools for which they received little recompense.

Over time, therefore, external EDA vendors started to supply portions of the puzzle, starting with schematic capture and then moving into mapping and packing (via logic synthesis as discussed in Chapters 9 and 19). Having said this, the FPGA vendors still typically provide internally developed, less sophisticated (compared to the state-of-the-art) versions of tools like

1877: America. Thomas Watson devises a "thumper" to alert users of incoming telephone calls.

1878: America.
First public long-
distance telephone lines
between Boston and
Providence become
operational.

schematic capture as part of their basic tool suite, and they also maintain a Vulcan Death Grip on their crown jewels (the place-and-route software).

For many engineers today, driving a design using schematic capture at the gate-level of abstraction is but a distant memory. In some cases, FPGA vendors offer little support for this type of flow for their latest devices to the extent that they only provide schematic libraries for older component generations. However, schematic capture does still find a role with some older engineers and also with folks who need to make minor functional changes to legacy designs. Furthermore, graphical entry mechanisms that are descended from early schematic capture packages still find a place in modern design flows, as is discussed in the next chapter.

HDL-Based Design Flows

Schematic-based flows grind to a halt

Toward the end of the 1980s, as designs grew in size and complexity, schematic-based ASIC flows began to run out of steam. Visualizing, capturing, debugging, understanding, and maintaining a design at the gate level of abstraction became increasingly difficult and inefficient when juggling 5,000 or more gates and reams of schematic pages.

In addition to the fact that capturing a large design at the gate level of abstraction is prone to error, it is also extremely time-consuming. Thus, some EDA vendors started to develop design tools and flows based on the use of *hardware description languages*, or HDLs.

EDA is pronounced by spelling it out as "E-D-A."

HDL is pronounced by spelling it out as "H-D-L."

The advent of HDL-based flows

The idea behind a hardware description language is, perhaps not surprisingly, that you can use it to describe hardware. In a wider context, the term *hardware* is used to refer to any of the physical portions of an electronics system, including the ICs, printed circuit boards, cabinets, cables, and even the nuts and bolts holding the system together. In the context of an HDL, however, "hardware" refers only to the electronic portions (components and wires) of ICs and printed circuit boards. (The HDL may also be used to provide limited representations of the cables and connectors linking circuit boards together.)

In the early days of electronics, almost anyone who created an EDA tool created his or her own HDL to go with it. Some of these were analog HDLs in that they were intended to represent circuits in the analog domain, while others were focused on representing digital functionality. For the purposes of this book, we are interested in HDLs only in the context of designing digital ICs in the form of ASICs and FPGAs.

Different levels of abstraction

Some of the more popular digital HDLs are introduced later in this chapter. For the nonce, however, let's focus more on how a generic digital HDL is used as part of a design flow. The first thing to note is that the functionality of a digital circuit can be represented at different levels of abstraction and that different HDLs support these levels of abstraction to a greater or lesser extent (figure 9-1).

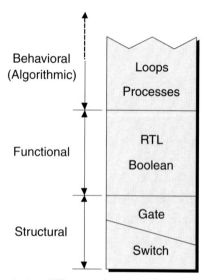

Figure 9-1. Different levels of abstraction.

Largely self-taught, George Boole made significant contributions in several areas of mathematics, but was immortalized for two works published in 1847 and 1854 in which he represented logical expressions in a mathematical form now known as Boolean algebra.

In 1938, Claude Shannon published an article based on his master's thesis at MIT, in which he showed how Boole's concepts could be used to represent the functions of switches in electronic circuits.

The lowest level of abstraction for a digital HDL would be the *switch level*, which refers to the ability to describe the circuit as a netlist of transistor switches. A slightly higher level of abstraction would be the *gate level*, which refers to the ability to describe the circuit as a netlist of primitive logic gates and functions. Thus, the early gate-level netlist formats generated by schematic capture packages as discussed in the previous chapter were in fact rudimentary HDLs.

Both switch-level and gate-level netlists may be classed as *structural* representations. It should be noted, however, that "structural" can have different connotations because it may also be used to refer to a hierarchical block-level netlist in which each block may have its contents specified using any of the levels of abstraction shown in Figure 9-1.

The next level of HDL sophistication is the ability to support *functional* representations, which covers a range of constructs. At the lower end is the capability to describe a function using Boolean equations. For example, assuming that we had already declared a set of signals called Y, SELECT, DATA-A, and DATA-B, we could capture the functionality of a simple 2:1 multiplexer using the following Boolean equation:

```
Y = (SELECT & DATA-A) | (!SELECT & DATA-B);
```

Note that this is a generic syntax that does not favor any particular HDL and is used only for the purposes of this example. (As we discussed in chapter 3, the "&" character represents a logical AND, the "|" character represents an OR, and the "!" character represents a NOT.)

The functional level of abstraction also encompasses *register transfer level (RTL)* representations. The term *RTL* covers a multitude of manifestations, but the easiest way to wrap one's brain around the underlying concept is to consider a design formed from a collection of registers linked by combinational logic. These registers are often controlled by a common clock signal, so assuming that we have already declared two signals called CLOCK and CONTROL, along with a set of registers called REGA, REGB, REGC, and REGD, then an RTL-type statement might look something like the following:

RTL is pronounced by spelling it out as "R-T-L."

```
when CLOCK rises
  if CONTROL == "1"
    then REGA = REGB & REGC;
    else REGA = REGB | REGD;
  end if;
end when;
```

In this case, symbols like *when, rises, if, then, else,* and the like are keywords whose semantics are defined by the owners of the HDL. Once again, this is a generic syntax that does not favor any particular HDL and is used only for the purposes of this example.

The highest level of abstraction sported by traditional HDLs is known as *behavioral*, which refers to the ability to describe the behavior of a circuit using abstract constructs like loops and processes. This also encompasses using algorithmic elements like adders and multipliers in equations; for example:

```
Y = (DATA-A + DATA-B) * DATA-C;
```

We should note that there is also a *system* level of abstraction (not shown in figure 9-1) that features constructs intended for system-level design applications, but we'll worry about this level a little later.

FSM is pronounced by spelling it out as "F-S-M."

Many of the early digital HDLs supported only structural representations in the form of switch or gate-level netlists. Others such as ABEL, CUPL, and PALASM were used to capture the required functionality for PLD devices. These languages (which were introduced in chapter 3) supported different levels of functional abstraction, such as Boolean equations, text-based truth tables, and text-based *finite state machine (FSM)* descriptions.

The next generation of HDLs, which were predominantly targeted toward logic simulation, supported more sophisticated levels of abstraction such as RTL and some behavioral constructs. It was these HDLs that formed the core of the first true HDL-based design flows as discussed below.

A *simple (early) HDL-based ASIC flow*

The key feature of HDL-based ASIC design flows is their use of *logic synthesis* technology, which began to appear on the market around the mid-1980s. These tools could accept an RTL representation of a design along with a set of timing constraints. In this case, the timing constraints were presented in a side-file containing statements along the lines of "the maximum delay from input X to output Y should be no greater than N nanoseconds" (the actual format would be a little drier and more boring).

The logic synthesis application automatically converted the RTL representation into a mixture of registers and Boolean equations, performed a variety of minimizations and optimizations (including optimizing for area and timing), and then generated a gate-level netlist that would (or at least, should) meet the original timing constraints (Figure 9-2).

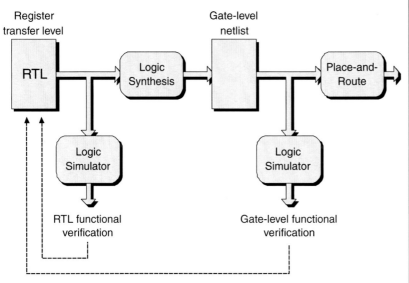

Figure 9-2. Simple HDL-based ASIC flow.

There were a number of advantages to this new type of flow. First of all, the productivity of the design engineers rose dramatically because it was a lot easier to specify, understand,

1878: England.
Sir Joseph Wilson Swan demonstrates a true incandescent light bulb.

1878: England.
William Crookes invents
his version of a cathode
ray tube called the
Crookes' Tube.

discuss, and debug the required functionality of the design at the RTL level of abstraction as opposed to working with reams of gate-level schematics. Also, logic simulators could run designs described in RTL much more quickly than their gate-level counterparts.

One slight glitch was that logic simulators could work with designs specified at high levels of abstraction that included behavioral constructs, but early synthesis tools could only accept functional representations up to the level of RTL. Thus, design engineers were obliged to work with a *synthesizable subset* of their HDL of choice.

Once the synthesis tool had generated a gate-level netlist, the flow became very similar to the schematic-based ASIC flows discussed in the previous chapter. The gate-level netlist could be simulated to ensure its functional validity, and it could also be used to perform timing analysis based on estimated values for tracks and other circuit elements. The netlist could then be used to drive the place-and-route software, following which a more accurate timing analysis could be performed using extracted resistance and linefeed capacitance values.

A simple (early) HDL-based FPGA flow

It took some time for HDL-based flows to flourish within the ASIC community. Meanwhile, design engineers were still coming to grips with the concept of FPGAs. Thus, it wasn't until the very early 1990s that HDL-based flows featuring logic synthesis technology became fully available in the FPGA world (Figure 9-3).

As before, once the synthesis tool had generated a gate-level netlist, the flow became very similar to the schematic-based FPGA flows discussed in the previous chapter. The gate-level netlist could be simulated to ensure its functional validity, and it could also be used to perform timing analysis based on estimated values for tracks and other circuit elements. The netlist could then be used to drive the FPGA's mapping, packing, and place-and-route software, following

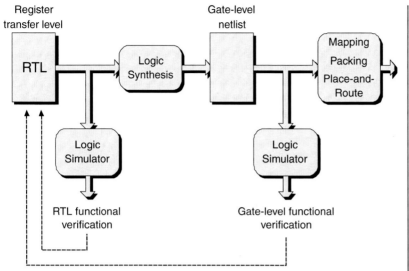

Figure 9-3. Simple HDL-based FPGA flow.

which a more accurate timing report could be generated using real-world (physical) values.

Architecturally aware FPGA flows

The main problem besetting the original HDL-based FPGA flows was that their logic synthesis technologies were derived from the ASIC world. Thus, these tools "thought" in terms of primitive logic gates and registers. In turn, this meant that they output gate-level netlists, and it was left to the FPGA vendor to perform the mapping, packing, and place-and-route functions.

Sometime around 1994, synthesis tools were equipped with knowledge about different FPGA architectures. This meant that they could perform mapping—and some level of packing—functions internally and output a LUT/CLB-level netlist. This netlist would subsequently be passed to the FPGA vendor's place-and-route software. The main advantage of this approach was that these synthesis tools had a better idea about timing estimations and area utilization, which allowed them to generate a better *quality of results (QoR)*. In real terms, FPGA designs generated by architecturally aware synthesis tools were

QoR is pronounced by spelling it out as "Q-o-R."

1878: Ireland.
Denis Redmond
demonstrates capturing
an image using
selenium photocells.

15 to 20 percent faster than their counterparts created using traditional (gate-level) synthesis offerings.

Logic versus physically aware synthesis

We're jumping a little bit ahead of ourselves here, but this is as good a place as any to briefly introduce this topic. The original logic synthesis tools were designed for use with the multimicron ASIC technologies of the mid-1980s. In these devices, the delays associated with the logic gates far outweighed the delays associated with the tracks connecting those gates together. In addition to being relatively small in terms of gate-count (by today's standards), these designs featured relatively low clock frequencies and correspondingly loose design constraints. The combination of all of these factors meant that early logic synthesis tools could employ relatively simple algorithms to estimate the track delays, but that these estimations would be close enough to the real (post-place-and-route) values that the device would work.

Over the years, ASIC designs increased in size (number of gates) and complexity. At the same time, the dimensions of the structures on the silicon chip were shrinking with two important results:

- Delay effects became more complex in general.
- The delays associated with tracks began to outweigh the delays associated with gates.

By the mid-1990s, ASIC designs were orders of magnitude larger—and their delay effects were significantly more sophisticated—than those for which the original logic synthesis tools had been designed. The result was that the estimated delays used by the logic synthesis tool had little relation to the final post-place-and-route delays. In turn, this meant that achieving *timing closure* (tweaking the design to make it achieve its original performance goals) became increasingly difficult and time-consuming.

For this reason, ASIC flows started to see the use of *physically aware synthesis* somewhere around 1996. The ways in which physically aware synthesis performs its magic are discussed in more detail in chapter 19. For the moment, we need only note that, during the course of performing its machinations, the physically aware synthesis engine makes initial placement decisions for the logic gates and functions. Based on these placements, the tool can generate more accurate timing estimations.

Ultimately, the physically aware synthesis tool outputs a placed (but not routed) gate-level netlist. The ASIC's physical implementation (place-and-route) tools use this initial placement information as a starting point from which to perform local (fine-grained) placement optimizations followed by detailed routing. The end result is that the estimated delays used by the physically aware synthesis application more closely correspond to the post-place-and-route delays. In turn, this means that achieving timing closure becomes a less taxing process.

"But what of FPGAs," you cry. Well, these devices were also increasing in size and complexity throughout the 1990s. By the end of the millennium, FPGA designers were running into significant problems with regard to timing closure. Thus, around 2000, EDA vendors started to provide FPGA-centric, physically aware synthesis offerings that could output a mapped, packed, and placed LUT/CLB-level netlist. In this case, the FPGA's physical implementation (place-and-route) tools use this initial placement information as a starting point from which to perform local (fine-grained) placement optimizations followed by detailed routing.

Graphical design entry lives on

When the first HDL-based flows appeared on the scene, many folks assumed that graphical design entry and visualization tools, such as schematic capture systems, were poised to exit the stage forever. Indeed, for some time, many design engineers prided themselves on using text editors like VI

In an expert's hands, the VI editor (pronounced by spelling it out as "V-I") was (and still is) an extremely powerful tool, but it can be very frustrating for new users.

1879: America
Thomas Alva Edison
invents an incandescent
light bulb (a year after
Sir Joseph Wilson Swan
in England).

(from Visual Interface) or EMACS as their only design entry mechanism.

But a picture tells a thousand words, as they say, and graphical entry techniques remain popular at a variety of levels. For example, it is extremely common to use a block-level schematic editor to capture the design as a collection of high-level blocks that are connected together. The system might then be used to automatically create a skeleton HDL framework with all of the block names and inputs and outputs declared. Alternatively, the user might create a skeleton framework in HDL, and the system might use this to create a block-level schematic automatically.

From the user's viewpoint, "pushing" down into one of these schematic blocks might automatically open an HDL editor. This could be a pure text-and-command–based editor like VI, or it might be a more sophisticated HDL-specific editor featuring the ability to show language keywords in different colors, automatically complete statements, and so forth.

Furthermore, when pushing down into a schematic block, modern design systems often give you a choice between entering and viewing the contents of that block as another, lower-level block-level schematic, raw HDL code, a graphical state diagram (used to represent an FSM), a graphical flowchart, and so forth. In the case of the graphical representations like state diagrams and flowcharts, these can subsequently be used to generate their RTL equivalents automatically (Figure 9-4).

Furthermore, it is common to have a tabular file containing information relating to the device's external inputs and outputs. In this case, both the top-level block diagram and the tabular file will (hopefully) be directly linked to the same data and will simply provide different views of that data. Making a change in any view will update the central data and be reflected immediately in all of the views.

Graphical State Diagram

Textual HDL

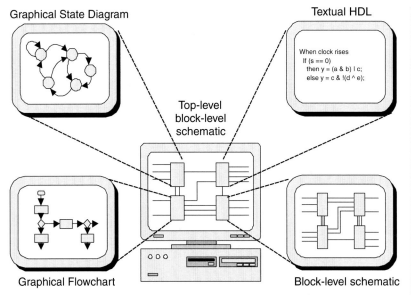

```
When clock rises
  If (s == 0)
    then y = (a & b) I c;
    else y = c & !(d ^ e);
```

Top-level
block-level
schematic

Graphical Flowchart

Block-level schematic

Figure 9-4. Mixed-level design capture environment.

A positive plethora of HDLs

Life would be so simple if there were only a single HDL to worry about, but no one said that living was going to be easy. As previously noted, in the early days of digital IC electronics design (circa the 1970s), anyone who created an HDL-based design tool typically felt moved to create his or her own language to accompany it. Not surprisingly, the result was a morass of confusion (you had to be there to fully appreciate the dreadfulness of the situation). What was needed was an industry-standard HDL that could be used by multiple EDA tools and vendors, but where was such a gem to be found?

Verilog HDL

Sometime around the mid-1980s, Phil Moorby (one of the original members of the team that created the famous HILO logic simulator) designed a new HDL called Verilog. In 1985, the company he was working for, Gateway Design Automation, introduced this language to the market along with an accompanying logic simulator called Verilog-XL.

PLI is pronounced by spelling it out as "P-L-I."

API is pronounced by spelling it out as "A-P-I."

One very cool concept that accompanied Verilog and Verilog-XL was the Verilog *programming language interface (PLI)*. The more generic name for this sort of thing is *application programming interface (API)*. An API is a library of software functions that allow external software programs to pass data into an application and access data from that application. Thus, the Verilog PLI is an API that allows users to extend the functionality of the Verilog language and simulator.

As one simple example, let's assume that an engineer is designing a circuit that makes use of an existing module to perform a mathematical function such as an FFT. A Verilog representation of this function might take a long time to simulate, which would be a pain if all the engineer really wanted to do was verify the new portion of the circuit. In this case, the engineer might create a model of this function in the C programming language, which would simulate, say, 1,000 times faster than its Verilog equivalent. This model would incorporate PLI constructs, allowing it to be linked into the simulation environment. The model could subsequently be accessed from the Verilog description of the rest of the circuit by means of a PLI call providing a bidirectional link to pass data back and forth between the main circuit (represented in Verilog) and the FFT (captured in C).

Yet one more really useful feature associated with Verilog and Verilog-XL was the ability to have timing information specified in an external text file known as a *standard delay format (SDF)* file. This allowed tools like post-place-and-route timing analysis packages to generate SDF files that could be used by the simulator to provide more accurate results.

As a language, the original Verilog was reasonably strong at the structural (switch and gate) level of abstraction (especially with regard to delay modeling capability); it was very strong at the functional (Boolean equation and RTL) level of abstraction; and it supported some behavioral (algorithmic) constructs (Figure 9-5).

FFT is pronounced by spelling it out as "F-F-T."

SDF is pronounced by spelling it out as "S-D-F."

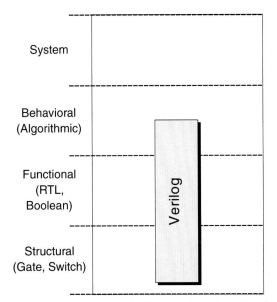

Figure 9-5. Levels of abstraction (Verilog).

1879: England. William Crookes postulates that cathode rays may be negative charged particles.

In 1989, Gateway Design Automation along with Verilog (the HDL) and Verilog-XL (the simulator) were acquired by Cadence Design Systems. The most likely scenario at that time was for Verilog to remain as just another proprietary HDL. However, with a move that took the industry by surprise, Cadence put the Verilog HDL, Verilog PLI, and Verilog SDF specifications into the public domain in 1990.

This was a very daring move because it meant that anybody could develop a Verilog simulator, thereby becoming a potential competitor to Cadence. The reason for Cadence's largesse was that the VHDL language (introduced later in this section) was starting to gain a significant following. The upside of placing Verilog in the public domain was that a wide variety of companies developing HDL-based tools, such as logic synthesis applications, now felt comfortable using Verilog as their language of choice.

Having a single design representation that could be used by simulation, synthesis, and other tools made everyone's life a lot easier. It is important to remember, however, that Verilog was originally conceived with simulation in mind; applications like synthesis were something of an afterthought. This means that when creating a Verilog representation to be used for both simulation and synthesis, one is restricted to using a *synthesizable subset* of the language (which is loosely defined as whatever collection of language constructs your particular logic synthesis package understands and supports).

The formal definition of Verilog is encapsulated in a document known as the *language reference manual (LRM)*, which details the syntax and semantics of the language. In this context, the term *syntax* refers to the grammar of the language—such as the ordering of the words and symbols in relation to each other—while the term *semantics* refers to the underlying meaning of the words and symbols and the relationships between the things they denote ... phew!

In an ideal world, an LRM would define things so rigorously that there would be no chance of any misinterpretation. In the real world, however, there were some ambiguities with respect to the Verilog LRM. Admittedly, these were corner-case conditions along the lines of "if a control signal on this register goes inactive at the same time as the clock signal triggers, which signal will be evaluated by the simulator first?" But the end result was that different Verilog simulators might generate different results, which is always somewhat disconcerting to the end user.

Verilog quickly became very popular. The problem was that different companies started to extend the language in different directions. In order to curtail this sort of thing, a nonprofit body called *Open Verilog International (OVI)* was established in 1991. With representatives from all of the major EDA vendors of the time, OVI's mandate was to manage and standardize Verilog HDL and the Verilog PLI.

The popularity of Verilog continued to rise exponentially, with the result that OVI eventually asked the IEEE to form a working committee to establish Verilog as an IEEE standard. Known as IEEE 1364, this committee was formed in 1993. May 1995 saw the first official IEEE Verilog release, which is

LRM is pronounced by spelling it out as "L-R-M."

OVI is pronounced by spelling it out as "O-V-I."

formally known as IEEE 1364-1995, and whose unofficial designation has come to be Verilog 95.

Minor modifications were made to this standard in 2001; hence, it is often referred to as the Verilog 2001 (or Verilog 2K1) release. At the time of this writing, the IEEE 1364 committee is working feverishly on a forthcoming Verilog 2005 offering, while the design world holds its breath in dread anticipation (see also the section on "Superlog and System-Verilog" later in this chapter).

VHDL and VITAL

In 1980, the U.S. *Department of Defense (DoD)* launched the *very high speed integrated circuit (VHSIC)* program, whose primary objective was to advance the state of the art in digital IC technology.

> Don't ask me how VHSIC is pronounced (it's been a long day).

This program sought to address, among other things, the fact that it was difficult to reproduce ICs (and circuit boards) over the long life cycles of military equipment because the function of the parts wasn't documented in a rigorous fashion. Furthermore, different components forming a system were often designed and verified using diverse and incompatible simulation languages and design tools.

In order to address these issues, a project to develop a new hardware description language called *VHSIC HDL* (or *VHDL* for short) was launched in 1981. One unique feature of this process was that industry was involved from a very early stage. In 1983, a team comprising Intermetrics, IBM, and Texas Instruments was awarded a contract to develop VHDL, the first official release of which occurred in 1985.

> VHDL is pronounced by spelling it out as "V-H-D-L."

Also of interest is the fact that in order to encourage acceptance by the industry, the DoD subsequently donated all rights to the VHDL language definition to the IEEE in 1986. After making some modifications to address a few known problems, VHDL was released as official standard IEEE 1076 in 1987. The language was further extended in a 1993 release and again in 1999.

Initially, VHDL didn't have an equivalent to Verilog's PLI. Today, different simulators have their own ways of doing this sort of thing, such as ModelSim's foreign language interface (FLI). We can but hope that these diverse offerings will eventually converge on a common standard.

As a language, VHDL is very strong at the functional (Boolean equation and RTL) and behavioral (algorithmic) levels of abstraction, and it also supports some system-level design constructs. However, VHDL is a little weak when it comes to the structural (switch and gate) level of abstraction, especially with regard to its delay modeling capability.

It quickly became apparent that VHDL had insufficient timing accuracy to be used as a sign-off simulator. For this reason, the VITAL initiative was launched at the *Design Automation Conference (DAC)* in 1992. *VHDL Initiative toward ASIC Libraries (VITAL)* was an effort to enhance VHDL's abilities for modeling timing in ASIC and FPGA design environments. The end result encompassed both a library of ASIC/FPGA primitive functions and an associated method for back-annotating delay information into these library models, where this delay mechanism was based on the same underlying tabular format used by Verilog (Figure 9-6).

DAC may be pronounced to rhyme with "sack," or it may be spelled out as "D-A-C."

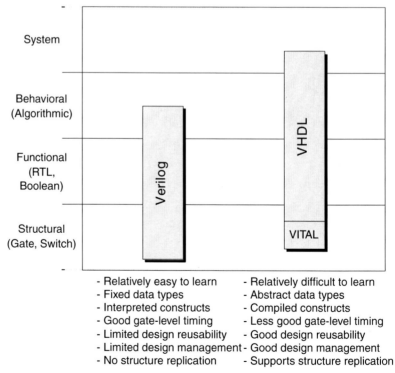

Figure 9-6. Levels of abstraction (Verilog versus VHDL).

Mixed-language designs

Once upon a time, it was fairly common for an entire design to be captured using a single HDL (Verilog or VHDL). As designs increased in size and complexity, however, it became more common for different portions of the design to be created by different teams. These teams might be based in different companies or even reside in different countries, and it was not uncommon for the different groups to be using different design languages.

Another consideration was the increasing use of legacy design blocks or third-party IP, where the latter refers to a design team purchasing a predefined function from an external supplier. As a general rule of thumb related to Murphy's Law, if you were using one language, then the IP you wanted was probably available only in the other language.

The early 1990s saw a period known as the HDL Wars, in which the proponents of one language (Verilog or VHDL) stridently predicted the imminent demise of the other … but the years passed and both languages retained strong followings. The end result was that EDA vendors began to support mixed-language design environments featuring logic simulators, logic synthesis applications, and other tools that could work with designs composed from a mixture of Verilog and VHDL blocks (or modules, depending on your language roots).

Murphy's Law—if anything can go wrong, it will—is attributed to Capt. Edward Murphy, an engineer working at Edwards Air Force Base in 1949.

UDL/I

As previously noted, Verilog was originally designed with simulation in mind. Similarly, VHDL was created as a design documentation and specification language that took simulation into account. As a result one can use both of these languages to describe constructs that can be simulated, but not synthesized.

In order to address these problems, the *Japan Electronic Industry Development Association (JEIDA)* introduced its own HDL, the *unified design language for integrated circuits (UDL/I)* in 1990.

1880: America. Alexander Graham Bell patents an optical telephone system called the Photophone.

The key advantage of UDL/I was that it was designed from the ground up with both simulation and synthesis in mind. The UDL/I environment includes a simulator and a synthesis tool and is available for free (including the source code). However, by the time UDL/I arrived on the scene, Verilog and VHDL already held the high ground, and UDL/I never really managed to attract much interest outside of Japan.

Superlog and SystemVerilog

In 1997, things started to get complicated because that's when a company called Co-Design Automation was formed. Working away furiously, the folks at Co-Design developed a "Verilog on steroids" called Superlog.

Superlog was an amazing beast that combined the simplicity of Verilog with the power of the C programming language. It also included things like temporal logic, sophisticated design verification capabilities, a dynamic API, and the concept of *assertions* that are key to the formal verification strategy known as *model checking*. (VHDL already had a simple assert construct, but the original Verilog had nothing to boast about in this area.)

The two main problems with Superlog were (a) it was essentially another proprietary language, and (b) it was so much more sophisticated than Verilog 95 (and later Verilog 2001) that getting other EDA vendors to enhance their tools to support it would have been a major feat.

Meanwhile, while everyone was scratching their heads wondering what the future held, the OVI group linked up with their equivalent VHDL organization called VHDL International to form a new body called Accellera. The mission of this new organization was to focus on identifying new standards and formats, to develop these standards and formats, and to foster the adoption of new methodologies based on these standards and formats.

In the summer of 2002, Accellera released the specification for a hybrid language called SystemVerilog 3.0 (don't even ask me about 1.0 and 2.0). The great advantage to this language was that it was an incremental enhancement to the existing Verilog, rather than the death-defying leap represented by a full-up Superlog implementation. Actually, SystemVerilog 3.0 featured many of Superlog's language constructs donated by Co-Design. It included things like the assertion and extended synthesis capabilities that everyone wanted and, being an Accellera standard, it was well placed to quickly gain widespread adoption.

The current state of play (at the time of this writing) is that Co-Design was acquired by Synopsys in the fall of 2002. Synopsys maintained the policy of donating language constructs from Superlog to SystemVerilog, but no one is really talking about Superlog as an independent language anymore. After a little pushing and pulling, all of the mainstream EDA vendors officially endorsed SystemVerilog and augmented their tools to accept various subsets of the language, depending on their particular application areas and requirements. System-Verilog 3.1 hit the streets in the summer of 2003, followed by a 3.1a release (to add a few enhancements and fix some annoying problems) around the beginning of 2004. Meanwhile, the IEEE is set to release the next version of Verilog in 2005. In order to avert a potential schism between Verilog 2005 and SystemVerilog, Accellera has promised to donate their SystemVerilog copyright to the IEEE by the summer of 2004.

SystemC

And then we have SystemC, which some design engineers love and others hate with a passion. SystemC—discussed in more detail in chapter 11—can be used to describe designs at the RTL level of abstraction.[1] These descriptions can subsequently be simulated 5 to 10 times faster than their Verilog or VHDL counterparts, and synthesis tools are available to convert the SystemC RTL into gate-level netlists.

---

[1] SystemC can support higher levels of abstraction than RTL, but those levels are outside the scope of this chapter; instead, they are discussed in more detail in chapter 11.

1880: France. Pierre and Jacques Currie discover piezoelectricity.

1881:
Alan Marquand invents a graphical technique of representing logic problems.

One big argument for SystemC is that it provides a more natural environment for hardware/software codesign and co-verification. One big argument against it is that the majority of design engineers are very familiar with Verilog or VHDL, but they are not familiar with the object-orientated aspects of SystemC. Another consideration is that the majority of today's synthesis offerings represent hundreds of engineer years of development in translating Verilog or VHDL into gate-level netlists. By comparison, there are far fewer SystemC-based synthesis tools, and those that are available tend to be somewhat less sophisticated than their more traditional counterparts.

In reality, SystemC is more applicable to a system-level versus an RTL design environment. Having said this, SystemC seems to be gaining a lot of momentum in Asia and Europe, and the debate on SystemC versus SystemVerilog versus VHDL will doubtless be with us for quite some time.

Points to ponder

Be afraid, be very afraid

Most software engineers throw up their hands in horror when they look at another programmer's code, and they invariably start a diatribe as to the lack of comments, consistency, whatever … you name it, and they aren't happy about it.

They don't know how lucky they are because the RTL source code for a design often sets new standards for awfulness! Sad to relate, the majority of designs described in RTL are almost unintelligible to another designer. In an ideal world, the RTL description of a design should read like a book, starting with a "table of contents" (an explanation of the design's structure), having a logical flow partitioned into "chapters" (logical breaks in the design), and having lots of "commentary" (comments explaining the structure and operation of the design).

It's also important to note that coding style can impact performance (this typically affects FPGAs more than ASICs). One reason for this is that, although they might be logically equivalent, different RTL statements can yield different results. Also, tools are part of the equation because different tools can yield different results.

The various FPGA vendors and EDA vendors are in a position to provide their customers with reams of information on particular coding styles and considerations with regard to their chips and tools, respectively. However, the following points are reasonably generic and will apply to most situations.

Serial versus parallel multiplexers

When creating RTL code, it is useful to understand what your synthesis tool is going to do in certain circumstances. For example, every time you use an if-then-else statement, the result will be a 2:1 multiplexer. This becomes interesting in the case of nested if-then-else statements, which will be synthesized into a priority structure. For example, assume that we have already declared signals Y, A, B, C, D, and SEL (for select) and that we use them to create a nested if-then-else (Figure 9-7).

```
if      SEL == 00" then Y = A;
elseif SEL == 01" then Y = B;
elseif SEL == 10" then Y = C;
else                    Y = D;
end if;
```

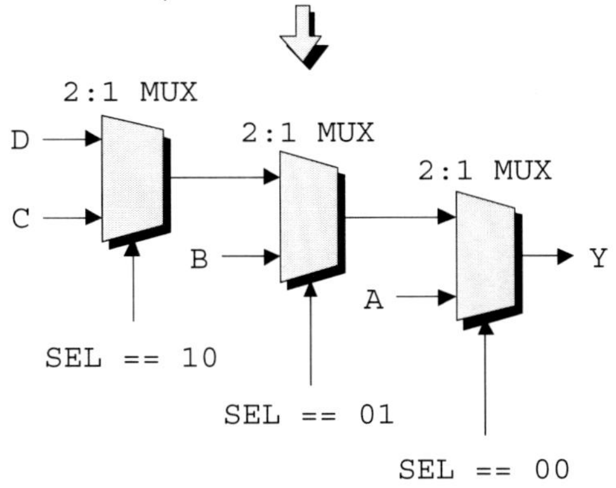

Figure 9-7. Synthesizing nested if-then-else statements.

1883: America. William Hammer and Thomas Alva Edison discover the "Edison Effect".

1884: Germany.
Paul Gottleib Nipkow
uses spinning disks to
scan, transmit, and
reproduce images.

As before, the syntax used here is a generic one that doesn't really reflect any of the mainstream languages. In this case, the innermost if-then-else will be the fastest path, while the outermost if-then-else will be the critical signal (in terms of timing). Having said this, in some FPGAs all of the paths through this structure will be faster than using a case statement. Speaking of which, a case statement implementation of the above will result in a 4:1 multiplexer, in which all of the timing paths associated with the inputs will be (relatively) equal (Figure 9-8).

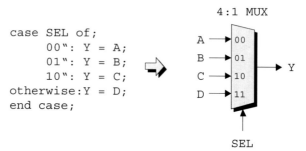

```
case SEL of;
    00": Y = A;
    01": Y = B;
    10": Y = C;
otherwise:Y = D;
end case;
```

Figure 9-8. Synthesizing a case statement.

Beware of latch inference

Generally speaking, it's a good idea to avoid the use of latches in FPGA designs unless you really need them. One other thing to watch out for: If you use an *if-then-else* statement, but neglect to complete the "else" portion, then most synthesis tools will infer a latch.

Use constants wisely

Adders are the most used of the more complex operators in a typical design. In certain cases, ASIC designers sometimes employ special versions using combinations of half-adders and full-adders. This may work very efficiently in the case of a gate array device, for example, but it will typically result in a very bad FPGA implementation.

When using an adder with constants, a little thought goes a long way. For example, "A + 2" can be implemented more efficiently as "A + 1 with carry-in," while "A – 2" would be better implemented as "A – 1 with carry-in."

Similarly, when using multipliers, "A * 2" can be implemented much more efficiently as "A SHL 1" (which translates to "A shifted left by one bit"), while "A * 3" would be better implemented as "(A SHL 1) + A."

In fact, a little algebra also goes a long way in FPGAs. For example, replacing "A * 9" with "(A SHL 3) + A" results in at least a 40 percent reduction in area.

Consider resource sharing

Resource sharing is an optimization technique that uses a single functional block (such as an adder or comparator) to implement several operators in the HDL code.

If you do not use resource sharing, then each RTL operation is built using its own logic. This results in better performance, but it uses more logic gates, which equates to silicon real estate. If you do decide to use resource sharing, the result will be to reduce the gate-count, but you will typically take a hit in performance. For example, consider the statement illustrated in Figure 9-9.

Note that frequency values shown in Figure 9-9 are of interest only for the purposes of this comparison, because these values will vary according to the particular FPGA architecture, and they will change as new process nodes come online.

The following operators can be shared with other instances of the same operator or with related operators on the same line:

```
*
+  -
>  <  >=  <=
```

For example, a + operator can be shared with instances of other + operators or with – operators, while a * operator can be shared only with other * operators.

1886:
Reverend Charles Lutwidge Dodgson (Lewis Carrol) publishes a diagrammatic technique for logic representation in *The Game of Logic.*

For not-so-technical readers, the circles with ">" symbols indicate comparators (circuits that compare two numbers to determine which is the larger); the circles with "+" symbols indicate adders; and the wedge-shaped blocks are 2:1 multiplexers that select between their inputs based on the value of the control signals coming out of the comparators.

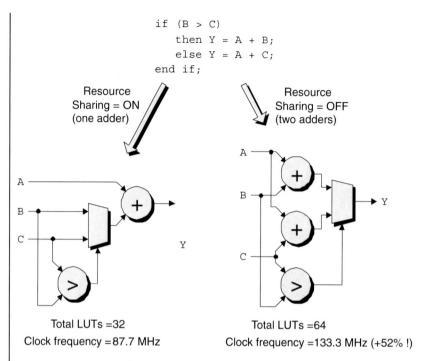

```
if (B > C)
    then Y = A + B;
    else Y = A + C;
end if;
```

Resource Sharing = ON (one adder)

Resource Sharing = OFF (two adders)

Total LUTs =32
Clock frequency =87.7 MHz

Total LUTs =64
Clock frequency =133.3 MHz (+52% !)

Figure 9-9. Resource sharing.

If nothing else, it's a good idea to check whether or not your synthesis application has resource sharing enabled or disabled by default. And one final point is that resource sharing in ASICs can alleviate routing congestion, but it may actually cause routing problems in FPGAs.

Last but not least

Internal tri-state buses are slow in most FPGAs and should be avoided unless you are 100 percent confident that you know what you're doing. If at all possible, use tri-state buffers only at the top-most level of the design. If you do wish to use internal tri-state buffers, then in the case of FPGA families that don't support these gates, the majority of today's synthesis tools provide automatic tri-state-to-multiplexer conversion (this basically involves converting the tri-state buffers specified in the RTL into corresponding LUT/CLB-based logic.)

Also, bidirectional buffers can cause timing loop problems, so if you use them, make sure that any false paths are clearly marked.

Silicon Virtual Prototyping for FPGAs

Just what is an SVP?

Before we leap headfirst into the concept of silicon virtual prototyping for FPGAs, it's probably worth reminding ourselves how the *silicon virtual prototype (SVP)* concept originated in the ASIC world, some of the alternative SVP manifestations one might see in that world, and some of the problems associated with those manifestations.

As high-end ASIC devices containing tens of millions of logic gates appeared on the scene, capacity and complexity issues associated with these megadesigns caused design flows to become a little wobbly around the edges.

The problem is that, with traditional flows, many design issues do not become apparent until accurate timing analysis can be performed following extraction of realistic physical values (capacitance, resistance, and sometimes inductance), based on the results from place-and-route. This requires the engineers to go all of the way through the flow (including synthesis and place-and-route) before they discover a major problem that would have been better detected and resolved earlier in the process.

This is extremely irritating, and the end result often involves numerous time-consuming iterations that can so delay a design that it completely misses its time-to-market window. (In many cases there is only room in the market for the winner, and there's no such thing as second place!)

One solution is to create an SVP, which is a representation of the design that can be generated relatively quickly, but which (hopefully) contains sufficient information to allow the

SVP is pronounced by spelling it out as "S-V-P."

1887: England.
J. Thomson discovers
the electron.

designers to identify and address a large proportion of potential problems before they undergo the time-consuming portions of the design flow. In theory, the time taken to iterate a design using an SVP can be measured in hours, as opposed to days or weeks using conventional design flows.

ASIC-based SVP approaches

As was discussed in the previous chapter, the role of logic synthesis is to accept an RTL representation of a design along with a set of timing constraints. The logic synthesis application automatically converts this RTL representation into a mixture of registers and Boolean equations, performs a variety of minimizations and optimizations (including optimizing for area and timing), and then generates a gate-level netlist that hopefully meets the original timing constraints.

Conventional logic synthesis solutions operate in the gate-size versus delay plane, which means they are constantly making trade-offs with regard to the size of gates and the delays associated with those gates. Due to their underlying modus operandi, these tools perform tremendous amounts of compute-intensive, time-consuming evaluations. Even worse, many of the optimization decisions performed by the synthesis tool are often rendered meaningless when the design is handed over to the physical implementation (place-and-route) portion of the flow.

Gate-level SVPs (from fast-and-dirty synthesis)

One key aspect of an SVP is the ability to generate it quickly and easily. The majority of current ASIC SVPs are based on the use of a gate-level netlist representation of the design that is subsequently placed using a rough-and-ready placement algorithm. Unfortunately, conventional synthesis tools consume too much time and computational resources to meet the speed demands of prototyping. Thus, some ASIC-based SVP flows make use of a fast-and-dirty synthesis engine (Figure 10-1).

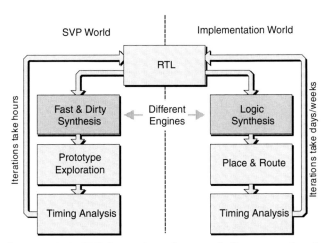

Figure 10-1. SVP based on fast-and-dirty synthesis.

1887: England. William Crookes demonstrates that cathode rays travel in a straight line.

This fast-and-dirty engine is typically based on completely different algorithms from the main synthesis application, for example, direct RTL mapping. Thus, the ensuing gate-level netlist used to form the SVP is not as accurate a representation of the design's final implementation as one might hope for.

In turn, this means that once the SVP has been used to perform RTL exploration and timing analysis, engineers still have to perform a full-up logic synthesis (or physically aware synthesis) step using a completely different synthesis engine in order to generate the real netlist to be passed on to the physical implementation (place-and-route) tools.

So, the big problem with this SVP-based approach is that the prototyping tools and their methodologies are separate and distinct from the implementation tools and their methodologies. This leads to unpredictability of design convergence due to lack of correlation, which can result in time-consuming back-end–to–front-end iterations, which sort of defeats the whole purpose of using an SVP in the first place!

Gate-level SVPs (from gain-based synthesis)

As opposed to conventional logic synthesis that is based in the gate-size versus delay plane, a concept known as *gain-based*

1887: Germany. Heinrich Hertz demonstrates the transmission, reception, and reflection of radio waves.

synthesis[1] is a kettle of fish of a different color (I never metaphor I didn't like).

This form of synthesis is derived from ideas put forward by Ivan Sutherland, Bob Sproull, and David Harris in their 1999 book *Logical Effort: Designing Fast CMOS Circuits*.[2] In this case, the synthesis engine uses logical effort concepts to establish a fixed-timing plane, and the physical implementation (place-and-route) tools subsequently work within this plane.

This means that all timing optimizations are completed and all circuit delays are determined and frozen by the end of the synthesis step. When the placement engine performs its task, it uses a size-driven algorithm in which all of the cells are dynamically sized to meet their timing budgets based on the actual loads they see. Following placement, a load-driven routing engine is used to tune the width and spacing of the tracks so as to maintain the original timing budgets and to ensure signal integrity.

One interesting point with regard to the gain-based approach is that the amount of computer memory and computational effort required to perform this type of synthesis are a fraction of that demanded by conventional synthesis tools. This means that a gain-based synthesis engine claims an order of magnitude increase in capacity over conventional synthesis approaches.

Another interesting point is that the gain-based synthesis engine automatically uses up any slack in path delays. This means that the smallest possible sizes are used for each gate that will just meet the timing budget. Thus, the resulting implementation occupies the smallest amount of silicon real estate, which significantly reduces congestion, power consumption, and noise problems.

---

[1] At the time of this writing, one of the chief proponents of gain-based synthesis is Magma Design Automation (www.magma-da.com).

[2] Ivan Sutherland is internationally renowned for his pioneering work on logic design.

"But," you cry, "what does all of this have to do with SVPs?" Well, the speed and capacity inherent to gain-based synthesis means that the same synthesis engine can be used for both prototyping and implementation (Figure 10-2).

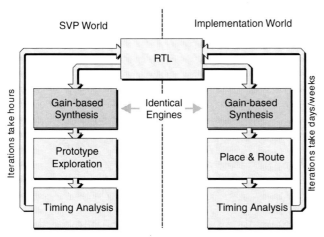

Figure 10-2. SVP based on gain-based synthesis.

Basing both the prototyping and implementation environments on the same algorithms, tools, and methodologies provides high correlation and predictable design convergence and significantly reduces time-consuming back-end–to–front-end iterations.

Cluster-level SVPs

As discussed earlier, the majority of today's SVPs are based on full-blown gate-level netlist representations of the design. Even though these representations may be generated using fast-and-dirty synthesis, they can still contain millions upon millions of logic gates, which can strain the capacity of the SVP's placement and analysis engines.

One solution is to use the concept of *clustering* as a basis for the SVP's placement decisions and track-delay estimations. In this case the cells (gates and registers) generated by fast-and-dirty or gain-based synthesis are automatically gathered into groups called *clusters*. Each cluster typically consists of tens to

1888: America. First coin-operated public telephone invented.

1889: America.
Almon Brown Strowger
invents the first
automatic telephone
exchange.

hundreds of cells, which means that they are small enough to preserve overall placement quality; however, the number of clusters is orders of magnitude smaller than the number of cells, providing extremely significant run-time improvements.

The actual number of cells may vary from cluster to cluster so as to keep the areas of the clusters as uniform as possible. In order to streamline computational complexity and capacity requirements, optimization and analysis are performed on the clustered data. Furthermore, in cases where two clusters are linked by multiple wires (which is a common occurrence), these wires are considered to be a single "weighted" wire for the purposes of estimating routing resource utilization, which has an effect on cluster placement.

RTL-based SVPs

A well-accepted engineering rule of thumb states that detecting, isolating, and resolving a problem at any stage of the design, implementation, or deployment process costs 10 times more than addressing the same problem at the previous stage in the process. In the case of digital ICs, there are three major breakpoints in the design flow with respect to analyzing area, timing, and so forth. (Figure 10-3).

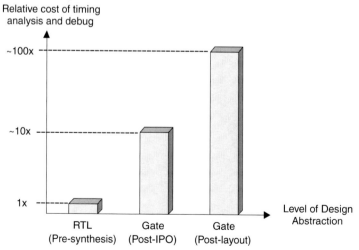

Figure 10-3. Major breakpoints with respect to analyzing area, timing, and so forth.

The term *timing closure* refers to analyzing a design or architecture to detect and correct any problematic timing paths. Irrespective of the level it is performed at, timing closure is an iterative process, which means that the analyze-detect-correct steps typically need to be run a number of times in order to achieve convergence.

With regard to the levels of abstraction shown in Figure 10-3, postlayout timing analysis is the most accurate by far, but it is extremely expensive with regard to cost and time. Iterating at the postlayout level is a painful proposition, and design teams try very hard to avoid making changes at this level.

In the case of conventional flows, the first breakpoint for relatively accurate timing analysis occurs at the gate level following synthesis and *in-place optimization (IPO)*. The problem is that getting to this post-IPO breakpoint using conventional flows requires the use of physically aware synthesis to provide a placed gate-level netlist. This approach is therefore extremely compute-intensive and time-consuming, and large blocks can take days to go through the full physical synthesis and timing analysis process. Not only does this stretch out the design and timing closure process, but it also ties up expensive EDA tools that could be being used for implementing chips rather than analyzing their timing.

> IPO means that, after the placement algorithm has performed its initial pass, it is possible to make certain "tweaks" (optimizations), such as changing the size of cells based on updated estimates of the length of the tracks they will see.

One alternative is to use a gate-level SVP as discussed above; but, once again, these representations have their own problems, including requiring the use of some form of compute-intensive and time-consuming synthesis and placement.

Another approach is to work with an RTL-based SVP,[3] which allows engineers to quickly identify and address paths that will cause downstream timing problems. In order to wrap one's brain around how this works, it's first necessary to understand that there's a related application that takes the logical

---

[3] At the time of this writing, one of the chief proponents of RTL-based SVPs is InTime Software (www.intimesw.com).

LEF stands for "logical exchange format," where this file details the logical functionality of the cells in the library.

Similarly, DEF stands for "design exchange format," where this file details the physical aspects of the cells in the library, such as their resistance and capacitance values and their physical dimensions.

and physical (LEF and DEF) definition files associated with an ASIC cell library and generates a corresponding *design kit database* to be used by the RTL-based SVP (Figure 10-4).

LEF

DEF

Design Kit Generator

Design Kit

Figure 10-4. Generating a design kit.

It's important to note that such a design kit is not a library of characterized gates, but is instead a database of characterized logical functions (such as counters, XOR trees, etc.). The design kit generator captures the behavior of these logical functions, including timing and area estimations.

The RTL-based SVP generator and analysis engine subsequently accepts the RTL code for the design, the time constraints associated with the design block (in industry-standard SDF format), and the design kit associated with the target cell library. As the SVP generator reads in the RTL, it converts it into a netlist of entities called *work functions*. Each work function is an abstraction that directly maps onto an equivalent function in the design kit.

Once the RTL has been converted into a netlist of work functions, the SVP generator performs identical logical operations to those that are typically performed at the gate level, including common subexpression elimination, constant propagation, loop unraveling, the removal of all redundant functional computations, and so forth.

The SVP generator and analysis engine uses the resulting minimal irredundant network of work functions to perform a "virtual placement" of these functions. This placement is then used to generate accurate area estimates, which are subsequently used to generate accurate time estimates. In conjunction with the design kit, the SVP generator and analysis engine understands how the various synthesis engines will weight various factors and modify their implementation strategies (such as swapping counter realizations) in order to meet the specified timing constraints. All of these factors are taken into account when performing the analysis.

Proponents of RTL-based SVPs claim a 40-fold speed increase as compared to generating a post-IPO, pre-place-and-route gate-level netlist using a physically aware synthesis approach. In an example 4.5-million-gate design circa 2003, this equated to a 2.5-hour iteration to generate and analyze an RTL-based SVP as compared to 99 hours to generate and analyze a post-IPO gate-level netlist.

Of course the big question is, just how accurate are RTL-based SVPs? The supporters of this form of SVP claim that its timing analysis results typically correlate to post-IPO delays with an error of 20 percent or less (worst-case errors may rise to 30 percent). Although this may sound pretty dire, the latest generation of synthesis tools is capable of closing timing on RTL that is within 20 to 30 percent of the desired timing (it's the paths that are off by 80, 150, 200 percent, and higher that cause problems). Thus, RTL-based SVPs are accurate enough to allow design engineers to generate RTL code that can subsequently be fully implemented by the downstream syntheses and layout engines.

I know, I know. We've digressed again, although you have to admit that this is all interesting stuff! But now let's return to pondering FPGAs.

FPGA-based SVPs

Not surprisingly, multimillion-gate FPGA designs are hitting the same problems that befell ASICs, including the fact that it takes ages to place, route, and perform timing analysis on the little rascals.

One particularly painful aspect of this process is that, although the original RTL representation of the design is

1890: America Census is performed using Herman Hollerith's punched cards and automatic tabulating machines.

1890s:
John Venn proposes logic representation using circles and eclipses.

almost invariably hierarchical,[4] the FPGA's place-and-route tools typically end up working on flattened representations of the design. This means that even if you make the smallest of changes to a single block of RTL code and resynthesize only that block, you end up having to rerun place-and-route on the entire design. In turn, this means that you can grow to be old and gray before you finally get to achieve timing closure on your design.

In order to address these problems, some EDA vendors have started to provide tools that support the concept of an FPGA SVP by providing a mixture of floor planning and pre-place-and-route timing analysis. This is coupled with the ability to perform place-and-route on individual design blocks, which dramatically speeds up the implementation process.[5]

This form of SVP commences with a graphical top-down view of the target FPGA device showing all of the internal logical resources, such as LUTs, registers, slices, CLBs, embedded RAMs, multipliers, and so forth.

Following the logic synthesis step (using the synthesizer of your choice), the SVP generator loads the ensuing hierarchical LUT/CLB-level netlist, along with any associated timing and physical constraints, and automatically creates an initial floor plan. This auto-generated floor plan shows a collection of square and/or rectangular blocks, each of which corresponds to a top-level module in the design. Furthermore, if any of these top-level modules itself contains submodules, then these are shown as embedded blocks in the floor plan (and so on down through the hierarchy).

The SVP generator performs its own initial placement of the resources (LUTs, registers, RAMs, multipliers, etc.) used by each block. These resources are also shown in the top-down view of the device, along with graphical representations

---

[4] By "hierarchical," we mean that the top level of the design is typically formed from a number of functional modules, which may themselves call submodules and so forth.

[5] At the time of this writing, one of the chief proponents of FPGA SVPs, in the form described here, is Hier Design (www.hierdesign.com).

as to the amount of routing resources required to link the various blocks together.

Interactive manipulation

The initial placement of the design in the SVP allows it to provide accurate timing estimations on a block-by-block basis prior to running place-and-route. If any potential problem areas are detected, you can interactively modify the floor plan in order to address them.

The simplest form of manipulation is to reshape the rectangular blocks in the floor plan by pulling their sides to make them taller, thinner, shorter, or fatter. Alternatively, you can create more complex outlines such as "L," "U," and "T" shapes (pretty much any contour you can form out of squares and rectangles).

Next, you can move the blocks around. When you grab a block and start to drag it across the face of the device, the system will provide a graphical indication as to whether or not there are the necessary resources required to implement that block at its current location (you can only drop the block in an area where there are sufficient resources). Furthermore, as you manipulate a block by reshaping it or moving it around, the system dynamically displays the utilization of resources (LUTs, registers, RAMs, multipliers, etc.) inside that block relative to the total amount of each resource type currently encompassed by that block.

You can also split existing blocks into two or more subblocks, which you can then manipulate independently. Alternatively, you can merge two or more blocks into a single block. Also, in some cases (say, areas of control logic), you might wish to pull one or more subblocks out of their parent blocks and move them up to the top level of the design, at which point you can reshape them, merge them together, move them around, and so forth. Much of this reflects a different philosophy of how one might use an ASIC floor-planning tool. In the case of an ASIC, for example, if you have two

1892: America. First automatic telephone switchboard comes into service.

1894: Germany. Heinrich Hertz discovers that radio waves travel at the speed of light and can be refracted and polarised.

blocks with lots of interconnect between them, you would typically place them side by side. By comparison, in the case of an FPGA, merging the blocks (thereby allowing the place-and-route tools to do a much better job of optimization using local versus global routing resources) might provide a more efficacious solution.

Furthermore, you aren't limited to manipulating blocks only as described in the original RTL hierarchy. You can actually manipulate individual FPGA resources like LUTs, registers, slices, CLBs, and the like. This includes dragging them around and repositioning them within their current hierarchical block, dragging them from one hierarchical block to another, creating new blocks, and dragging groups of LUTs from one or more existing blocks into this new block, and so forth.

Where things start to get really clever is that, if you go back to make changes to your original RTL and resynthesize those modules, then when you reimport the resulting LUT/CLB-level netlist(s), the SVP generator sorts everything out for you and loads the right logic into the appropriate floor-plan blocks. (How do they do it? I don't have a clue!)

Incremental place-and-route

As soon as you are ready to rock and roll, you can select one or more floor-plan blocks and kick off the FPGA vendor's place-and-route software. Each block is treated as an individual entity, so once you've laid out a block, it will remain untouched unless you decide you want to change it. This has a number of advantages. First of all, place-and-route run times for individual blocks are extremely small compared to the traditional times associated with full-up multimillion-gate designs.

And even if you add up the place-and-route times for running all of the blocks individually, the total elapsed time is much less than it would be if one were performing place-and-route on the design in its entirety. This is because the complexity (and associated run times) of place-and-route increases in a nonlinear manner as the size of the block being processed increases. Furthermore, once you've run place-and-route on all of the blocks, you can make changes to individual blocks and rerun place-and-route only on those blocks without affecting the rest of the chip.

An additional advantage associated with this SVP approach is that it lends itself to creating and preserving IP. That is, once a block has undergone place-and-route, you can lock it down and export it as a new structural LUT/CLB-level netlist along with its associated physical and timing constraints. This block can subsequently be used in other designs (its placement is relative, which means that it can be dragged around the chip and relocated as discussed above).

RTL-based FPGA SVPs

In an ideal world, it would be nice to be able to work with RTL-based FPGA SVPs. The various FPGA and EDA vendors do provide RTL-level floor-planning tools with varying degrees of sophistication. At the time of this writing, however, there is no FPGA equivalent to the state-of-the-art in RTL-based ASIC SVP technology (but we will doubtless see such a beast in the not-so-distant future).

1894: Italy. Guglielmo Marconi invents wireless telegraphy.

C/C++ etc.–Based Design Flows

Problems with traditional HDL-based flows

With regard to the traditional HDL-based flows introduced in chapter 9, a design commences with an original concept, whose high-level definition is determined by system architects and system designers. It is at this stage that *macroarchitecture* decisions are made, such as partitioning the design into hardware and software components (see also chapter 13).

The resulting specification is then handed over to the hardware design engineers, who commence their portion of the development process by performing *microarchitecture definition* tasks such as detailing control structures, bus structures, and primary data path elements. These microarchitecture definitions, which are often performed in brainstorming sessions on a whiteboard, may include performing certain operations in parallel versus sequential, pipelining portions of the design versus nonpipelining, sharing common resources (for example, two operations sharing a single multiplier, versus using dedicated resources) and so forth.

Eventually, the design intent is captured by writing RTL VHDL/Verilog. Following verification via simulation, this RTL is then synthesized down to a structural netlist suitable for use by the target technology's place-and-route applications (Figure 11-1).

At the time of this writing, these VHDL or Verilog-based flows account for around 95 percent of all ASIC and FPGA designs; however, there are a number of problems associated with these flows:

Note that this chapter focuses on C/C++ flows in the context of generic digital designs. Considerations such as quantization (commencing with floating-point representations which are subsequently coerced into their fixed-point counterparts) are covered in the discussions on DSP-centric designs in chapter 12.

In the case of an FPGA target, the LUT/CLB-level netlist may be presented in EDIF, VHDL, or Verilog depending on the FPGA vendor.

With regards to physically aware synthesis-based flows, EDIF remains the "netlist of choice." In this case, the placement information may be incorporated in the EDIF itself or presented in an external "constraints" side-file.

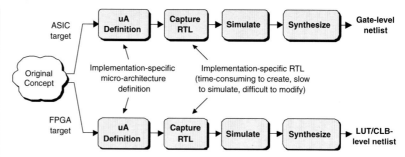

Figure 11-1. Traditional (simplified) HDL-based flows.

- *Capturing the RTL is time-consuming:* Even though Verilog and VHDL are intended to represent hardware, it is still time-consuming to use these languages to capture the functionality of a design.
- *Verifying RTL is time-consuming:* Using simulation to verify large designs represented in RTL is computationally expensive and time-consuming.
- *Evaluating alternative implementations is difficult:* Modifying and reverifying RTL to perform a series of what-if evaluations of alternative microarchitecture implementations is difficult and time-consuming. This means that the number of evaluations the design team can perform may be limited, which can result in a less-than-optimal implementation.
- *Accommodating specification changes is difficult:* If any changes to the specification are made during the course of the project, folding these changes into the RTL and performing any necessary reverification can be painful and time-consuming. This is a significant consideration in certain application areas, such as wireless projects, because broadcast standards and protocols are constantly evolving and changing.
- *The RTL is implementation specific:* Realizing a design in an FPGA typically requires a different RTL coding style from that used for an ASIC implementation (see also the discussions in Chapters 7, 9, and 18). This

means that it can be extremely difficult to retarget a complex design represented in RTL from one implementation technology to another. This is of concern when one is migrating an existing ASIC design into an FPGA equivalent or creating an FPGA design to be used as a prototype for a future ASIC implementation.

One way to view this is that all of the implementation intelligence associated with the design is hard-coded into the RTL, which therefore becomes implementation specific. It's important to understand that this implementation specificity goes beyond the coarse ASIC-versus-FPGA boundary, which dictates that RTL intended for an FPGA implementation is not suitable for an optimal ASIC realization, and vice versa. Even assuming a single target device architecture, the way in which a set of algorithms is used to process data may require a number of different micro-architecture implementations, depending on the target application areas.

Actually, to be scrupulously fair, we should probably note that the same RTL *may* be used to drive both ASIC and FPGA implementations. The reason for doing this is to avoid the risk of introducing a functional bug into the RTL when retargeting the code, but there is typically a penalty to be paid. That is, if code originally targeted toward an FPGA implementation is subsequently used to drive an ASIC implementation, the resulting ASIC will typically require more silicon real estate and have higher power consumption as compared to using RTL created with an ASIC architecture in mind. Similarly, if code originally targeted toward an ASIC implementation is subsequently used to drive an FPGA implementation, the ensuing FPGA will typically take a significant performance hit as compared to using RTL created with an FPGA architecture in mind.

■ *RTL is less than ideal for hardware-software codesign:* System-on-chip (SoC) devices are generally under-

1895: America. Dial telephones go into Milwaukee's city hall.

RTOS is pronounced "R-tos." That is, by spelling out the "R" followed by "TOS" to rhyme with "boss."

Real-time systems are those in which the correctness of a computation or action depends not only on *how* it is performed but also *when* it is performed.

stood to be those that include microprocessor cores. Irrespective of whether these designs are to be realized using ASICs or FPGAs, today's SoCs are exhibiting an ever-increasing amount of software content. When coupled with increased design reuse on the hardware side, in many cases it is necessary to verify the software and hardware concurrently so as to completely validate such things as the system diagnostics, RTOS, device drivers, and embedded application software. Generally speaking, it can be painful verifying (simulating) the hardware represented in VHDL or Verilog in conjunction with the software represented in C/C++ or assembly language.

One approach that addresses the issues enumerated above is to perform the initial design capture at a higher level of abstraction than can be achieved with RTL VHDL/Verilog. The first such level is to use some form of C/C++, but as usual nothing is simple because there are a variety of alternatives, including SystemC, augmented C/C++, and pure C/C++.

C versus C++ and concurrent versus sequential

Before we leap into the fray, we should tie down a couple of points to ensure that we're all marching in step to the same beat. First, there is a wide variety of programming languages available, but—excepting specialist application areas—the most commonly used by far are traditional C and its object-oriented offspring C++. For our purposes here, we will refer to these collectively as C/C++.

The next point of import is that, by default, statements in languages like C/C++ are executed *sequentially*. For example, assuming that we have already declared three integer variables called *a*, *b*, and *c*, then the following statements

```
a = 6;    /* Statement in C/C++ program */
b = 2;    /* Statement in C/C++ program */
c = 9;    /* Statement in C/C++ program */
```

would, perhaps not surprisingly, occur one after the other. However, this has certain implications; for example, if we now assume that the following statements occur sometime later in the program

```
a = b;   /* Statement in C/C++ program */
b = a;   /* Statement in C/C++ program */
```

1895: Germany. Wilhelm Kohnrad Roentgen discovers X-rays.

then *a* (which initially contained 6) will be loaded with the value currently stored in *b* (which is 2). Next, *b* (which initially contained 2) will be loaded with the value currently stored in *a* (which is now 2), so both *a* and *b* will end up containing the same value.

The sequential nature of programming languages is the way in which software engineers think. However, hardware design engineers have quite a different view of the world. Let's assume that a piece of hardware contains two registers called *a* and *b* that are driven by a common clock signal. Let's further assume that these registers have previously been loaded with values of 6 and 2, respectively. Finally, let's assume that at some point in the HDL code, we see the following statements:

```
a = b;   /* Statement in VHDL/Verilog Code */
b = a;   /* Statement in VHDL/Verilog Code */
```

As usual, the above syntax doesn't actually represent VHDL or Verilog; it's just a generic syntax used only for the purposes of this example. Generally speaking, hardware engineers would expect both of these statements to be executed *concurrently* (at the same time). This means that *a* (which initially contained 6) will be loaded with the value stored in *b* (which was 2) while—at the same time—*b* (which initially contained 2) will be loaded with the value stored in *a* (which was 6). The end result is that the initial contents of a and b will be exchanged.

As usual, of course, the above is something of a simplification. However, it's fair to say that HDL statements will execute concurrently by default, unless sequential behavior is forced by means of techniques like blocking assignments. Thus, by default, RTL-based logic simulators will execute the statements shown above in this concurrent manner; similarly RTL-based logic synthesis tools will generate hardware that handles these two activities simultaneously. By comparison, unless explicitly directed to do otherwise (by means of the techniques introduced later in this chapter), C/C++ statements will execute sequentially.

SystemC-based flows

What is SystemC (and where did it come from)?

SystemC is "managed" by the *Open SystemC Initiative (OSCI)*. This is an independent not-for-profit organization composed of companies, universities, and individuals dedicated to promoting SystemC as an open-source standard for system-level design.

The code for SystemC—along with an integrated simulator and design environment—is available from www.systemc.org.

Before we come to consider SystemC-based flows, it is probably a good idea to briefly summarize just what SystemC is, because there is typically some confusion on this point (not the least in the mind of the author).

SystemC 1.0

One of the underlying concepts behind SystemC is that it is an open-source environment to which everyone contributes. As an example, consider Linux, which was rough around the edges at first. Based on contributions from different folks, however, Linux eventually became a real *operating system* (OS) with the potential to challenge Microsoft.

In this spirit, a relatively undocumented SystemC 1.0 was let loose to roam wild and free circa 2000. SystemC 1.0 was a C++ class library that facilitated the representation of notions such as concurrency (things happening at the same time), timing, and I/O pins. By means of this class library, engineers could capture designs at the RTL level of abstraction.

One advantage of this early incarnation was that it facilitated hardware/software codesign environments. Another was that SystemC representations at the RTL level of abstraction might simulate 5 to 10 times faster than their VHDL and Verilog counterparts.[1] On the downside, it was harder and more time-consuming to capture an RTL-level design in SystemC 1.0 than with VHDL or Verilog. Furthermore, there was a scarcity of design tools that could synthesize SystemC 1.0 representations into netlist-level equivalents with any degree of sophistication.

SystemC 2.0

Later, in 2002, SystemC 2.0 arrived on the scene. This augmented the 1.0 release with some high-level modeling constructs such as FIFOs (a form of memory that can accept and subsequently make available a series of words of data and that operates on a first-in first-out principle). The 2.0 release also included a variety of behavioral, algorithmic, and system-level modeling capabilities, such as the concepts of transactions and channels (which are used to describe the communication of data between blocks at an abstract level).

In order to gain a little perspective on all of this, let's first consider a typical scenario of how things would have worked using the original SystemC 1.0. As a simple example, let's assume that we have two functions called $f(x)$ and $g(x)$ that have to communicate with each other. (Figure 11-2).

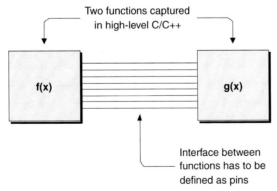

Figure 11-2. Interfacing in SystemC 1.0.

1895: Russia. Alexander Popov (also spelled Popoff) constructs a receiver for natural electricity waves and tries to detect thunderstorms.

---

[1] This is design-dependent; in reality, some SystemC RTL-level simulation run times are at parity with their HDL counterparts.

1897: England. Guglielmo Marconi transmits a Morse code message *"let it be so"* across the Bristol Channel.

In this case, the interface between the blocks would have to be defined at the pin level. The real problem with this approach occurs when you are in the early stages of a design, because you are already defining implementation details such as bus widths. This makes things difficult to change if you wish to experiment with different what-if architectural scenarios. This aspect of things became much easier with SystemC 2.0, which allowed abstract interfaces to be declared between the blocks (Figure 11-3).

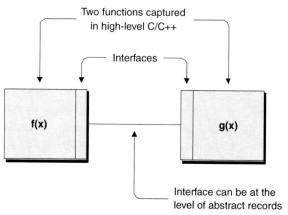

Figure 11-3. Interfacing in SystemC 2.0.

Now, the interfacing between the blocks can be performed at the level of abstract records on the basis that, in the early stages of the design cycle, we don't really care how data gets from point a to point b, just that it does get there somehow.

These abstract interfaces facilitate performing architectural evaluation early in the design cycle. Once the architecture starts to firm up, you can start refining the interface by using high-level constructs such as a FIFO to which one would assign attributes like width and depth and characteristics like blocking write, nonblocking read, and how to behave when empty or full. Still later, this logical interface can be replaced by a completely specified (pin-level) interface that binds the functional blocks together at a more physical level.

Levels of abstraction

Truth to tell, this is where things start to become a little fuzzy around the edges, not the least because one runs into different definitions depending on to whom one is talking. As a first pass, however, we might take a stab at capturing the different levels of SystemC abstraction, as shown in Figure 11-4.

1897: England. Marconi establishes the first Marconi-station at the Needles (Isle of Wight, England), sending a signal over 22 km to the English coast.

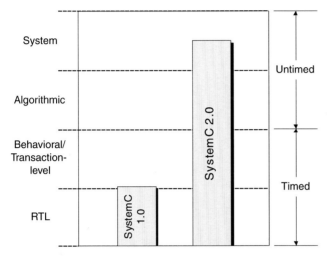

Figure 11-4. Levels of SystemC abstraction.

This is why things become confusing, because SystemC can mean all things to all people. To some it's a replacement for RTL VHDL/Verilog, while to others it's a single language that can be used for system-level specification, algorithmic and architectural analysis, behavioral design, and testbenches for use in verification.

One area of confusion comes when you start to talk about behavioral synthesis. This encompasses certain aspects of both the algorithmic and transactional levels (in the latter case, however, you have to be careful as to how to define your transactions).

SystemC-based design-flow alternatives

This is a tricky one because one might go various ways here. For example, many of today's designs begin life as com-

1901:
Hubert Booth invents
the first vacuum
cleaner.

plex algorithms. In this case, it is very common to start off by creating a C or C++ representation. This representation can be used to validate the algorithms by compiling it into a form that can be run (simulated) 1,000 or more times faster than an RTL equivalent.

In the case of the HDL-based flows discussed in chapter 9, this C/C++ representation of the algorithms would then be hand-translated into RTL VHDL/Verilog. The C/C++ representation will typically continue to be used as a *golden model*, which means it can be linked into the RTL simulator and run in parallel with the RTL simulation. The results from the C/C++ and RTL models can be compared so as to ensure that they are functionally equivalent.

Alternatively, in one flavor of a SystemC-based flow, the original C/C++ model could be incrementally modified by adding timing, concurrency, pin definitions, and so forth to transform it to a level at which it would be amenable to SystemC-based RTL or behavioral synthesis.

In another flavor of a SystemC-based flow, the design might be initially captured in SystemC using system, algorithmic, or transaction-level constructs that could be used for verification at a high level of abstraction. This representation could then be incrementally modified to bring it *down* to a level at which it would be amenable to SystemC-based RTL or behavioral synthesis.

Irrespective of the actual route by which one might get there, let's assume that we are in possession of a SystemC representation of a design that is suitable for SystemC-based behavioral or RTL synthesis. In this case, there are two main design-flow alternatives, which are (1) to translate the System C into RTL VHDL/Verilog automatically and then to use conventional RTL synthesis technology, or (2) to use SystemC-based synthesis to generate an implementation-level netlist directly.

There are two schools of thought here. One says that synthesizing the SystemC directly into the implementation-level netlist offers the cleanest, fastest, and most efficient route.

However, another view is that it's better to translate the SystemC into RTL VHDL/Verilog first because RTL is the way design engineers really visualize their world; that this level is a natural staging point for integrating design blocks (including third-party IP) originating from multiple sources; and that Verilog/VHDL synthesis technology is extremely mature and powerful (as compared to SystemC-based synthesis technology).

But we digress. Both of these flows can be applied to ASIC or FPGA targets (Figure 11-5).

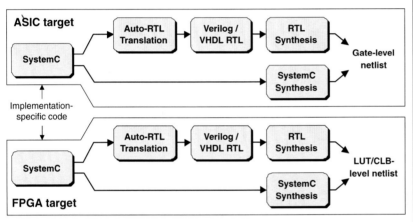

Figure 11-5. Alternative SystemC flows.

The first SystemC synthesis applications were predominantly geared toward ASIC flows, so they didn't do a very good job at inferring FPGA-specific entities such as embedded RAMs, embedded multipliers, and so forth. More recent incarnations do a much better job of this, but the level of sophistication exhibited by different tools is a moving target, so the prospective user is strongly advised to perform some in-depth evaluations before slapping a bundle of cash onto the bargaining table.

Note that figure 11-5 shows the use of implementation-specific SystemC to drive the ASIC versus FPGA flows. As soon as you start coding at the RTL level and adding timing

1901:
Marconi sends a radio signal across the Atlantic.

concepts, be it in VHDL, Verilog, or SystemC, then achieving an optimal implementation requires that the code be written with a specific target architecture in mind.

Once again, having said this, the same SystemC *can* be used to drive both ASIC and FPGA flows, but there is typically a penalty to be paid. If SystemC code originally targeted toward an FPGA implementation is subsequently used to drive an ASIC flow, the resulting ASIC will typically require more silicon real estate and have higher power consumption as compared to using code created with an ASIC architecture in mind. Similarly, if code originally targeted toward an ASIC implementation is subsequently used to drive an FPGA flow, the ensuing FPGA will typically take a significant performance hit as compared to using code created with an FPGA architecture in mind. This is primarily a result of hard-coding the microarchitecture definition in the source.

Love it or loath it

Depending on whom you are talking to, folks either love SystemC or they loath it. Most would agree that SystemC 2.0 is very promising and that there's no other language that provides the same capabilities (some of these capabilities are being added into SystemVerilog, but not all of them).

On the downside, a lot of design engineers are reasonably proficient at writing C, but most of them are significantly less familiar with the object-oriented aspects of C++. So requiring them to use SystemC means giving them more power on the one hand, while thrusting them into a world they don't like or understand on the other. It's also true that while SystemC can be very useful for verification and high-level system modeling, in some respects it's still relatively immature toolwise with regard to actual implementation flows.

One school of thought says that, although SystemC is difficult to write by hand and also difficult to synthesize, which makes it a somewhat clumsy specification language, it does provide a powerful framework for simulation across languages and levels of abstraction.

At the time of this writing, a number of companies that were strong supporters of SystemC in the United States have grown somewhat less vocal over the last few years. On the other hand, SystemC is gaining some ground in Europe and Asia. What does the future hold? Wait a few years, and I'll be happy to tell you!

Augmented C/C++-based flows

What do we mean by augmented C/C++?

There are two ways in which standard C/C++ can be augmented to extend its capabilities and the things it can be used to represent. The first is to include special comments, known by some as *commented directives* or *pragmas*, into the pure C/C++ code. These comments can subsequently be recognized and interpreted by parsers, precompilers, compilers, and other tools and used to add constructs to the code or modify the way in which it is processed.[2] One significant drawback to this approach is that simulation requires the use of proprietary C/C++ compilers as opposed to using standard off-the-shelf compilers. This limits the options customers have and is only viable if standards are developed for multiple EDA vendors to leverage.

The other way in which C/C++ can be augmented is to add special keywords and statements into the language. This is a very popular technique, and there are a veritable plethora of such language variants roaming wild and free around the world, each tailored toward a different application area. One downside of this approach is that, once again, it requires proprietary C/C++ compilers; otherwise, tools such as simulators that have not been enhanced to understand these new keywords and statements will crash and burn. A common solution

1902:
Robert Bosch invents the first spark plug.

---

[2] One example of this form of C/C++ augmentation is demonstrated by 0-In Design Automation (www.0in.com) for use with its *assertion-based verification* (ABV) technology. Another example of particular relevance here is Future Design Automation (www.future-da.com), which employs this technique with its C/C++ to RTL synthesis engine.

1902:
Transpacific cable links
Canada and Australia.

to this problem is to wrap standard *#ifdef* directives around the new keywords and statements such that a precompiler can be used to discard them as required (this is somewhat inelegant, but it works).

In the case of capturing the functionality of hardware for ASIC and FPGA designs, it is necessary to augment standard C/C++ with special statements to support such concepts as *clocks, pins, concurrency, synchronization,* and *resource sharing.*[3]

Assuming that you have an initial model represented in pure C/C++, the first step would be to augment it with clock statements, along with interface statements used to define the input and output pins. You could then use an appropriate synthesis tool to generate an implementation (as discussed below). However, because C/C++ is by nature sequential, the resulting hardware can be horribly slow and inefficient if the synthesis tool is not capable of locating potential parallelisms and exploiting them.

For example, assume that we have the following statements in a C/C++ representation of the design:

```
a = 6;        /* Standard C/C++ statement */
b = 2;        /* Standard C/C++ statement */
c = 9;        /* Standard C/C++ statement */
d = a + b;    /* Standard C/C++ statement */
  :
etc
```

By default, each = sign is assumed by the synthesis application to represent one clock cycle. Thus, if the above code were left as is, the augmented C/C++ synthesis tool would generate hardware that loaded variable (register) *a* with 6 on the first clock, then *b* with 2 on the next clock, then *c* with 9 on the next clock, and so forth. Thus, by hardware standards, this would run horribly slowly.

---

[3] A big player in this form of C/C++ augmentation for ASIC and FPGA design capture, simulation, and synthesis is Celoxica (www.celoxica.com) with its Handel-C language.

Of course, most synthesis tools would be capable of locating and exploiting the potential parallelisms in the above example, but they might well miss more complex cases that require human consideration and intervention. For the purposes of these discussions, however, we shall continue to work with this simple test case. The point is that an augmented C/C++ language will have keywords like "parallel" (or "par") and "sequential" (or "seq") that will instruct the downstream synthesis application as to which statements should be executed in parallel, and so forth. For example:

```
parallel;      /* Augmented C/C++ statement */
a = 6;         /* Standard  C/C++ statement */
b = 2;         /* Standard  C/C++ statement */
c = 9;         /* Standard  C/C++ statement */
sequential;    /* Augmented C/C++ statement */
d = a + b;     /* Standard  C/C++ statement */
  :
  etc
```

In this case, the *parallel* statement instructs the synthesis tool that the following statements can be implemented concurrently, while the *sequential* statement implies that the preceding operations must occur prior to any subsequent actions taking place. Of course, these parallel and sequential statements can be nested as required.

Things become more complex in the case of loops, depending on whether the designer wishes to unravel them partially or fully. Just to give a point of reference, we might visualize a loop as being something like *"for i = 1 to 10 in increments of 1 do xxxx, yyyy, and zzzz"*. *In some cases, it may be possible to simply associate a* parallel or sequential statement with the loop, but if more subtlety is required, the designer may be obliged to completely rewrite these constructs.

It may also be necessary to add *"share"* statements if resource sharing is required, and *"channel"* statements to share signals between expressions, and the list goes on.

As was previously noted, tools such as simulators that have not been enhanced to understand these new keywords and statements will "crash-and-burn" when presented with this representation. One solution is to "wrap" standard "#ifdef" directives around the new keywords and statements such that a precompiler can be used to discard them as required. However, this means that the simulator and synthesis engines will be working on different views of the design, which is typically not a good idea. The other solution is to use a proprietary simulator, but this may not have the power, capacity, or capabilities of your existing simulation technology.

1902:
US Navy installs
radiotelephones aboard
ships.

Augmented C/C++ design-flow alternatives

As usual, one might go various ways here. As we previously discussed, in the case of a design that begins life as a suite of algorithms, it is very common to start off by creating a C or C++ representation. Following verification, this C/C++ model can be incrementally modified by adding statements for clocks, pins, concurrency, synchronization, and resource sharing so as to make the model suitable for the appropriate synthesis utility. Alternatively, the design might be captured using the augmented C/C++ language from the get-go.

Irrespective of the actual route we might take to get there, let's assume that we are in possession of an augmented C/C++ representation of a design that is suitable for synthesis. Once again, there are two main design-flow alternatives, which are (1) to translate the augmented C/C++ into Verilog or VHDL at the RTL level of abstraction automatically and to then use conventional RTL synthesis technology, or (2) to use an appropriate augmented C/C++ synthesis engine.

And, once again, one school of thought says that synthesizing the augmented C/C++ directly into the implementation- level netlist offers the cleanest, fastest, and most efficient route. Others say that the RTL Verilog/VHDL level is the natural staging post for design integration and that today's RTL synthesis technology is extremely mature and powerful.

Both of these flows can be applied to ASIC or FPGA targets (Figure 11-6). The first augmented C/C++ synthesis applications were predominantly geared toward ASIC flows. This meant that these early incarnations didn't do a tremendous job when it came to inferring FPGA-specific entities such as embedded RAMs, embedded multipliers, and so forth. More recent versions of these tools do a much better job at this, but, as usual, the prospective user is strongly advised to perform some in-depth evaluations before handing over any hard-earned cash.

Note that figure 11-6 shows the use of implementation-specific code to drive the ASIC versus FPGA flows because

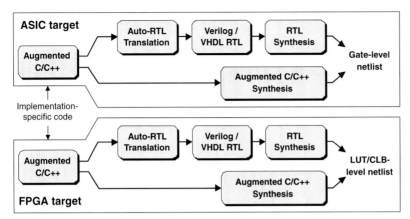

Figure 11-6. Alternative augmented C/C++ flows.

1904: England. John Ambrose Fleming invents the vacuum tube diode rectifier.

achieving an optimal implementation requires that the code be written with a specific target architecture in mind. In reality, the same code can be used to drive both ASIC and FPGA flows, but there is usually a penalty to be paid (see the discussions on SystemC for more details).

Pure C/C++-based flows

Last, but not least, we come to pure C/C++-based flows.[4] In reality, the term *pure* C/C++ actually refers to industry-standard C/C++ that is minimally augmented with SystemC data types to allow specific bit widths to be associated with variables and constants.

Although relatively new, pure C/C++-based flows offer a number of advantages as compared to other C-based flows and traditional Verilog-/VHDL-based flows:

- *Creating pure C/C++ is fast and efficient:* Pure untimed C/C++ representations are more compact and easier to

---

[4] At the time of this writing, perhaps the best example of a pure C/C++ based flow is provided by *Precision C Synthesis* from Mentor (www.mentor.com). Also of interest is the *SPARK* C-to-VHDL synthesis tool developed at the Center for Embedded Computer Systems, University of California, San Diego and Irvine (www.cecs.uci.edu/~spark).

1904:
First practical
photoelectric cell is
developed.

create and understand than equivalent SystemC and augmented C/C++ representations (and they are much more compact than their RTL equivalents, requiring perhaps 1/10th to 1/100th of the code).

■ *Verifying C/C++ is fast and efficient:* A pure untimed C/C++ representation will simulate significantly faster than a timed SystemC or augmented C/C++ model and 100 to 10,000 times faster than an equivalent RTL representation. In fact, pure C/C++ models are already widely created and used by system designers for algorithm and system validation.

■ *Evaluating alternative implementations is fast and efficient:* Modifying and reverifying pure untimed C/C++ to perform a series of what-if evaluations of alternative microarchitecture implementations is fast and efficient. This facilitates the design team's ability to arrive at fundamentally superior microarchitecture solutions. In turn, this can result in significantly smaller and faster designs as compared to flows based on traditional hand-coded RTL methods.

■ *Accommodating specification changes is relatively easy:* If any changes to the specification are made during the course of the project, it's relatively easy to implement and evaluate these changes in a pure untimed C/C++ representation, thereby allowing the changes to be folded into the resulting implementation.

Furthermore, as noted earlier in this chapter, one of the most significant problems associated with existing SystemC and augmented C/C++-based design flows is that the implementation intelligence associated with the design has to be hard-coded into the model, which therefore becomes implementation specific.

A key aspect associated with a pure untimed C/C++-based design flow is that the code presented to the synthesis engine is just what someone would write if he or she didn't have any preconceived hardware implementation or target device architecture in mind. This means that the C/C++ code that system designers write today is an ideal input to this form of synthesis. The only modification typically required to use a pure C/C++ model with the synthesis engine is to add a single special comment to the source code to indicate the top of the functional portion of the design (anything conceptually above this point is considered to form part of the testbench).

As opposed to adding intelligence to the source code (thereby locking it into a target implementation), all of the intelligence is provided by the user controlling and guiding the synthesis engine itself (Figure 11-7).

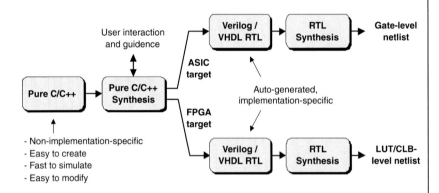

Figure 11-7. A pure untimed C/C++-based design flow.

Once the synthesis engine has parsed the source code, the user can use it to perform microarchitecture trade-offs and evaluate their effects in terms of size and speed. The synthesis engine analyzes the code, identifies its various constructs and operators, along with their associated data and memory dependencies, and automatically provides for parallelism wherever possible. The engine also provides a graphical interface that allows the user to specify how different elements should be handled. For example, the interface allows the user to associate ports with registers or RAM blocks; it identifies constructs like loops and allows the user to specify on an

1904:
First ultraviolet lamps are introduced.

1904:
Telephone answering
machine is invented.

individual basis whether they should be fully unraveled, par-tially unraveled, or left alone; it allows the user to specify whether or not loops and other constructs should be pipe-lined; it allows the user to perform resource sharing on specific entities; and so forth.

These evaluations are performed on the fly, and the syn-thesis engine reports total size/area and latency in terms of clock cycles and I/O delays (or throughput time/cycles in the case of pipelined designs). The user-defined configuration associated with each what-if scenario can be named, saved, and reused as required (it would be almost impossible to per-form these trade-offs in a timely manner using a conventional hand-coded RTL-based flow).

The fact that the pure untimed C/C++ source code used by the synthesis engine is not required to contain any imple-mentation intelligence and that all such intelligence is supplied by controlling the engine itself means that the same source code can be easily retargeted to alternative microarchi-tectures and different implementation technologies.

Once the user's evaluations are completed, clicking the "Go" button causes the synthesis engine to generate corre-sponding RTL VHDL. This code can subsequently be used by conventional logic synthesis or physically aware synthesis applications to generate the netlist used to drive the down-stream implementation (place-and-route, etc.) tools.

As usual, it would be possible to synthesize the pure unti-med C/C++ directly into a gate-level netlist (this alternative is not shown in figure 11-7). However, generating the inter-mediate RTL provides a comfort zone for the engineers by allowing them to check that they are satisfied with the imple-mentation decisions that have been made during the course of the C/C++ to RTL translation.

Furthermore, generating intermediate RTL is useful because this is the level of abstraction where hardware design engineers generally stitch together the various functional blocks forming their designs. Large portions of today's designs are typically presented in the form of IP blocks represented in RTL. This means that the intermediate RTL step shown in figure 11-7 is a useful point in the design flow for integrating and verifying the entire hardware system. The design engineers can then take full advantage of their existing RTL synthesis technology, which is mature, robust, and well understood.

Different levels of synthesis abstraction

The fundamental difference between the various C/C++-based flows presented in this chapter is the level of synthesis abstraction each can support. For example, although SystemC offers significant system-level, algorithmic, and transaction-level modeling capabilities, its synthesizable subset is at a relatively low level of abstraction. Similarly, although augmented C/C++ representations are closer to pure C/C++ than are their SystemC counterparts, which means that they simulate much more quickly, their synthesizable subset remains significantly lower than would be ideal.

This lack of synthesis abstraction causes the timed SystemC and augmented C/C++ representations to be implementation specific. In turn, this makes them difficult to create and modify and significantly reduces their flexibility with regard to performing what-if evaluations and retargeting them toward alternative implementation technologies (Figure 11-8).

By comparison, the latest generation of pure untimed C/C++ synthesis technology supports a high level of synthesis abstraction. Non-implementation-specific C/C++ models are very compact and can be quickly and easily created and modified. By means of the synthesis engine itself, the user can quickly and easily perform what-if evaluations and retarget the design toward alternative implementation technologies. The end result is that a pure C/C++-based design flow can dramatically speed implementation and increase design flexibility as compared to other C/C++-based flows.

Before anyone starts to pen irate letters claiming the author is anti-SystemC, it should be reiterated that the discussions presented here are focused on the use of the various flavors of C/C++ in the context of FPGA implementation flows.

In this case, the tool-chain used to progress SystemC representations through to actual implementations is relatively immature and unsophisticated.

When it comes to system-level modeling and verification applications, however, SystemC can be extremely efficacious (many observers see SystemC and SystemVerilog being used in conjunction with each other, with SystemC being employed for the initial system-level design representation, and then SystemVerilog being used to "flesh out" the implementation-level details).

Similarly, if one is coming from a software background and is working on embedded software applications and hardware/software co-design and co-verification, then SystemC is considered by many to be "the bees knees" as it were.

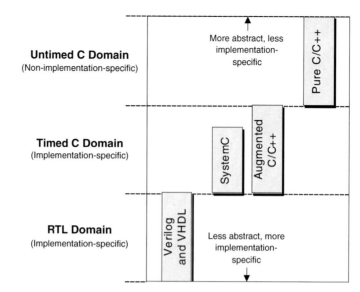

Figure 11-8. Different levels of C/C++ synthesis abstraction.

Mixed-language design and verification environments

Last, but not least, we should note that a number of EDA companies can provide mixed-level design and verification environments that can support the cosimulation of models specified at multiple levels of abstraction.

In some cases, this may simply involve linking a C/C++ model to a Verilog simulator via its *programming language interface (PLI)* or to a VHDL simulator via its *foreign language interface (FLI)*. Alternatively, one might find a SystemC environment with the ability to accept blocks represented in Verilog or VHDL.

And then there are very sophisticated environments that start off with a graphical block-based editor showing the design's major functional units, where the contents of each block can be represented using the following:

- VHDL
- Verilog
- SystemVerilog
- SystemC

One point that we haven't really considered is that, when you create a representation of your design in one of the flavors of C/C++ discussed here, you often create a testbench in the same language.

Such a testbench typically employs language constructs that aren't understood by any of the downstream tools like C/C++ to RTL generators. So in the past, you typically had to hand-translate the testbench from your C/C++ representation into an RTL equivalent for use with your VHDL/Verilog simulator.

- Handel-C
- Pure C/C++

The top-level design might be in a traditional HDL that calls submodules in the various HDLs and in one or more flavors of C/C++. Alternatively, the top-level design might be in one of the flavors of C/C++ that calls submodules in the various languages.

In this type of environment, the VHDL, Verilog, and SystemVerilog representations are usually handled by a single-kernel simulation engine. This engine is then cosimulated with appropriate engines for the various flavors of C/C++. Furthermore, this type of environment will incorporate source-code debuggers that support the various flavors of C/C++; it will allow testbenches to be created using any of the languages; and supporting tools like graphical waveform displays will be capable of displaying signals and variables associated with any of the language blocks.[5]

In reality, the various mixed-language design and verification environment solution combinations and permutations change on an almost weekly basis, so you need to take a good look at what's out there before you leap into the fray.

One advantage of a mixed-language design and verification environment is that you can continue to use your original C/C++ testbench to drive the downstream version of your design in VHDL/Verilog at the RTL and gate levels of abstraction. (You may need to "tweak" a few things, but that's a lot better than rewriting everything from the ground up.)

---

[5] A good example of a mixed-language simulation and verification environment of this type that is focused on FPGA—and, to a lesser extent, ASIC—designs is offered by Aldec Inc. (www.aldec.com). Another good example is *ModelSim*® from Mentor; this includes native SystemC support, thereby allowing single-kernel simulation between VHDL, Verilog, and SystemC.

DSP-Based Design Flows

Introducing DSP

The term *digital signal processing*, or *DSP*, refers to the branch of electronics concerned with the representation and manipulation of signals in digital form. This form of processing includes compression, decompression, modulation, error correction, filtering, and otherwise manipulating audio (voice, music, etc.), video, image, and silimar data for such applications as telecommunications, radar, and image processing (including medical imaging).

In many cases, the data to be processed starts out as a signal in the real (analog) world. This analog signal is periodically sampled, with each sample being converted into a digital equivalent by means of an *analog-to-digital (A/D)* converter (Figure 12-1).

Analog input signal	A/D
Digital input samples	DSP
Modified output samples	D/A
Analog output signal	

Analog domain ← → **Digital domain** ← → **Analog domain**

Figure 12-1. What is DSP?

These samples are then processed in the digital domain. In many cases, the processed digital samples are subsequently

DSP is pronounced by spelling it out as "D-S-P."

Analog is spelled "analogue" in England (and it is also pronounced with a really cool accent over there).

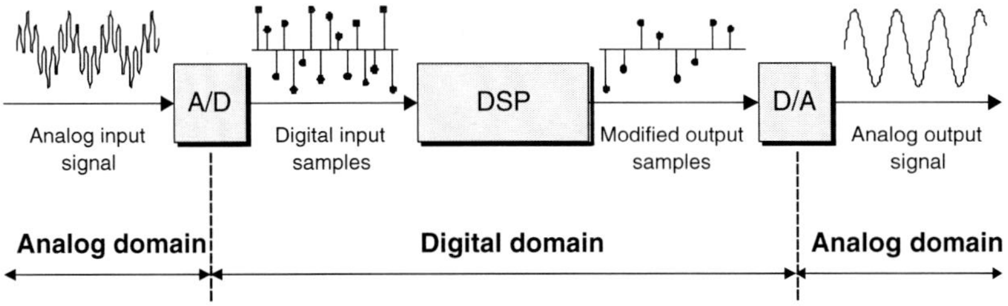

Analog-to-digital (A/D) converters may also be referred to as ADCs.

Digital-to-analog (D/A) converters may also be referred to as DACs.

The term CODEC is often bandied around by folks working in the DSP arena.

This sometimes stands for COmpressor/ DECompressor; that is, something that compresses and decompresses data.

In telecommunications, however, it typically stands for COder/ DECoder; that is, something that encodes and decodes a signal.

CODECs can be implemented in software, hardware, or as a mixture of both.

converted into an analog equivalent by means of a *digital-to-analog (D/A)* converter.

DSP occurs all over the place—in cell phones and telephone systems; CD, DVD, and MP3 players; cable desktop boxes; wireless and medical equipment; electronic vision systems; … the list goes on. This means that the overall DSP market is huge; in fact, some estimates put it at $10 billion in 2003!

Alternative DSP implementations

Pick a device, any device, but don't let me see which one

As usual, nothing is simple because DSP tasks can be implemented in a number of different ways:

- A *general-purpose microprocessor (μP)*: This may also be referred to as a central processing unit (CPU) or a microprocessor unit (MPU). The processor can perform DSP by running an appropriate DSP algorithm.
- A *digital signal processor (DSP)*: This is a special form of microprocessor chip (or core, as discussed below) that has been designed to perform DSP tasks much faster and more efficiently than can be achieved by means of a general-purpose microprocessor.
- *Dedicated ASIC hardware*: For the purposes of these discussions, we will assume that this refers to a custom hardware implementation that executes the DSP task. However, we should also note that the DSP task could be implemented in software by including a microprocessor or DSP core on the ASIC.
- *Dedicated FPGA hardware*: For the purposes of these discussions, we will assume that this refers to a custom hardware implementation that executes the DSP task. Once again, however, we should also note that the DSP functionality could be implemented in software by means of an embedded microprocessor

core on the FPGA (at the time of this writing, dedicated DSP hard cores do not exist for FPGAs).

1904:
Telephone answering
machine is invented.

System-level evaluation and algorithmic verification

Irrespective of the final implementation technology (µP, DSP, ASIC, FPGA), if one is creating a product that is to be based on a new DSP algorithm, it is common practice to first perform system-level evaluation and algorithmic verification using an appropriate environment (we consider this in more detail later in this chapter).

Although this book attempts to avoid focusing on companies and products as far as possible, it would be rather coy of us not to mention that—at the time of this writing—the de facto industry standard for DSP algorithmic verification is MATLAB[®1] from The MathWorks (www.mathworks.com).[2]

For the purposes of these discussions, therefore, we shall refer to MATLAB as necessary. However, it should be noted that there are a number of other very powerful tools and environments available to DSP developers. For example, Simulink[®] from The MathWorks has a certain following; the Signal Processing Worksystem (SPW) environment from CoWare[3] (www.coware.com) is very popular, especially in telecom markets; and tools from Elanix (www.elanix.com) also find favor with many designers.

Software running on a DSP core

Let's assume that our new DSP algorithm is to be implemented using a microprocessor or DSP chip (or core). In this case, the flow might be as shown in Figure 12-2.

---

[1] MATLAB and Simulink are registered trademarks of The MathWorks Inc.

[2] It should be noted that MATLAB and Simulink can be used for a wide range of tasks, including control system design and analysis, image processing, financial modeling, and so forth.

[3] EDA is a fast-moving beast. For example, SPW came under the auspices of Cadence when I first started penning this chapter, but it fell under the purview of CoWare by the time I was half-way through!

1906: America.
First radio program of
voice and music is
broadcast.

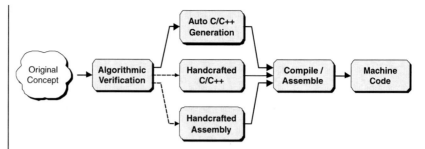

Figure 12-2. A simple design flow for a software DSP realization.

The process commences with someone having an idea for a new algorithm or suite of algorithms. This new concept typically undergoes verification using tools such as MATLAB as discussed above. In some cases, one might leap directly from the concept into handcrafting C/C++ (or assembly language).

Once the algorithms have been verified, they have to be regenerated in C/C++ or in assembly language. MATLAB can be used to generate C/C++ tuned for the target DSP core automatically, but in some cases, design teams may prefer to perform this translation step by hand because they feel that they can achieve a more optimal representation this way. As yet another alternative, one might first auto-generate C/C++ code from the algorithmic verification environment, analyze and profile this code to determine any performance bottlenecks, and then recode the most critical portions by hand. (This is a good example of the old 80:20 rule, in which you spend 80 percent of your time working on the most critical 20 percent of the design.)

Once you have your C/C++ (or assembly language) representation, you compile it (or assemble it) into the machine code that will ultimately be executed by the microprocessor or DSP core.

This type of implementation is very flexible because any desired changes can be addressed relatively quickly and easily by simply modifying and recompiling the source code. However, this also results in the slowest performance for the DSP algorithm because microprocessor and DSP chips are both classed as *Turing machines*. This means that their primary role in life is to process instructions, so both of these devices operate as follows:

- Fetch an instruction.
- Decode the instruction.
- Fetch a piece of data.
- Perform an operation on the data.
- Store the result somewhere.
- ⋮
- Fetch another instruction and start all over again.

Of course, the DSP algorithm actually runs on hardware in the form of the microprocessor or DSP, but we consider this to be a software implementation because the actual (physical) manifestation of the algorithm is the program that is executed on the chip.

Dedicated DSP hardware

There are myriad ways in which one might implement a DSP algorithm in an ASIC or FPGA—the latter option being the focus of this chapter, of course. But before we hurl ourselves into the mire, let's first consider how different architectures can affect the speed and area (in terms of silicon real estate) of the implementation.

DSP algorithms typically require huge numbers of multiplications and additions. As a really simple example, let's assume that we have a new DSP algorithm that contains an expression something like the following:

```
Y = (A * B) + (C * D) + (E * F) + (G * H);
```

As usual, this is a generic syntax that does not favor any particular HDL and is used only for the purposes of these discussions. Of course, this would be a minuscule element in a horrendously complex algorithm. But, at the end of the day, DSP algorithms tend to contain a lot of this type of thing.

In 1937, while still a graduate student, the eccentric English genius Alan Turing wrote his ground-breaking paper "On Computable Numbers with an Application to the Entscheidungsproblem." Since Turing did not have access to a real computer (not unreasonably as they did not exist at the time), he invented his own as an abstract "paper exercise." This theoretical model, which became known as a *Turing machine*, subsequently inspired many "thought experiments."

For the nontechnical reader, each of the variable names (A, B, C, etc.) in this equation is assumed to represent a bus (group) of binary signals. Also, when you multiply two binary values of the same width together, the result is twice the width (so if A and B are each 16 bits wide, the result of multiplying them together will be 32 bits wide).

1906:
Dunwoody and Pickard build a crystal-and-cat-whisker-radio.

The point is that we can exploit the parallelism inherent in hardware to perform DSP functions much more quickly than can be achieved by means of software running on a DSP core. For example, suppose that all of the multiplications were performed in parallel (simultaneously) followed by two stages of additions (Figure 12-3).

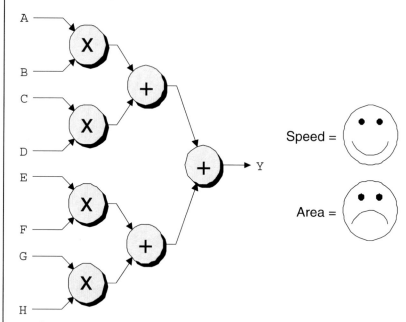

Figure 12-3. A parallel implementation of the function.

Remembering that multipliers are relatively large and complex and that adders are sort of large, this implementation will be very fast, but will consume a correspondingly large amount of chip resources.

As an alternative, we might employ *resource sharing* (sharing some of the multipliers and adders between multiple operations) and opt for a solution that is a mixture of parallel and serial (Figure 12-4).

This solution requires the addition of four 2:1 multiplexers and a register (remember that each of these will be the same multibit width as their respective signal paths). However, multiplexers and registers consume much less area than the

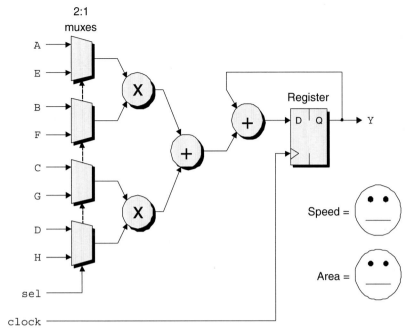

Figure 12-4. An in-between implementation of the function.

two multipliers and adder that are no longer required as compared to our initial solution.

On the downside, this approach is slower, because we must first perform the (A * B) and (C * D) multiplications, add the results together, add this total to the existing contents of the register (which will have been initialized to contain zero), and store the result in the register. Next, we must perform the (E * F) and (G * H) multiplications, add these results together, add this total to the existing contents of the register (which currently contains the results from the first set of multiplications and additions), and store this result in the register.

As yet another alternative, we might decide to use a fully serial solution (Figure 12-5).

This latter implementation is very efficient in terms of area because it requires only a single multiplier and a single adder. This is the slowest implementation, however, because we must first perform the (A * B) multiplication, add the result to the existing contents of the register (which will have been initial-

The process of trading off different datapath and control implementations is commonly known as micro-architecture exploration (see also chapter 11 for more discussions on this point).

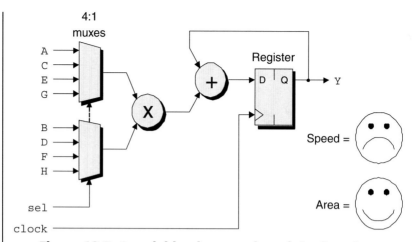

Figure 12-5. A serial implementation of the function.

ized to contain zero), and store the total in the register. Next, we must perform the (C * D) multiplication, add this result to the existing contents of the register, and store this new total in the register. And so forth for the remaining multiplication operations. (Note that when we say "this is the slowest implementation," we are referring to these hardware solutions, but even the slowest hardware implementation remains much, much faster than a software equivalent running on a microprocessor or DSP.)

DSP-related embedded FPGA resources

As previously discussed in chapter 4, some functions like multipliers are inherently slow if they are implemented by connecting a large number of programmable logic blocks together inside an FPGA. Because these functions are required by a lot of applications, many FPGAs incorporate special hard-wired multiplier blocks. (These are typically located in close proximity to embedded RAM blocks because these functions are often used in conjunction with each other.)

Similarly, some FPGAs offer dedicated adder blocks. One operation that is very common in DSP-type applications is called a multiply-and-accumulate. As its name would suggest, this function multiplies two numbers together and adds the result into a running total stored in an accumulator (register). Hence, it is commonly referred to as a MAC, which stands for *multiply, add,* and *accumulate* (Figure 12-6).

1907: America.
Lee de Forest creates a three-element vacuum tube amplifier (the triode).

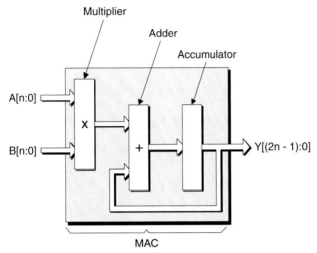

Figure 12-6. The functions forming a MAC.

Note that the multiplier, adder, and register portions of the serial implementation of our function shown in figure 12-5 offer a classic example of a MAC. If the FPGA you are working with supplies only embedded multipliers, you would be obliged to implement this function by combining the multiplier with an adder formed from a number of programmable logic blocks, while the result would be stored in a block RAM or in a number of distributed RAMs. Life becomes a little easier if the FPGA also provides embedded adders, and some FPGAs provide entire MACs as embedded functions.

FPGA-centric design flows for DSPs

Arrgggh! I'm quivering with fear (but let's call it anticipation) as I'm poised to pen these words. This is because, at the time of this writing, the idea of using FPGAs to perform DSP is still relatively new. Thus, there really are no definitive design flows or methodologies here—everyone seems to have his or her unique way of doing things, and whichever option you choose, you'll almost certainly end up breaking new ground one way or another.

Domain-specific languages

The way of the world is that electronic designs increase in size and complexity over time. In order to manage this problem while maintaining—or, more usually, increasing—productivity, it is necessary to keep raising the level of abstraction used to capture the design's functionality and verify its intent.

For this reason the gate-level schematics discussed in chapter 8 were superceded by the RTL representations in VHDL and Verilog discussed in chapter 9. Similarly, the drive toward C-based flows as discussed in chapter 11 is powered by the desire to capture complex concepts quickly and easily while facilitating architectural analysis and exploration.

In the case of specialist areas such as DSPs, system architects and design engineers can achieve a dramatic improvement in productivity by means of *domain-specific languages (DSLs)*, which provide more concise ways of representing specific tasks than do general-purpose languages such as C/C++ and SystemC.

One such language is MATLAB, which allows DSP designers to represent a signal transformation, such as an FFT, that can potentially take up an entire FPGA, using a single line of code[4] along the lines of

```
y = fft(x);
```

Actually, the term *MATLAB* refers both to a language and an algorithmic-level simulation environment. In order to avoid confusion, it is common to talk about M-code (meaning "MATLAB code") and M-files (files containing MATLAB code). Some engineers in the trenches occasionally refer to

DSL is pronounced by spelling it out as "D-S-L."

FFT is pronounced by spelling it out as "F-F-T."

The input stimulus to a MATLAB simulation might come from one or more mathematical functions such as a sine-wave generator, or it might be provided in the form of real-world data (for example, an audio or video file).

---

[4] Note that the semicolon shown in this example MATLAB statement is optional. If present, it serves to suppress the output display.

the "M language," but this is not argot favored by the folks at The MathWorks.

In addition to sophisticated transformation operators like the FFT shown above, there are also much simpler transformations like adders, subtractors, multipliers, logical operators, matrix arithmetic, and so forth. The more complex transformations like an FFT can be formed from these fundamental entities if required. The output from each transformation can be used as the input to one or more downstream transformations, and so forth, until the entire system has been represented at this high level of abstraction.

One important point is that such a system-level representation does not initially imply a hardware or software implementation. In the case of DSP core, for example, it could be that the entire function is implemented in software as discussed earlier in this chapter. Alternatively, the system architects could partition the design such that some functions are implemented in software, while other performance-critical tasks are implemented in hardware using dedicated ASIC or FPGA fabric. In this case, one typically needs to have access to a hardware or software codesign environment (see also chapter 13). For the purposes of these discussions, however, we shall assume pure hardware implementations.

System-level design and simulation environments

System-level design and simulation environments are conceptually at a higher level than DSLs. One well-known example of this genre is Simulink from The MathWorks. Depending on whom one is talking to, there may be a perception that Simulink is simply a graphical user interface to MATLAB. In reality, however, it is an independent dynamic modeling application that works *with* MATLAB.

If you are using Simulink, you typically commence the design process by creating a graphical block diagram of your system showing a schematic of functional blocks and the connections between them. Each of these blocks may be user-

M-files can contain scripts (actions to be performed) or transformations or a mixture of both. Also M-files can call other M-files in a hierarchical manner.

The primary (top-level) M-file typically contains a script that defines the simulation run. This script might prompt the user for information like the values of filter coefficients that are to be used, the name of an input stimulus file, and so forth, and then call other M-files and pass them these user-defined values as required.

First developed in 1962, FORTRAN (whose name was derived from its original use: formula translation) was one of the earliest high-level programming languages.

defined, or they may originate in one of the libraries supplied with Simulink (these include DSP, communications, and control function block sets). In the case of a user-defined block, you can "push" into that block and represent its contents as a new graphical block diagram. You can also create blocks containing MATLAB functions, M-code, C/C++, FORTRAN … the list goes on.

Once you've captured the design's intent, you use Simulink to simulate and verify its functionality. As with MATLAB, the input stimulus to a Simulink simulation might come from one or more mathematical functions, such as sine-wave generators, or it might be provided in the form of real-world data such as audio or video files. In many cases, it comes as a mixture of both; for example, real-world data might be augmented with pseudorandom noise supplied by a Simulink block.

The point here is that there's no hard-and-fast rule. Some DSP designers prefer to use MATLAB as their starting point, while others opt for Simulink (this latter case is much rarer in the scheme of things). Some folks say that this preference depends on the user's background (software DSP development versus ASIC/FPGA designs), but others say that this is a load of tosh. And it really doesn't matter, because, if the truth is told, the reasons behind who does what in this regard pale into insignificance compared to the horrors that are to come.

Floating-point versus fixed-point representations

Irrespective as to whether one opts for Simulink or MATLAB (or a similar environment from another vendor) as a starting point, the first-pass model of the system is almost invariably described using *floating-point* representations. In the context of the decimal number system, this refers to numbers like 1.235×10^3 (that is, a fractional number raised to some power of 10). In the context of applications like MATLAB, equivalent binary values are represented inside the computer using the IEEE standard for double-precision floating-point numbers.

Floating-point numbers of this type have the advantage of providing extremely accurate values across a tremendous dynamic range. However, implementing floating-point calculations of this type in dedicated FPGA or ASIC hardware requires a humongous amount of silicon resources, and the result is painfully slow (in hardware terms). Thus, at some stage, the design will be migrated over to use *fixed-point* representations, which refers to numbers having a fixed number of bits to represent their integer and fractional portions. This process is commonly referred to as *quantization.*

This is totally system/algorithm dependent, and it may take some considerable amount of experimentation to determine the optimum balance between using the fewest number of bits to represent a set of values (thereby decreasing the amount of silicon resources required and speeding the calculations), while maintaining sufficient accuracy to perform the task in hand. (One can think of this trade-off in terms of how much noise the designer is willing to accept for a given number of bits.) In some cases, designers may spend days deciding "should we use 14, 15, or 16 bits to represent these particular values?" And, just to increase the fun, it may be best to vary the number of bits used to represent values at different locations in the system/algorithm.

Things start to get really fun in that the conversion from floating-point to fixed-point representations may take place upstream in the system/algorithmic design and verification environment, or downstream in the C/C++ code. This is shown in more detail in the "System/algorithmic level to C/C++" section below. Suffice it to say that if one is working in a MATLAB environment, these conversions can be performed by passing the floating-point signals through special transformation functions called *quantizers.* Alternatively, if one is working in a Simulink environment, the conversions can be performed by running the floating-point signals through special fixed-point blocks.

1907:
Lee de Forest begins regular radio music broadcasts.

1908:
Charles Fredrick Cross
invents cellophane.

System/algorithmic level to RTL (manual translation)

At the time of this writing, many DSP design teams commence by performing their system-level evaluations and algorithmic validation in MATLAB (or the equivalent) using floating-point representations. (It is also very common to include an intermediate step in which a fixed-point C/C++ model is created for use in rapid simulation/validation.) At this point, many design teams bounce directly into hand-coding fixed-point RTL equivalents of the design in VHDL or Verilog (figure 12-7a). Alternatively, they may first transition the floating-point representations into their fixed-point counterparts at the system/algorithmic level, and then hand-code the RTL in VHDL or Verilog (Figure 12-7b).

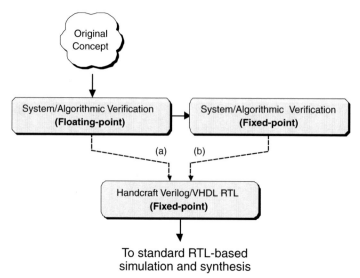

Figure 12-7. Manual RTL generation.

There are, of course, a number of problems with this flow, not the least being that there is a significant conceptual and representational divide between the system architects working at the system/algorithmic level and the hardware design engineers working with RTL representations in VHDL or Verilog.

Because the system/algorithmic and RTL domains are so different, manual translation from one to the other is time-consuming and prone to error. There is also the fact that the resulting RTL is implementation specific because realizing the optimal design in an FPGA requires a different RTL coding style from that used for an optimal ASIC implementation.

Another consideration is that manually modifying and reverifying RTL to perform a series of what-if evaluations of alternative microarchitecture implementations is extremely time-consuming (such evaluations may include performing certain operations in parallel versus sequential, pipelining portions of the design versus nonpipelining, sharing common resources—for example, two operations sharing a single multiplier—versus using dedicated resources, etc.)

Similarly, if any changes are made to the original specification during the course of the project, it's relatively easy to implement and evaluate these changes in the system-/algorithmic-level representations, but subsequently folding these changes into the RTL by hand can be painful and time-consuming.

Of course, once an RTL representation of the design has been created, we can assume the use of the downstream logic-synthesis-based flows that were introduced in chapter 9.

System/algorithmic level to RTL (automatic-generation)

As was noted in the previous section, performing system-/algorithmic-level-to -RTL translation manually is time-consuming and prone to error. There are alternatives, however, because some system-/algorithmic-level design environments offer direct VHDL or Verilog RTL code generation (Figure 12-8).

As usual, the system-/algorithmic-level design would commence by using floating-point representations. In one version of the flow, the system/algorithmic environment is used to migrate these representations into their fixed-point counter-

1909:
General Electric introduces the world's first electrical toaster.

1909:
Leo Baekeland patterns
an artificial plastic that
he calls Bakelite.

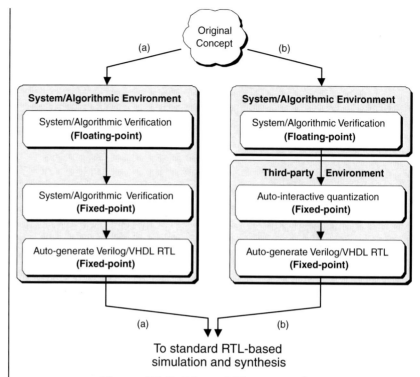

Figure 12-8. Direct RTL generation.

parts and then to generate the equivalent RTL in VHDL or
Verilog automatically (Figure 12-8a)[5].

Alternatively, a third-party environment might be used to
take the floating-point system-/algorithmic-level representation, autointeractively quantize it into its fixed-point
counterpart, and then automatically generate the equivalent
RTL in VHDL or Verilog (figure 12-8b)[6].

As before, once an RTL representation of the design has
been created, we can assume the use of the downstream logic-
synthesis-based flows that were introduced in chapter 9.

---

[5] A good example of this type of environment is offered by Elanix Inc.
(www.elanix.com).

[6] An example of this type of environment is offered by AccellChip Inc.
(www.accelchip.com), whose environment can accept floating-point
MATLAB M-files, output their fixed-point equivalents for verification,
and then use these new M-files to auto-generate RTL.

System/algorithmic level to C/C++ etc.

Due to the problems associated with exploring the design at the RTL level, there is an increasing trend to use a stepping-stone approach. This involves transitioning from the system-/algorithmic-level domain into to some sort of C/C++ representation, which itself is subsequently migrated into an RTL equivalent. One reason this is attractive is that the majority of DSP design teams already generate a C/C++ model for use as a golden (reference) model, in which case this sort of comes for free as far as the downstream RTL design engineer is concerned.

Of course, the first thing to decide is when and where in the flow one should transition from floating-point to fixed-point representations (Figure 12-9).

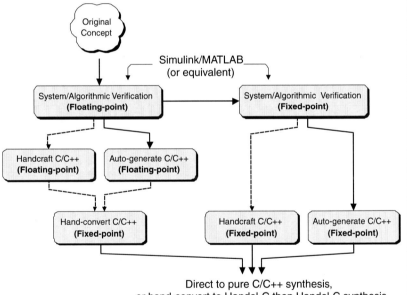

Figure 12-9. Migrating from floating point to fixed point.

It is somewhat difficult to qualify the relative effort associated with alternative paths through these flows. As a rule of thumb, one might make the following points:

a) Manual MATLAB to C/C++ translation is relatively easy, being in the order of hours to days (automatic translation is typically used only for simulation or DSP code generation depending on how critical the performance is).

b) Manual exploration of quantization effects is relatively easy, especially for experienced system designers (auto-interactive quantization is used less frequently). Also, many designers rely on noise analysis to guide them in this process.

c) Manual MATLAB or C/C++ to RTL translation is relatively hard, being in the order of weeks to months. Automation in this area provides a lot of value assuming it is possible to achieve sufficient quality of results.

Frighteningly enough, Figure 12-9 shows only a subset of the various potential flows. For example, in the case of the handcrafted options, as opposed to first hand-coding the

d) MATLAB/Simulink-based automated flows that rely on IP core generation are typically not well suited to designs that include substantial original content.

C/C++ and then gradually transmogrifying this representation into Handel-C or SystemC, one could hand-code directly into these languages.

However, the main thing to remember is that once we have a fixed-point representation in one of the flavors of C/C++, we can assume the use of the downstream C/C++ flows introduced in chapter 11 (one flow of particular interest in this area is the pure untimed C/C++ approach used by Precision C from Mentor).

Block-level IP environments

Nothing is simple in this world because there is always just one more way to do things. As an example, one might create a library of DSP functional blocks at the system/algorithmic level of abstraction along with a one-to-one equivalent library of blocks at the RTL level of abstraction in VHDL or Verilog.

The idea here is that you could then capture and verify your design using a hierarchy of functional blocks specified at the system/algorithmic level of abstraction. Once you were happy with your design, you could then generate a structural netlist instantiating the RTL-level blocks, and use this to drive downstream simulation and synthesis tools. (These blocks would have to be parameterized at all levels of abstraction so as to allow you to specify such things as bus widths and so forth.)

As an alternative, the larger FPGA vendors typically offer IP core generators (in this context, the term *core* is considered to refer to a block that performs a specific logical function; it does not refer to a microprocessor or DSP core). In several cases, these core generators have been integrated into system-/algorithmic-level environments. This means that you can create a design based on a collection of these blocks in the system-/algorithmic-level environment, specify any parameters associated with these blocks, and perform your system-/algorithmic-level verification.

Later, when you're ready to rock and roll, the core generator will automatically generate the hardware models corresponding to each of these blocks.[7] (The system-/algorithmic-level models and the hardware models ensuing from the core generator are bit identical and cycle identical.) In some cases the hardware blocks will be generated as synthesizable RTL in VHDL or Verilog. Alternatively, they may be presented as firm cores at the LUT/CLB level of abstraction, thereby making the maximum use of the targeted FPGA's internal resources.

One big drawback associated with this approach is that, by their very nature, IP blocks are based on hard-coded micro-architectures. This means that the ability to create highly tuned implementations to address specific design goals is somewhat diminished. The end result is that IP-based flows may achieve an implementation faster with less risk, but such an implementation may be less optimal in terms of area, performance, and power as compared to a custom hardware implementation.

Don't forget the testbench!

One point that the folks selling you DSP design tools often neglect to mention is the test bench. For example, let's assume that your flow involves taking your system-/algorithmic-level design and hand-translating it into RTL. In that case, you are going to have to do the same with your testbench. In many cases, this is a nontrivial task that can take days or weeks!

Or let's say that your flow is based on taking your floating-point system-/algorithmic-level design and hand-translating it into floating-point C/C++, at which point you will wish to verify this new representation. Then you might take your floating-point C/C++ and hand-translate it into fixed-point C/C++, at which point you will wish to verify this representation. And then you might take your fixed-point C/C++ and

1909:
Marconi shares Noble prize in physics for his contribution to telegraphy.

---

[7] A good example of this type of approach is the integration of Simulink with the *System Generator* utility from Xilinx (www.xilinx.com).

1909:
Radio distress signals save 1900 lives after two ships collide.

(hopefully) automatically synthesize an equivalent RTL representation, at which point … but you get my drift.[8]

The problem is that at each stage you are going to have to do the same thing with your testbench[9] (unless you do something cunning as discussed in the next (and last—hurray!) section.

Mixed DSP and VHDL/Verilog etc. environments

In the previous chapter, we noted that a number of EDA companies can provide mixed-level design and verification environments that can support the cosimulation of models specified at multiple levels of abstraction. For example, one might start off with a graphical block-based editor showing the design's major functional units, where the contents of each block can be represented using

- VHDL
- Verilog
- SystemVerilog
- SystemC
- Handel-C
- Pure C/C++

In this case, the top-level design might be in a traditional HDL that calls submodules represented in the various HDLs and in one or more flavors of C/C++. Alternatively, the top-level design might be in one of the flavors of C/C++ that calls submodules in the other languages.

More recently, integrations between system-/algorithmic-level and implementation-level environments have become available. The way in which this works depends on who is doing what and what that person is trying to do (sorry, I don't

[8] Don't laugh, because I personally know of one HUGE system house that does things in just this way!

[9] With regards to the C/C++ to RTL stage of the process, even if you have a C/C++ to RTL synthesis engine, your testbench will typically contain language constructs are aren't amenable to synthesis, which means that you're back to doing things by hand.

mean to be cryptic). For example, a system architect working at the system/algorithmic level (e.g., in MATLAB) might decide to replace one or more blocks with equivalent representations in VHDL or Verilog at the RTL level of abstraction. Alternatively, a design engineer working in VHDL or Verilog at the RTL level of abstraction might decide to call one or more blocks at the system/algorithmic level of abstraction.

Both of these cases require cosimulation between the system-/algorithmic-level environment and the VHDL/Verilog environment, the main difference being who calls whom. Of course, this sounds easy if you say it quickly, but there is a whole host of considerations to be addressed, such as synchronizing the concept of time between the two domains and specifying how different signal types are translated as they pass from one domain to the other (and back again).

This really is a case of treating any canned demonstration with a healthy amount of suspicion. If you are planning on doing this sort of thing, you need to sit down with the vendor's engineer and work your own example through from beginning to end. Call me an old cynic if you will, but my advice is to let their engineer guide you, while keeping your hands firmly on the keyboard and mouse. (You'd be amazed how much activity can go on in just a few seconds should you turn your head in response to the age-old question, "Good grief! Did you see what just flew by the window?")

1910: America. First installation of teleprinters on postal lines between New York City and Boston.

Embedded Processor-Based Design Flows

Introduction

For the purposes of this book, we are concerned only with electronic systems that include one or more FPGAs on the *printed circuit board (PCB)*. The vast majority of such systems also make use of a general-purpose microprocessor, or μP, to perform a variety of control and data-processing applications.[1] This is often referred to as the *central processing unit (CPU)* or *microprocessor unit (MPU)*.

Until recently, the CPU and its peripherals typically appeared in the form of discrete chips on the circuit board. There are an almost infinite number of possible scenarios here, but the two main ones involve the way in which the CPU is connected to its memory (Figure 13-1).

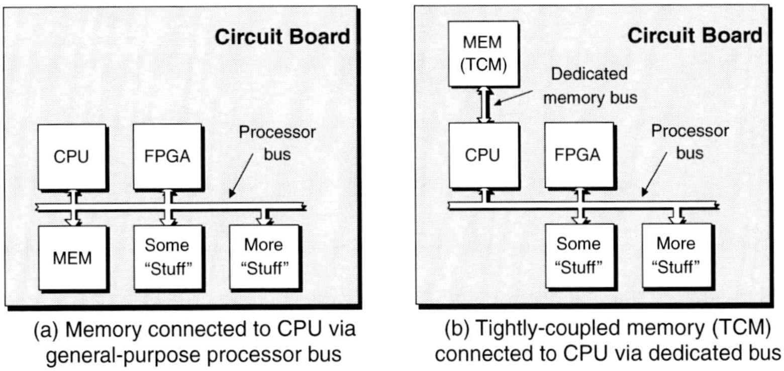

PCB is pronounced by spelling it out as "P-C-B."

CPU and MPU are pronounced by spelling them out as "C-P-U" and "M-P-U," respectively.

(a) Memory connected to CPU via general-purpose processor bus

(b) Tightly-coupled memory (TCM) connected to CPU via dedicated bus

Figure 13-1. Two scenarios at the circuit board level.

---

[1] Alternatively, one might use a *microcontroller (μC)* device, which combines a CPU core with selected peripherals and specialized inputs and outputs.

1910:
First electric washing machines are introduced.

In both of these scenarios, the CPU is connected to an FPGA and some other stuff via a general-purpose processor bus. (By "stuff" we predominantly mean peripheral devices such as counter timers, interrupt controllers, communications devices, etc.)

In some cases, the main *memory (MEM)* will also be connected to the CPU by means of the main processor bus, as shown in figure 13-1a (actually, this connection will be via a special peripheral called a memory controller, which is not shown here because we're trying to keep things simple). Alternatively, the memory may be connected directly to the CPU by means of a dedicated memory bus, as shown in Figure 13-1b).

The point is that presenting the CPU and its various peripheral devices in the form of dedicated chips on the circuit board costs money and occupies real estate. It also impacts the reliability of the board because every solder joint (connection point) is a potential failure mechanism.

One alternative is to embed the CPU along with some of its peripherals in the FPGA itself (Figure 13-2).[2]

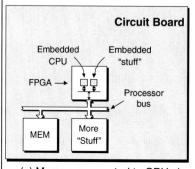

(a) Memory connected to CPU via general-purpose processor bus

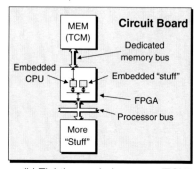

(b) Tightly-coupled memory (TCM) connected to CPU via dedicated bus

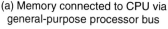

Figure 13-2. Two scenarios at the FPGA level.

---

2 Another alternative would be to embed a microprocessor core in an ASIC, but that's a tale for another book!

It is common for a relatively small amount of memory used by the CPU to be included locally in the FPGA. At the time of this writing, however, it is rare for all of the CPU's memory to be included in the FPGA.

Creating an FPGA design of this type brings a whole slew of new problems to the table. First of all, the system architects have to decide which functions will be implemented in software (as instructions to be executed by the CPU) and which functions will be implemented in hardware (using the main FPGA fabric). Next, the design environment must support the concept of coverification, in which the hardware and embedded software portions of the system can be verified together to ensure that everything works as it should. Both of these topics are considered in more detail later in this chapter.

Hard versus soft cores

Hard cores

A hard microprocessor core is one that is implemented as a dedicated, predefined (hardwired) block (these cores are only available in certain device families). Each of the main FPGA vendors has opted for a particular processor type to implement its hard cores. For example, Altera offer embedded ARM processors, QuickLogic have opted for MIPS-based solutions, and Xilinx sports PowerPC cores.

Of course, each vendor will be delighted to explain at great length why its implementation is far superior to any of the others (the problem of deciding which one actually is better is only compounded by the fact that different processors may be better suited to different tasks).

As noted in chapter 4, there are two main approaches for integrating such cores into the FPGA. The first is to locate it in a strip to the side of the main FPGA fabric (Figure 13-3).

In this scenario, all of the components are typically formed on the same silicon chip, although they could also be formed on two chips and packaged as a *multichip module* (MCM).

In addition to the microprocessor core itself, each FPGA vendor also supports an associated processor bus. For example, Altera and QuickLogic both support the *AMBA* bus from ARM (this is an open specification that can be downloaded from www.arm.com free of any charges).

By comparison, Xilinx embedded cores make use of the *CoreConnect* bus from IBM.

CoreConnect has two flavors. The main 64-bit bus is known as the *processor local bus (PLB)*. This can be used in conjunction with one or more 32-bit *on-chip peripheral busses (OPBs)*.

MCM is pronounced by spelling it out as "M-C-M."

1910: France.
George Claude
introduces neon lamps.

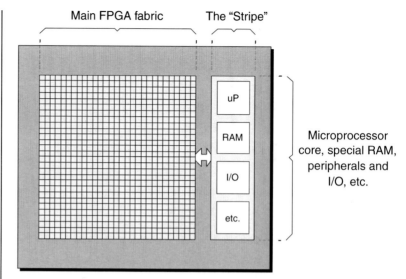

Figure 13-3. Bird's-eye view of chip with embedded core outside of the main fabric.

One advantage of this implementation is that the main FPGA fabric is identical for devices with and without the embedded microprocessor core, which can make things easier for the design tools used by the engineers. The other advantage is that the FPGA vendor can bundle a whole load of additional functions in the strip to complement the microprocessor core, such as memory and special peripherals.[3]

An alternative is to embed one or more microprocessor cores directly into the main FPGA fabric. One, two, and even four core implementations are currently available at the time of this writing (Figure 13-4).

In this case, the design tools have to be able to take account of the presence of these blocks in the fabric; any memory used by the core is formed from embedded RAM blocks, and any peripheral functions are formed from groups of general-purpose programmable logic blocks. Proponents of this scheme can argue that there are inherent speed advan-

---

[3] This approach is favored by vendors such as Altera (www.altera.com) and QuickLogic (www.quicklogic.com).

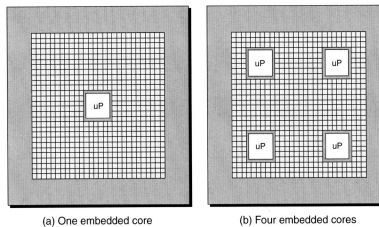

(a) One embedded core   (b) Four embedded cores

Figure 13-4. Bird's-eye view of chips with embedded cores inside the main fabric.

tages to be gained from having the microprocessor core in intimate proximity to the main FPGA fabric.[4]

Soft microprocessor cores

As opposed to embedding a microprocessor physically into the fabric of the chip, it is possible to configure a group of programmable logic blocks to act as a microprocessor. These are typically called "soft cores," but they may be more precisely categorized as either soft or firm, depending on the way in which the microprocessor's functionality is mapped onto the logic blocks. For example, if the core is provided in the form of an RTL netlist that will be synthesized with the other logic, then this truly is a soft implementation. Alternatively, if the core is presented in the form of a placed and routed block of LUTs/CLBs, then this would typically be considered a firm implementation.

In both of these cases, all of the peripheral devices like counter timers, interrupt controllers, memory controllers, communications functions, and so forth are also implemented as

One tool of interest in the soft core arena is LisaTek from CoWare Inc. (www.coware.com). Using a special language you define a required instruction set and microarchitecture (resources, pipelining, cycle timing) associated with a desired micro-processor. LisaTek takes this definition generates the corresponding RTL for your soft core, along with associated software tools such as a C compiler, assembler, linker, and instruction set simulator (ISS).

---

[4] This approach is favored by Xilinx (www.xilinx.com), who also provide a multitude of peripherals in the form of soft IP cores.

The Nios is based on a SPARC architecture using the concept of register windows, while the MicroBlaze is based on a classical RISC architecture.

IDE is pronounced by spelling it out as "I-D-E."

Depending on who you are talking to and the FPGA or RTOS vendor in question, the 'D' in IDE can stand for "design" or "development."

QuickLogic offer a 9-bit soft microcontroller that goes under the catchy name of Q90C1xx. (Having a 9-bit data word can be useful for certain communication functions.)

soft or firm cores (the FPGA vendors are typically able to supply a large library of such cores).

Soft cores are slower and simpler than their hard-core counterparts (of course they are still incredibly fast in human terms). However, in addition to being practically free, they also have the advantages that you only have to implement a core if you need it and that you can instantiate as many cores as you require until you run out of resources in the form of programmable logic blocks.

Once again, each of the main FPGA vendors has opted for a particular processor type to implement its soft cores. For example, Altera offers the *Nios*, while Xilinx sports the *Micro-Blaze*. The Nios has both 16-bit and 32-bit architectural variants, which operate on 16-bit or 32-bit chunks of data, respectively (both variants share the same 16-bit-wide instruction set). By comparison, the MicroBlaze is a true 32-bit machine (that is, it has 32-bit-wide instruction words and performs its magic on 32-bit chunks of data). Once again, each of the vendors will be more than happy to tell you why its soft core rules and how its competitors' offerings fail to make the grade (sorry, you're on your own here).

One cool thing about the *integrated development environment (IDE)* fielded by Xilinx is that it treats the PowerPC hard core and the MicroBlaze soft core identically. This includes both processors being based on the same CoreConnect processor bus and sharing common soft peripheral IP cores. All of this makes it relatively easy to migrate from one processor to the other.

Also of interest is the fact that Xilinx offers a small 8-bit soft core called the *PicoBlaze*, which can be implemented using only 150 logic cells (give or take a handful). By comparison, the MicroBlaze requires around 1,000 logic cells[5]

---

[5] For the purposes of these discussions, a logic cell can be assumed to contain a 4-input LUT, a register element, and various other bits and pieces like multiplexers and fast carry logic.

(which is still extremely reasonable for a 32-bit processor implementation, especially when one is playing with FPGAs that can contain 70,000[6] or more such cells.)

Partitioning a design into its hardware and software components

As noted in chapter 4, almost any portion of an electronic design can be realized in hardware (using logic gates and registers, etc.) or software (as instructions to be executed on a microprocessor). One of the main partitioning criteria is how fast you wish the various functions to perform their tasks:

- *Picosecond and nanosecond logic:* This has to run insanely fast, which mandates that it be implemented in hardware (in the FPGA fabric).
- *Microsecond logic:* This is reasonably fast and can be implemented either in hardware or software (this type of logic is where you spend the bulk of your time deciding which way to go).
- *Millisecond logic:* This is the logic used to implement interfaces such as reading switch positions and flashing light-emitting diodes, or LEDs. It's a pain slowing the hardware down to implement this sort of function (using huge counters to generate delays, for example). Thus, it's often better to implement these tasks as microprocessor code (because processors give you lousy speed—compared to dedicated hardware—but fantastic complexity).

The trick is to solve every problem in the most cost-effective way. Certain functions belong in hardware, others cry out for a software realization, and some functions can go either way depending on how you feel you can best use the resources

Some cynics say that those aspects of the design that are well understood are implemented in hardware, while any portions of the design that are somewhat undefined at the beginning of the design process are often relegated to a software realization (on the basis that the software can be "tweaked" right up until the last minute).

---

[6] This 70,000 value was true when I ate my breakfast this morning, but it will doubtless have increased by the time you come to read this book.

(both chip-level resources and hardware/software engineers) available to you.

It is possible to envisage an "ideal" *electronic system level (ESL)* environment in which the system architects initially capture the design via a graphical interface as a collection of functional blocks that are connected together. Each of these blocks could then be provided with a system-/algorithmic-level SystemC representation, for example, and the entire design could be verified prior to any decisions being made as to which portions of the design were to be implemented in hardware and software.

When it comes to the partitioning process itself, we might dream of having the ability to tag each graphical block with the mouse and select a hardware or software option for its implementation. All we would then have to do would be to click the "Go" button, and the environment would take care of synthesizing the hardware, compiling the software, and pulling everything together.

And then we return to the real world with a resounding thud. Actually, a number of next-generation design environments show promise, and new tools and techniques are arriving on an almost daily basis. At the time of this writing, however, it is still very common for system architects to partition a design into its hardware and software portions by hand, and to then pass these top-level functions over to the appropriate engineers and hope for the best.

With regard to the software portion of the design, this might be something as simple as a state machine used to control a human-level interface (reading the state of switches and controlling display devices). Although the state machine itself may be quite tricky, this level of software is certainly not rocket science. At the other end of the spectrum, one might have incredibly complex software requirements, including

- System initialization routines and a hardware abstraction layer
- A hardware diagnostic test suite

RTOS is pronounced by spelling it out as "R-T-O-S."

Real-time systems are those in which the correctness of a computation or action depends not only on how it is performed but also when it is performed.

- A real-time operating system (RTOS)
- RTOS device drivers
- Any embedded application code

1911:
Dutch physicist Heike Kamerlingh Onnes discovers superconductivity.

This code will typically be captured in C/C++ and then compiled down to the machine instructions that will be run on the processor core (in extreme cases where one is attempting to squeeze the last drop of performance out of the design, certain routines may be handcrafted in assembly code).

At the same time, the hardware design engineers will typically be capturing their portions of the design at the RTL level of abstraction using VHDL or Verilog (or SystemVerilog).

Today's designs are so complex that their hardware and software portions have to be verified together. Unfortunately, wrapping one's brain around the plethora of coverification alternatives and intricacies can make a grown man (well, me actually) break down and weep.

Hardware versus software views of the world

One of the biggest problems to overcome when it comes to the coverification of the hardware and software portions of a design is the two totally different worldviews of their creators.

The hardware folks typically visualize their portion of the design as blocks of RTL representing such things as registers, logical functions, and the wires connecting them together. When hardware engineers are debugging their portion of the design, they think in terms of an editor showing their RTL source code, a logic simulator, and a graphical waveform display showing signals changing values at specific times. In a typical hardware design environment, clicking on a particular event in the waveform display will automatically locate the corresponding line of RTL code that caused this event to occur.

By comparison, the software guys and gals think in terms of C/C++ source code, of registers in the CPU (and in the

1912: America.
Dr Sidney Russell
invents the electric
blanket.

peripherals), and of the contents of various memory locations. When software engineers are debugging a program, they often wish to single-step through the code one line at a time and watch the values in the various registers changing. Or they might wish to set one or more breakpoints (this refers to placing markers at specific points in the code), run the program until they hit one of those breakpoints, and then pause to see what's going on. Alternatively, they might wish to specify certain conditions such as a register containing a particular value, then run the program until this condition is met, and once again pause to see what's happening.

When a software developer is writing application code such as a game, he or she has the luxury of being reasonably confident that the hardware (say, a home computer) is reasonably robust and bug-free. However, it's a different ball game when one is talking about a software engineer creating embedded applications intended to run on hardware that's being designed at the same time. When a problem occurs, it can be mega tricky determining if it was a fault in the software or if the hardware was to blame. The classic joke is a conversation between the two camps:

> **Software Engineer:** "I think I may have hit a hardware problem while running my embedded application."
> **Hardware Engineer:** "At what time did the error occur? Can you give me a test case that isolates the problem?"
> **Software Engineer:** "The error occurred at 9:30 this morning, and the test case is my application!"

In the case of today's state-of-the-art coverification environments, the hardware and software worlds are tightly coupled. This means that if the software engineers detect a potential hardware bug, identifying the particular line of code being executed will take the hardware engineers directly to the corresponding simulation time frame in the graphical waveform display. Similarly, if the hardware engineers detect a potential software bug (such as code requesting an illegal hardware transaction), they can use their interface to guide the software team to the corresponding line of source code. Unfortunately, this type of environment can cost a lot of money, so sometimes you have to opt for a less sophisticated solution.

Using an FPGA as its own development environment

Perhaps the simplest place to start is the scenario where the FPGA is used as its own development environment. The idea here is that you have an SRAM-based FPGA with an embedded processor (hard or soft) mounted on a development board that's connected to your computer. In addition to the FPGA, this development board will also have a memory device that will be used to store the software programs that are to be run by the embedded CPU (figure 13-5).

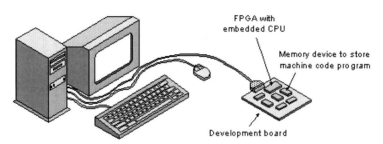

Figure 13-5. Using an FPGA as its own development environment.

Once the system architects have determined which portions of the design are to be implemented in hardware and software, the hardware engineers start to capture their RTL blocks and functions and synthesize them down to a LUT/CLB-level netlist. Meanwhile, the software engineers start to capture their C/C++ programs and routines and compile them down to machine code. Eventually, the LUT/CLB-level netlist will be loaded into the FPGA via a configuration file, the linked machine code image will be loaded into the memory device, and then you let the system run wild and free (Figure 13-6).

1912:
Feedback and heterodyne systems usher in modern radio reception.

1912:
The *Titanic* sends out radio distress signals when it collides with an iceberg and sinks on its maiden voyage.

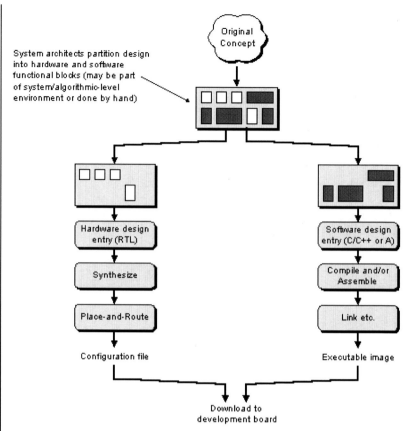

Figure 13-6. A (very) simple design flow.

Also, any of the machine code that is to be embedded in the FPGA's on-chip RAM blocks would actually be loaded via the configuration file.

Improving visibility in the design

The main problem with the scenario discussed in the previous section is lack of "visibility" as to what is happening in the hardware portion of the design. One way to mitigate this is to use a virtual logic analyzer to observe what's happening in the hardware (this is discussed in more detail in Chapter 16).

Things can be a little trickier when it comes to determining what's happening with the software. One point to remember is that—as discussed in chapter 5—an embedded CPU core will have its own dedicated JTAG boundary scan chain (Figure 13-7).

1913:
William D.Coolidge invents the hot-tungsten filament X-ray tube. This Coolidge Tube becomes the standard generator for medical X-rays.

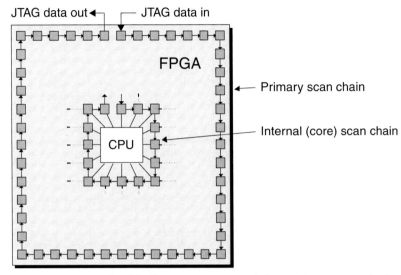

Figure 13-7. Embedded processor JTAG boundary scan chain.

This is true of both hard cores and the more sophisticated soft cores. In this case, the coverification environment can use the scan chain to monitor the activity on the buses and control signals connecting the CPU to the rest of the system. The CPU's internal registers can also be accessed via the JTAG port, thereby allowing an external debugger to take control of the device and single-step through instructions, set breakpoints, and so forth.

A few coverification alternatives

If you really want to get visibility into what's happening in the hardware portions of design, one approach is to use a logic simulator. In this case, the majority of the system will be modeled and simulated in VHDL or Verilog/SystemVerilog at the RTL level of abstraction. When it comes to the CPU core, however, there are various ways in which to represent this (Figure 13-8).

1914: America.
Traffic lights are used
for the first time (in
Cleveland, Ohio)

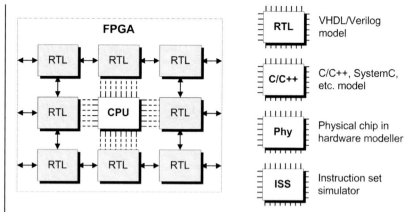

Figure 13-8. Alternative representations of the CPU.

Irrespective of the type of model used to represent the CPU, the embedded software (machine code) portion of the design will be loaded into some form of memory—either embedded memory in the FPGA or external memory devices—and the CPU model will then execute those machine code instructions.

Note that figure 13-8 shows a high-level representation of the contents of the FPGA only. If the machine code is to be stored in external memory devices, then these devices would also have to be part of the simulation. In fact, as a general rule of thumb, if the software talks to any stuff, then that stuff needs to be part of the coverification environment.

RTL (VHDL or Verilog)

Perhaps the simplest option here is when one has an RTL model of the CPU, in which case all of the activity takes place in the logic simulator. One disadvantage of this approach is that a CPU performs tremendous numbers of internal operations in order to perform the simplest task, which equates to incredibly slow simulation runs (you'll be lucky to be able to simulate 10 to 20 system clocks per second in real time).

The other disadvantage is that you have no visibility into what the software is doing at the source code level. All you'll be able to do is to observe logic values changing on wires and inside registers.

And there's always the fact that whoever supplies the real CPU doesn't want you to know how it works internally because that supplier may be using cunning proprietary tricks and wish to preserve their IP. In this case, you may well find it very difficult to lay your hands on an RTL model of the CPU at all.

C/C++, SystemC, etc.

As opposed to using an RTL model, it is very common to have access to some sort of C/C++ model of the CPU. (The proponents of SystemC have a vision of a world in which the CPU and the main peripheral devices all have SystemC models provided as standard for use in this type of design environment.)

The compiled version of this CPU model would be linked into the simulation via the *programming language interface (PLI)* in the case of a Verilog simulator or the *foreign language interface (FLI)*—or equivalent—in the case of a VHDL simulator.

The advantages of such a model are that it will run much faster than its RTL counterpart; that it can be delivered in compiled form, thereby preserving any secret IP; and that, at least in FPGA circles, such a model is usually provided for free (the FPGA vendors are trying to sell chips, not models).

One disadvantage of this approach is that the C/C++ model may not provide a 100 percent cycle-accurate representation of the CPU, which has the potential to cause problems if you aren't careful. But, once again, the main disadvantage of such a model is that its only purpose is to provide an engine to execute the machine code program, which means that you have no visibility into what the software is doing at the source code level. All you'll be able to do is observe logic values changing on wires and inside registers.

Way back in the mists of time, the Logic Modeling Corporation (LMC)—which was subsequently acquired by Synopsys—defined an interface for connecting behavioral models of hardware blocks to logic simulators. This is known as the SWIFT interface, and models—such as CPUs—that comply with this specification may be referred to as SWIFT models.

Physical chip in hardware modeler

Yet another possibility is to use a physical device to represent a hard CPU core. For example, if you are using a PowerPC core in a Xilinx FPGA, you can easily lay your hands on a real PowerPC chip. This chip can be installed in a box called a hardware modeler, which can then be linked into the logic simulation system.

The advantage of this approach is that you know the physical model (chip) is going to functionally match your hard core as closely as possible. Some disadvantages are that hardware modelers aren't cheap and they can be a pain to use.

The majority of hardware-modeler-based solutions don't support source-level debugging, which, once again, means that you have no visibility into what the software is doing at the source code level.[7] All you'll be able to do is to observe logic values changing on wires and inside registers.

Instruction set simulator

As previously noted, in certain cases, the role of the software portion of a design may be somewhat limited. For example, the software may be acting as a state machine used to control some interface. Alternatively, the software's role may be to initialize certain aspects of the hardware and then sit back and watch the hardware do all of the work. If this is the case, then a C/C++ model or a physical model is probably sufficient—at least as far as the hardware design engineer is concerned.

At the other extreme, the hardware portions of the design may exist mainly to act as an interface with the outside world. For example, the hardware may read in a packet of data and store it in the FPGA's memory, and then the CPU may perform huge amounts of complex processing on this data. In

ISS is pronounced by spelling it out as "I_S_S."

7 Actually, some hardware modelers do provide a certain amount of source-level debug capability, for example, Simpod Inc. (www.simpod.com) offers an interesting solution.

cases like these, it is necessary for the software engineer to have sophisticated source-level debugging capabilities. This requires the use of an *instruction set simulator (ISS)*, which provides a virtual representation of the CPU.

Although an ISS will almost certainly be created in C/C++, it will be architected very differently from the C/C++ models of the CPU discussed earlier in this section. This is because the ISS is created at a very high level of abstraction; it thinks in terms of transactions like "get me a word of data from location *x* in the memory," and it doesn't concern itself with details like how signals will behave in the real world. The easiest way to explain how this works is by means of an illustration (Figure 13-9).

1914:
Better triode improves radio reception.

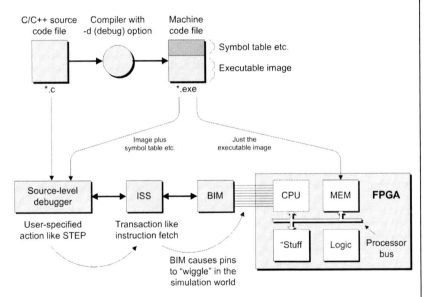

Figure 13-9. How an ISS fits into the picture.

First of all, the software engineers capture their program as C/C++ source code. This is then compiled using the -d (debug) option, which generates a symbol table and other debug-specific information along with the executable machine code image.

1914:
First trans-continental
telephone call.

When we come to perform the coverification, there are a number of pieces to the puzzle. At one end we have the source-level debugger, whose interface is used by the software engineer to talk to the environment. At the other end we have the logic simulator, which is simulating representations of the memory, stuff like peripheral devices, general-purpose logic, and so forth (for the sake of simplicity, this illustration assumes that all of the program memory resides in the FPGA itself).

In the case of the CPU, however, the logic simulator essentially sees a hole where this function should be. To be more precise, the simulator actually sees a set of inputs and outputs corresponding to the CPU. These inputs and outputs are connected to an entity called a *bus interface model (BIM)*, which acts as a translator between the simulator and the ISS.

Both the source code and the executable image (along with the symbol table and other debug-centric information) are loaded into the source-level debugger. At the same time, the executable image is loaded into the MEM block. When the user requests the source-level debugger to perform an action like stepping through a line of source code, it issues commands to the ISS. In turn, the ISS will execute high-level transactions such as an instruction fetch, or a memory read/write, or an I/O command. These transactions are passed to the BIM, which causes the appropriate pins to "wiggle" in the simulation world.

Similarly, when something connected to the processor bus in the FPGA attempts to talk to the CPU, it will cause the pins driving the BIM to "wriggle." The BIM will translate these low-level actions into high-level transactions that it passes to the ISS, which will in turn inform the source-level debugger what's happening. The source-level debugger will then display the state of the program variables, the CPU registers, and other information of this ilk.

There are a variety of incredibly sophisticated (often frighteningly expensive) environments of this type on the market.[8] Each has its own cunning tricks and capabilities, and some are more appropriate for ASIC designs than FPGAs or vice versa. As usual, however, this is a moving target, so you need to check around to see who is doing what before putting any of your precious money on the table.

A rather cunning design environment

As far as possible (and insofar as makes sense), this book attempts to steer away from discussing specific companies and products. But there's an exception to every rule, and this is it, because a company called Altium Ltd. (www.altium.com) has come up with a rather cunning FPGA design environment called Nexar that deserves mention.

It's difficult to know where to start, so let's kick off by saying that we're talking about a complete FPGA hardware/ software codesign and coverification environment for around $7,995.[9] This environment targets engineers designing things like simple controllers for domestic appliances like washing machines and is based on the fact that you can now purchase FPGAs containing more than 1 million system gates for around $20.[10]

Nexar includes a hardware development board that plugs into the back of your PC. This development board comes equipped with two daughter cards: one carrying a Xilinx FPGA and the other equipped with an Altera device. Nexar also features a number of soft microprocessor cores that replicate the functionality of industry-standard 8-bit devices like the 8051, Z80, and PIC microcontrollers (a range of 16-bit and 32-bit processor and DSP cores are planned for the future). Also

---

[8] For example, Seamless from Mentor (www.mentor.com), Incisive from Cadence (www.cadence.com), and XoC from Axis Systems (www.axissystems.com).

[9] This price was true circa November 2003.

[10] Again, this gate-count and price are circa November 2003.

1915:
First trans-atlantic radio
telephone conversation

included are a library of peripheral devices, a library of around 1,500 component blocks that range from simple gates to more complex functions such as counters, and a small RTOS.

By means of a schematic capture interface, the user places blocks representing the processors, peripherals, and various logic functions and wires them together. All of the blocks supplied with Nexar are provided royalty-free. These blocks have been presynthesized, so when you are ready to rock and roll, they can be directly downloaded into the FPGA on the development board. (If necessary, you can also create your own blocks and capture their contents in RTL. These will subsequently be processed by the synthesis engine bundled with Nexar.)

Clicking on a processor block allows you to enter the C/C++ source code program to be associated with that processor. This will subsequently be processed by one of the compilers bundled with Nexar.

The idea is that everything associated with the design—hardware and software—will be downloaded into the FPGA on the development board. In order to see what's happening in the hardware, you can include a variety of virtual instrument blocks in your schematic, including logic analyzers, frequency counters, frequency generators, and so forth. When it comes to the software, Nexar provides a source-level debugger that allows you to perform all of the usual tasks like setting breakpoints, specifying watch expressions, single-stepping, stepping over, stepping into, and so on.

What can I say? I've actually seen one of these little rascals performing its magic, and I was impressed. I really like the fact that this is essentially a turnkey solution, and you get everything (no costly add-ons required) in a package the size of a shoebox. And for the class of designs it is targeting, I personally think that Nexar is going to be a hard act to follow for some time to come.

Modular and Incremental Design

Handling things as one big chunk

In order to provide a basis for these discussions, let's consider an FPGA as containing a series of columns, each of which comprises large numbers of programmable logic blocks along with some blocks of RAM and other hard-wired elements such as multipliers or MACs (Figure 14-1).

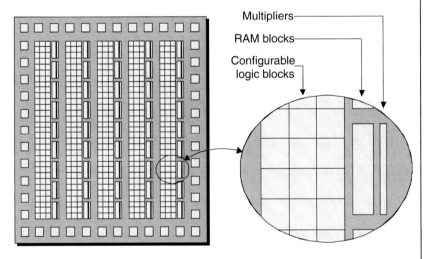

Multipliers ————

RAM blocks ————

Configurable logic blocks

Figure 14-1. A column-based architecture.

Of course, this illustration is a gross simplification, because a modern device can contain more columns than you can swing a stick at and each column can contain humongous amounts of programmable logic, and so forth.

1917:
Clarence Birdseye
preserves food by
means of freezing.

When we initially discussed the programming of SRAM-based FPGAs in chapter 5, we stated that we could visualize all of the SRAM configuration cells as comprising a single (long) shift register. For example, consider a simple bird's-eye view of the surface of the chip showing only the I/O pins/pads and the SRAM configuration cells (Figure 14-2).

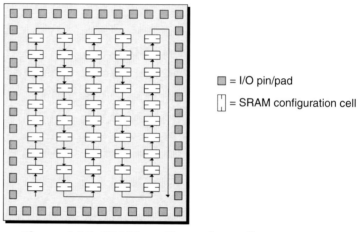

☐ = I/O pin/pad

⊟ = SRAM configuration cell

Figure 14-2. SRAM configuration cells presented as a single (long) register chain.

Once again, we can think of the SRAM configuration cells as a series of columns, each of which maps onto one of the columns of programmable logic shown in figure 14-1. This, too, is a grossly simplified representation because an FPGA can contain tens of millions of these configuration cells, but it will serve our purposes here. The ways in which the two ends of this register chain are made accessible to the outside world will depend on the selected programming mode (this is not relevant to these discussions).

In the early days of FPGA-based designs—circa the mid to late 1980s—devices were relatively small in terms of their logic capacity. One by-product of this was that a single design engineer was typically in charge of creating all of the RTL for the device. This RTL was subsequently synthesized, and the ensuing netlist was passed to the place-and-route software, which processed the design in its entirety.

The result was a monolithic configuration file that defined the function of the entire device and would be loaded as one big chunk. This obviously worked well with having the configuration cells presented as a single long register chain, so everyone was happy.

Partitioning things into smaller chunks

Over time, FPGAs have grown larger and more sophisticated, while the size and complexity of designs have increased by leaps and bounds. One way to address this is to partition the design into functional blocks and to give each block to one or more design engineers.

Each of these blocks can be synthesized in isolation. At the end of the day, however, all of the netlists associated with the blocks are gathered back together before being handed over to the place-and-route applications. Once again, place-and-route typically works on the design in its entirety, which can require an overnight run when you're talking about multimillion-gate designs.

Somewhere around 2002, some FPGA vendors started to offer larger devices in which the SRAM configuration cells are presented as multiple (relatively short) register chains (Ffigure 14-3).

The idea of presenting the device with these multiple chains may have been conceived with the concepts of modular and incremental design practices in mind. Alternatively, it may have come about for some mundane hardware-related reason, and then some bright spark said, "Just a minute, now that we have these multiple chains, what if we started to support the concepts of modular and incremental design?"

If I were a betting man, I'd probably put my money on the latter option, but let's be charitable and assume that someone somewhere actually knew what he or she was doing (hey, it could happen). However it came about, the end result of this architecture is that, along with associated software applications, it can support the concepts of modular and incremental design.

1917: Frank Conrad builds a radio station that eventually becomes KDKA (this call sign is still used to this day).

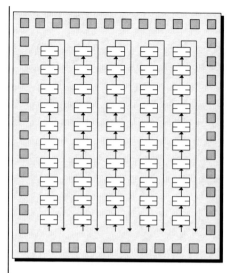

☐ = I/O pin/pad

☐ = SRAM configuration cell

Figure 14-3. SRAM configuration cells presented as multiple (relatively short) register chains.

Modular design

The terms "block-based" and "bottom-up" may also be associated with modular design.

Known as *team design* by some, this refers to the concept of partitioning a large design into functional blocks and giving each block, along with its associated timing constraints, to a different design engineer or group of engineers. The RTL for each model is captured and synthesized independently, and the final physical netlist is handed over to a system integrator.

Ultimately, each block (or a small group of blocks) will be assigned to a specific area in the device. The system integrator is responsible for "stitching" all of these areas together. In a way, this is similar to having a design split across multiple FPGAs, except that everything is in the same device.

The primary advantage of this scenario is that the netlist for each area can be run through the place-and-route applications in isolation (these tools will be given constraints restricting them to specific, predefined areas). This means that each team member can complete his or her portion of the design to the point that it fully meets its timing requirements after implementation, not just after synthesis.

Incremental design

This refers to the fact that, so long as you've tied down the interfaces between blocks/columns, you can modify the RTL associated with a particular block, resynthesize that block, and rerun place-and-route on that block in isolation. This is much, much faster than having to rerun place-and-route on the entire design.

Actually, the term *isolation* as used in the previous paragraph is possibly a tad misleading. It may be more appropriate to say that the incremental design tools "freeze" all of the unchanged blocks in place, and they only reimplement the changed block(s) in the context of the entire design. This provides an advantage over modular design in which the other blocks aren't present (of course, the modular and incremental design techniques may be used in conjunction with each other).

On the downside

One problem with the techniques described here is that they can lead to substantial waste of resources because, at the time of this writing, the finest resolution is that of an entire column, so if a particular functional block only occupies, say, 75 percent of the logic in that column, the remaining 25 percent will remain unused and go to waste. (FPGA vendors who support these architectures are talking about providing mechanisms to support finer resolutions in the future.)

Another problem is that the methodology described here is almost bound to result in "tall-and-thin" implementations for each functional block because you are essentially restricting the blocks to one or more vertical columns. This is obviously a pain in the case of those functional blocks that would benefit from a "short-and-fat" realization (spanning multiple columns and using small portions of those columns).

Perhaps the most significant problem with the early releases of tools and flows using these architectures to support

1918:
First radio link between UK and Australia.

1919:
People can dial their
own telephone
numbers.

modular and incremental design practices is that someone
(say the system integrator) is obliged to create a floor plan by
hand. This poor soul also has to define and place special interface blocks called *bus macros* that are used to link buses and
individual signals crossing from one block to another
(Figure 14-4).

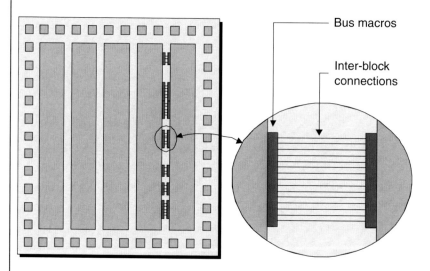

Figure 14-4. Placing bus macros.

The initial implementation of the tools made creating the
floor plan and defining and locating the bus macros awkward
to say the least. The rumor on the streets is that changes to
the software are in the offing that will greatly simplify this
process (on the bright side, it couldn't get any harder).

There's always another way

Way back in chapter 10, we introduced the concept of
FPGA-centric silicon virtual prototypes, or SVPs. At that
time, we noted that some EDA vendors have started to provide tools that support the concept of an FPGA SVP by
providing a mixture of floor planning and pre-place-and-route

timing analysis. This is coupled with the ability to perform place-and-route on individual design blocks, which dramatically speeds up the implementation process.[1]

The point is, if you go back and reread that chapter, you'll find that the implementation of an FPGA SVP described there fully supports the concepts of modular and incremental design without any of the problems associated with the techniques presented in this chapter. The only problem is that, being much more sophisticated, the tools from an EDA vendor will be substantially more expensive than the offerings from the FPGA vendors. As always, it's a case of "you pay your money and you make your choice."

1919:
Short-wave radio is invented.

---

[1] At the time of writing, one of the chief proponents of FPGA SVPs—in the form described in this book—is Hier Design (www.hierdesign.com).

High-Speed Design and Other PCB Considerations

Before we start

If you are desperately seeking information on FPGAs containing gigabit serial I/O transceivers, then you're in the wrong place, and you need to bounce over to chapter 21.

We were all so much younger then

In many respects, life was so much simpler for FPGA design engineers in the not-so-distant past (let's say 1990, just to stick a stake in the ground). In those halcyon days, no one gave much thought to the lot of the poor old layout designer tasked with creating the PCB.

Here's the way things went. First of all, even the highest-end FPGAs only had around 200 pins, which is relatively few by today's standards. If these pins were presented in a *pin grid array (PGA)* package, the pin pitch (the distance between pins) on these devices was around 1/10 inch (2.5 mm), which is absolutely huge by today's standards. Last but not least, signal delays through devices like FPGAs were massively large compared to the signal delays along circuit board tracks. All of these points led to a fairly simplistic design flow.

The process would commence with the system architects creating a very rough floor plan of the circuit board by hand—usually on a whiteboard or a scrap of paper. In fact, "floor plan" is probably too strong a term for what we're talking about, which was really more of a sketch showing the major components and the major connection paths between them (Figure 15-1).

PGA is pronounced by spelling it out as "P-G-A."

A PGA package has an array of pins presented across the bottom face of the device. The circuit board is created with a corresponding set of holes or vias. These devices are attached to the circuit board by pushing each pin through a corresponding hole or via in the board.

Circa 1990s, FPGAs presented in PGA packages were predominantly used for military applications; the norm for commercial applications was the *plastic quad flat package (PQFP)* with pins presented around the perimeter of the device.

1919:
The concept of flip-flop (memory) circuits is invented.

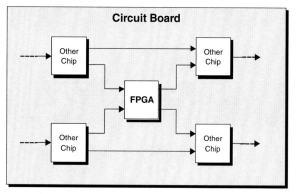

Figure 15-1. The system architects sketch a rough floor plan.

Based on this floor plan, the system architects would wave their hands around, make educated guesses about a whole range of things, and eventually pull some input-to-output timing constraints for the FPGA out of the air.

Armed with these timing constraints and a specification of the function the FPGA was to perform, the design engineer (remember, there was typically only one engineer per device) would wander off into the sunset to perform his or her machinations.

Generally speaking, it was relatively rare for design engineers to worry too much about FPGA pin assignments. To a large extent, they would let the place-and-route software run wild and free, and they would accept any pin assignments it decided upon.

Once the FPGA design, including the pin assignments, had been finalized, someone would be tasked with creating a graphical symbol of the device for use with schematic capture, along with a graphical representation of the device's physical footprint for use in the circuit board layout environment. These symbols would include details as to the signal names associated with the physical pins (and the physical locations of the pins in the case of the layout representation).

Meanwhile, the circuit board designer would have been working away in the background placing the other devices and, as far as possible, routing them. It was only after the

FPGA design had been finalized and the symbol created that the FPGA could be fully integrated into the circuit board environment and the routing completed. This meant that, at the end of the day, it was largely left up to the circuit board designer to make everything work.

The bad news was that when we said that the FPGA design had been "finalized," we really meant that hopefully it was getting close. In the real world, it was almost invariably the case that as soon as the circuit board designer had routed the final track, the FPGA engineer remembered a tweak that just had to be made. Implementing this tweak often ended up modifying the pin assignments, which left the circuit board designer feeling somewhat blue (it was not unknown for strong words to ensue).

The times they are a-changing

Frightening as it may seem, the simplistic flow discussed above persisted throughout most of the 1990s, but the size and complexity of today's FPGA devices means that this flow simply can't stand up under the strain.

At the time of this writing, we're talking about high-end FPGAs containing as many as 1,700 pins presented in *ball grid array (BGA)* packages with pin pitches of only 1 mm. Furthermore, today's ICs (including FPGAs) are as fast as lightning compared to their ancestors, which makes the delays associated with the circuit board tracks much more significant.

The bottom line is that it is no longer acceptable for the system architects to assign timing constraints to the FPGA in a fairly arbitrary manner and then leave it up to the circuit board designer to make things work at the end of the day. This scenario just won't fly. Instead, the process needs to start at the board level with the FPGA being treated as a black box (Figure 15-2).

In this case, the circuit board layout designer performs board-level timing based on a preliminary placement, and this information is used to calculate realistic constraints to feed to the FPGA design engineer. In the case of modern designs,

BGA is pronounced by spelling it out as "B-G-A."

A BGA package has an array of pads presented across the bottom face of the device. The circuit board is created with a corresponding set of pads. Each pad on the FPGA has a small ball of solder attached to it. These devices are attached to the circuit board by placing them in the correct location and then melting the solder balls to form good ball-to-pad connections.

1919:
Walter Schottky invents
the Tetrode (the first
multiple-grid vacuum
tube).

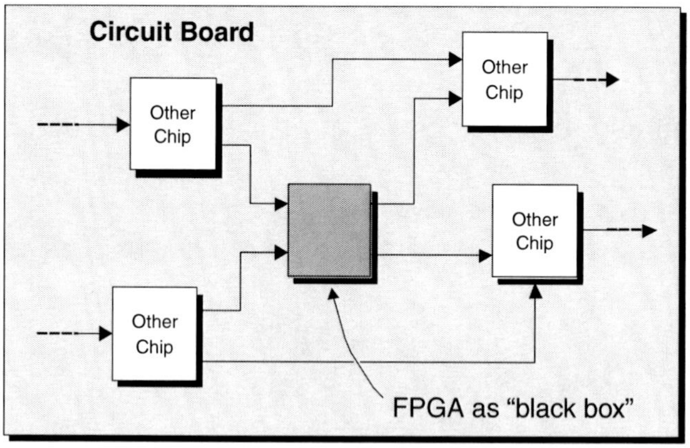

Figure 15-2. The circuit board designer performs preliminary placement.

there could be hundreds or thousands of such timing constraints, and it simply wouldn't be possible to generate and prioritize them without performing this board-level analysis.

But wait; we have to go farther than this. In order to ensure that the FPGA can be routed successfully, it's now the board designer who has to perform the initial assignments of signals to the FPGA's I/O pins. In order to do this, new tools are becoming available to the board designer. These tools provide a graphical representation of the physical footprint for the device along with an interactive interface that allows the user to declare signal names and associate them with specific device pins.[1]

These tools also provide for the auto-generation of the schematic symbol. In the case of devices with 1,000 or more pins, the tool can partition the symbol into multiple parts. One popular push-button option is to create these partitions based on the FPGA's I/O banks, but it's also possible to define partitions by hand on a pin-by-pin basis.

---

[1] At the time of this writing, a good example of the current state-of-the-art is the *BoardLink Pro* application from Mentor (www.mentor.com).

Once the circuit board designer has performed this up-front work, it's necessary to have some mechanism by which to transfer these pin assignments over to the FPGA design engineer, who will use them as physical constraints to guide the FPGA's place-and-route applications. In the real world, there may still be a number of iterations if the FPGA engineer finds it necessary to make modifications to the original pin assignments, but these tend to be minor compared to the horrors seen when using the flow of yesteryear as was introduced earlier in this chapter.

FPGA Xchange

Until recently, the passing of data back and forth between the board designer and the FPGA design engineer has typically involved a substantial amount of hands-on tweaking to get things to work. That is set to change because a new ASCII file format called *FPGA Xchange* is being defined by Mentor in conjunction with Altera, Xilinx, and the other major FPGA players.

This format will allow the circuit board tools and the FPGA tools to share common definitions of device aspects, such as how signal names have been assigned to physical device pins. This will allow board designers and FPGA engineers to pass data between their two domains quickly and easily.

For example, the board designer may create the original pin assignments and use the associated FPGA Xchange file to pass these as constraints to the FPGA engineer's place-and-route tools. The board designer can then proceed to layout the circuit board.

Meanwhile, the FPGA engineer may find it necessary to modify some of the pin assignments. These changes would be incorporated into the original FPGA Xchange file, which would subsequently be used by the board-level layout software to rip up any tracks associated with pins that had changed. These tracks can then be rerouted automatically or auto-interactively.

1921:
Albert hull invents the Magnetron (a microwave generator).

Other things to think about

High-speed designs

There is a common misconception that the term *high-speed design* means having a fast system clock. In reality, high-speed effects are associated with the speed of edges (the rate at which signals transition from logic 0 to logic 1, and vice versa). The faster the edge, the more significant are *signal inegrity (SI)* issues such as noise, crosstalk, and the like. Now it's certainly true that as the frequency of the system clock increases, the speed of edges also has to increase, but you can run into high-speed design problems with even a one megahertz clock if your signals have fast edge rates (and the vast majority of signals have fast edge rates these days).

SI is pronounced by spelling it out as "S-I."

SI analysis

One of the nice things about FPGAs is that the vendor has already dealt with the vast majority of SI issues inside the chip; however, it is becoming increasingly important to perform SI analysis at the board level. The best tools aren't cheap, but neither is creating a board that doesn't work. So you have to choose whether to perform the SI analysis or just roll the dice and see what happens.

SPICE versus IBIS

Performing SI analysis at the board level using SPICE models can be time-consuming. In the early 1990s, Intel created and promoted the *input/output buffer information specification (IBIS)*, which is a modeling format that describes the analog characteristics of drivers and receivers.

The reason for Intel's largesse was that they didn't want to give detailed SPICE models to customers because these models are at the transistor-capacitor-resistor level, and they can provide a lot of information that a component vendor might not wish its competitors to be aware of.

IBIS models are behavioral in nature and any process-related information is hidden. However, these models are only

SPICE, which is pronounced like the seasoning, stands for *simulation program with integrated circuit emphasis.* This analog simulation program was developed by the University of California at Berkeley and was made available for widespread use around the beginning of the 1970s.

accurate up to some maximum frequency, which can range from 500 megahertz to 1 gigahertz, depending on who you are talking to. After that point you are obliged to use a more accurate model such as SPICE.

Another problem is that the language has to be extended in order to accommodate new technologies. For example, IBIS has no mechanism to model the effects of pre-emphasis (see also chapter 21). However, the IBIS syntax is not inherently extensible, and augmenting the language via the various open forum committees is a long-winded process (by the time you get anything done, there's a new technology to worry about).

In late 2002, a proposal was made to augment the IBIS standard (this proposal was called BIRD75, where "BIRD" stands for *buffer information resolution document*). This proposal would allow calls to external models on a pin-by-pin basis. If adopted, this will allow extensibility, because the external models can be represented in languages such as SPICE, VHDL-AMS, Verilog-A, and so forth.

Startup power

Some FPGAs can have substantial power supply requirements due to high transient startup currents. Board-level designers need to check with the FPGA team to determine these requirements so as to ensure that the board can supply sufficient power to avoid any problems.

Use of internal termination impedances

Nearly all modern high-speed I/O standards require that the tracks on the circuit board have specific impedances and associated termination resistors (having the correct values eliminates the reflection and ringing effects that degrade SI and affect system performance).

Using termination resistors that are external to the device may necessitate additional layers in the board, resulting in higher costs and longer development times. In the case of FPGAs with hundreds or thousands of pins, it is almost impossible to place these termination resistors within reasonable

IBIS is prononuced "eye-bis."

DCI is pronounced by spelling it out as "D-C-I."

proximity to the device (distances greater than 1 cm from the pin cause problems). For these reasons, some FPGAs include *digitally controlled impedance (DCI)* capability.

Available on both inputs and outputs, DCI termination can be configured to support parallel and series termination schemes. These on-chip resistor values are completely user definable, and the digital implementation of this technology means that their values do not vary with changes in temperature or supply voltages.

A simple rule of thumb is that for any signals with rise/fall times of 500 picoseconds or less, external termination resistors cause discontinuities in the signal, in which case you should be using their on-chip counterparts.

Pushing data around in parallel versus serial

It is common in electronic systems to process groups of bits—called *words*—in parallel, where the width of the word depends on the system. In the case of 8-bit microprocessors and microcontrollers, for example, words are, perhaps not surprisingly, 8 bits wide.

In the days of yore, when device manufacturers agonized over every additional pin on an IC package, it was common for chips to include a function called a *universal asynchronous receiver transmitter (UART)*. Assuming 8-bit words, if the chip wished to send information to the outside world, the UART would convert an 8-bit byte of data from the internal bus into a series of pulses for transmission. Similarly, if the chip wished to access information from the outside world, the UART would accept that information as a series of pulses, collect it into an 8-bit byte, and place that byte on the internal bus. Thus, chips using this technique required only two pins to write and read data: *transmit data (TXD)* and *receive data (RXD)*.

As packaging technologies improved, increasing the number of pins became less of a burden, so it became more and more common to pass entire words of data around. In the case of an 8-bit system, this would require eight tracks on the circuit board and eight pins on each chip that was connected to this bus.

Over time, it became necessary to push more information around the system and to do so faster. Thus, bus widths increased to 16 bits, then 32 bits, then 64 bits, and so on. At the same time, clock speeds increased from integer multiples of megahertz, to tens of megahertz, to hundreds of megahertz, to thousands of megahertz, where 1,000 megahertz equates to 1 gigahertz.

As the speed of the system clock increases, it becomes more and more problematic to route wide buses around a circuit board with any hope of getting the signals where you want them to be, at the time you want them to be there, without running into all sorts of SI problems in the form of noise and crosstalk.[2] Thus, for the highest bandwidth applications, designers are turning back to serial data transmission in the form of gigabit transceivers. These are introduced in more detail in Chapter 21.

1921:
Canadian-American John Augustus Larson invents the polygraph lie detector.

---

[2] I know this is a long sentence, but that's appropriate because it's been a long day!

Observing Internal Nodes in an FPGA

Lack of visibility

One of the problems associated with debugging any chip design—be it an ASIC or an FPGA—is the lack of visibility as to what activity is taking place inside the device. Purely for the sake of these discussions, let's assume that we have a really simple pipelined design comprising a few registers and logic gates (Figure 16-1).

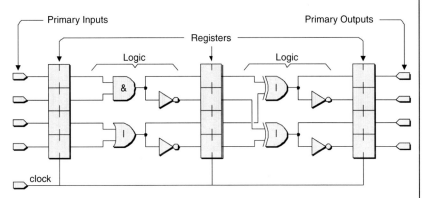

Figure 16-1. A very simple pipelined circuit.

Obviously, this is something of a nonsense circuit (you'd be amazed how tricky it can be to make something up that doesn't cloud the issue), but it will serve our purposes here.

The problem is that we only have access to the chip's primary inputs and outputs, so we can't see what's happening inside. This isn't particularly important when the design has been completed and verified, but it's a pain when we are trying

1921:
Czech author Karal
Capek coins the term
robot in his play *R.U.R.*

1921:
First use of quartz
crystals to keep radios
from wandering off
station.

to debug the chip to determine why it isn't doing what we expected it to.

One obvious solution would be to make the internal nodes visible by connecting them to primary output pins from the device (Figure 16-2).

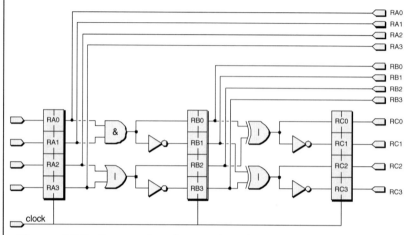

Figure 16-2. Connecting internal nodes to primary outputs.

The downside to this scheme is that most designs are "I/O limited," which means that the bottleneck is the number of primary I/O pins available on the package. In fact, even if we don't use any I/O pins to access internal nodes, many FPGA designs already leave a pile of internal resources unused because there aren't enough I/O pins available to convey all of the required control and data signals into and out of the device.

Multiplexing as a solution

One simple alternative is to multiplex the main outputs with the internal signals and bring them all out through the same set of output pins (Figure 16-3).

Of course, the select control would also require the use of some primary I/O pins. In the example shown here, the simplest case would be to bring the two select control signals directly to the outside world, which would therefore require

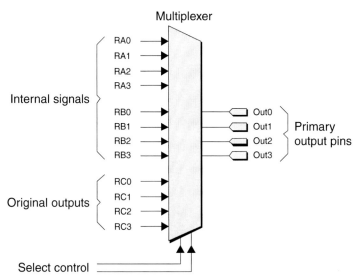

Figure 16-3. Multiplexing signals.

Internal signals: RA0, RA1, RA2, RA3, RB0, RB1, RB2, RB3

Original outputs: RC0, RC1, RC2, RC3

Multiplexer

Out0, Out1, Out2, Out3 — Primary output pins

Select control

1922:
First commercial broadcast ($100 for a 10-minute advert).

1923:
First neon advertising signs are introduced.

two I/O pins. Alternatively, we might use a small portion of logic to implement a simple state machine that required only a single I/O to act as a sort of clock to cycle between states, each of which would cause the multiplexer to select a different set of signals (see also the discussions on VirtualWires later in this chapter).

The main advantages of this scheme are that it offers great visibility and it's relatively fast. The main disadvantages are that it's relatively inflexible and time-consuming to implement because if you wish to change the internal pins that are being monitored, you have to modify the design's source code and then resynthesize it. Similarly, if you wish to change any trigger conditions that might be used by a state machine to determine which set of signals is to be selected by the multiplexer at any particular time, you once again have to change the design source code.

Another point to consider is that if, once you've debugged everything, you delete these test structures from your source code, you may introduce new problems into the design, not the least of which is that the routing will change, along with its associated delays.

Special debugging circuitry

Some FPGAs include special debugging circuitry that allows you to observe internal nodes. For example, FPGAs from Actel feature two special pins called PRA ("Probe A") and PRB ("Probe B"). By means of the embedded debugging circuitry combined with the use of a special debugging utility,[1] any internal signal can be connected to either of these pins, allowing the values on that node to be observed and analyzed.

The big advantage of this type of scheme is that you don't need to touch your source code. The disadvantage is that having only two probe pins might be considered a tad limiting when you have potentially hundreds of thousands of internal signals to worry about.

Virtual logic analyzers

Although the schemes discussed above are useful, it is often desirable to have access to more extensive logic analyzer instrumentation to allow for the tracing and debugging of groups of embedded signals, along with the ability to analyze signals in the context of other related signals or under specific triggering events.

OCI is prononcued by spelling it out as "O-C-I."

On-chip instrumentation (OCI) is an analysis approach that facilitates logic debugging by allowing the user to embed diagnostic IP blocks such as virtual logic analyzer applications into their designs. The idea here is to use some of the FPGA's resources to implement one or more virtual logic analyzer blocks that capture the activity of selected signals. This data will be stored in one or more of the FPGA's embedded RAM blocks, from whence it can be accessed by the external logic analyzer software by means of the device's JTAG port (Ffigure 16-4).

A portion of the virtual logic analyzer will be devoted to detecting trigger conditions on specific signals, where these

---

1 This used to be called *Actionprobe*®. Then a "new and improved" version called *Silicon Explorer* became available. At the time of this writing, *Silicon Explorer II* is the flavor of the day, and as for tomorrow …

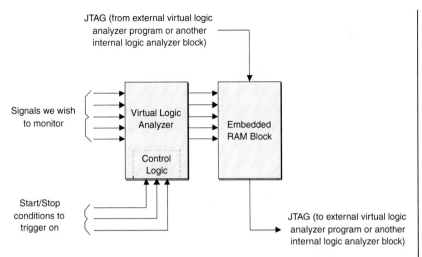

JTAG (from external virtual logic analyzer program or another internal logic analyzer block)

Signals we wish to monitor

Virtual Logic Analyzer

Embedded RAM Block

Control Logic

Start/Stop conditions to trigger on

JTAG (to external virtual logic analyzer program or another internal logic analyzer block)

Figure 16-4. Virtual logic analyzers.

1923: First photoelectric cell is introduced.

1923: First ship-to-ship communications (people on one ship can talk to people on another).

conditions will be used to start and stop the data capture on the signals being monitored.

Depending on the particular virtual logic analyzer implementation you are working with, you may or may not have to modify your design's source code to include this functionality. The big advantage of this type of scheme is that, even if you do have to include special macros in your source code, it's relatively easy to implement extremely sophisticated debugging capabilities in your design.

In some cases, the FPGA vendor will offer this sort of capability. Good examples of debugging tools of this ilk are *Chipscope*™ Pro from Xilinx (www.xilinx.com) and *SignalTap*® II from Altera (www.altera.com). Alternatively, if you are working with devices from a vendor who doesn't offer this capability, one option is to go to a third party such as First Silicon Solutions (www.fs2.com), which specializes in OCI and debugging for FPGA logic and embedded processors.

With regard to a virtual logic analyzer for use in tracing, analyzing, and debugging embedded signals in FPGAs, First Silicon Solutions boasts its *configurable logic analyzer module (CLAM)*. This little scallywag consists of an OCI block (available in both Verilog and VHDL) that is configured and

1925: America. Scientist, engineer, and politician Vannevar Bush designs an analogue computer called the Product Intergraph.

1925: First commercial picture/facsimile radio service across the USA.

synthesized as part of the design. This block (or blocks if you use more than one) is used in conjunction with control, probe, and display software that resides on your host PC.

VirtualWires

Sometime around the early 1990s, a company called Virtual Machine Works introduced a technology they called VirtualWires™. Originally intended as a technique for implementing massive multi-FPGA systems, VirtualWires provided a basis for a variety of FPGA-based emulation systems. One reason for mentioning it here is that it bears some similarities to the multiplexing solutions discussed earlier in this chapter. Another reason is that it's a really cool idea.

The problem

The starting point for the VirtualWires concept is that you have a large design that is too big to fit into a single FPGA, so you wish to split it across a number of devices. As a simple example, let's assume we have a design that equates to some number of system gates, but that the largest FPGA available offers only half this number of gates. Thus, an initial knee-jerk solution would almost certainly be to split the design across two devices (Figure 16-5).

Note that the logic in this figure is shown as comprising a number of subblocks labeled A through H. This is intended only to provide an aid in visualizing the way in which the logic might be partitioned across the devices.

The problem is that the chips typically won't have enough I/O pins to satisfy the requirements of the main inputs and outputs to the design along with the signals linking the two portions of the design. Prior to VirtualWires (or any similar concept), the only option was to further partition the design across more devices (Figure 16-6).

But now we have a new problem in that we are wasting huge amounts of each FPGA's logic resources, with the result that we are using way too many chips.

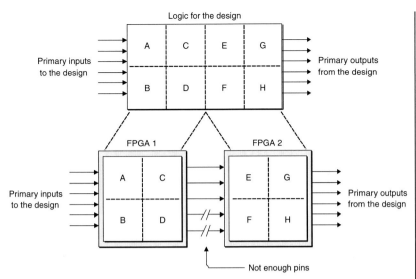

Figure 16-5. Not enough pins if we try to split the design across two devices.

1926: America. Dr. Julius Edgar Lilienfield from New York files a patent for what we would now recognize as an *npn* junction transistor being used in the role of an amplifier.

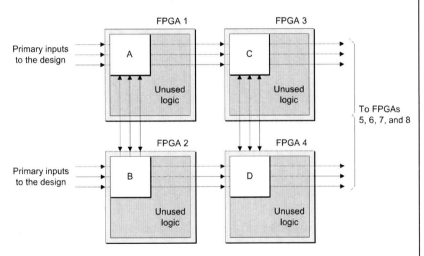

Figure 16-6. Lots of wasted FPGA logic resources when partitioning across multiple devices.

The VirtualWires solution

In order to see how the VirtualWires concept addresses our problem, let's first assume an extreme case in which we have access to some very strange FPGAs that can boast only three pins (two inputs and one output, where one of the inputs

1926: America.
First pop-up bread
toaster is introduced.

1926:
First commercial
picture/facsimile radio
service across the
Atlantic.

assumes the role of a clock). In this case, we would probably
end up using only a very small amount of each FPGA's logic
resources (Figure 16-7).

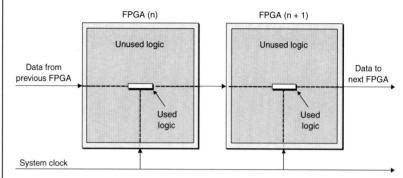

**Figure 16-7. An extreme case in which each
FPGA has only three pins.**

The idea behind VirtualWires is that, since we are wasting
so much of each device's internal resources anyway, we might
as well use some of these resources to implement some special
circuitry that allows our single data input to be latched into a
number of registers, each of which can be used to drive its
own block of logic. Similarly, the outputs from each of the
blocks of logic can be multiplexed together and registered
(Figure 16-8).

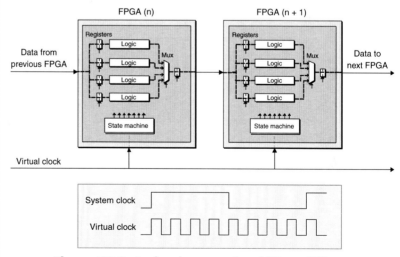

Figure 16-8. A simple example of VirtualWires.

Note that our original system clock has been superceded by a virtual clock, which subdivides each "beat" of the system clock into some number of "ticks." Also note that a state machine is implemented inside each FPGA. These state machines are used to enable and disable individual registers and also to control the multiplexers and so forth. (Of course, figure 16-8 is not to scale—the state machines and other VirtualWires structures would actually consume relatively little of the logic resources in each device compared to the number of logic blocks that are actually implementing the real design.)

On each tick of the virtual clock, the state machine inside each FPGA will enable a register driving one of the logic blocks, thereby allowing the data from the input pin to be loaded into that register. At the same time, the state machine will cause the multiplexer to select the output from one of the logic blocks, and it will store that data in a register driving the output pin, which in turn drives the input to the next FPGA in the chain.

In the real world, of course, our FPGAs will have hundreds or thousands of pins. Each input may be used to drive several blocks of logic, and each output will be driven by its own VirtualWires multiplexer that selects data from a number of blocks of logic. To cut a long story short, things will become much more complicated, but the underlying principle remains the same.

Last but not least, a key element to the VirtualWires concept is a compiler that takes the original design in the form of a gate-level netlist, partitions this design across multiple FPGAs, automatically creates the state machines and other VirtualWires-related structures, and then generates the configuration files that will be used to load each of the FPGAs.

1926:
John Logie Baird demonstrates an electromechanical TV system.

1927:
First five-electrode vacuum tube (the Pentrode) is introduced.

Intellectual Property

Sources of IP

Today's FPGA designs are so big and complex that it would be impractical to create every portion of the design from scratch. One solution is to reuse existing functional blocks for the boring stuff and spend the bulk of your time and resources creating the cunning new portions of the design that feature your "secret sauce" and that will differentiate your design from any competitor offerings.

Any existing functional blocks are typically referred to as IP. The three main sources of such IP are (1) internally created blocks from previous designs, (2) FPGA vendors, and (3) third-party IP providers. For the purposes of these discussions, we shall concentrate on the latter two categories.

IP is pronounced by spelling it out as "I-P."

Handcrafted IP

One scenario is that the IP provider has handcrafted an IP block starting with an RTL description (the provider might also have used an IP block/core generator application, as discussed later in this chapter). In this case, there are several ways in which the end user might purchase and use such a block (Figure 17-1).

IP at the unencrypted RTL level

In certain cases, FPGA designers can purchase IP at the RTL level as blocks of unencrypted source code. These blocks can then be integrated into the RTL code for the body of the design (Figure 17-1a). (Note that the IP provider would

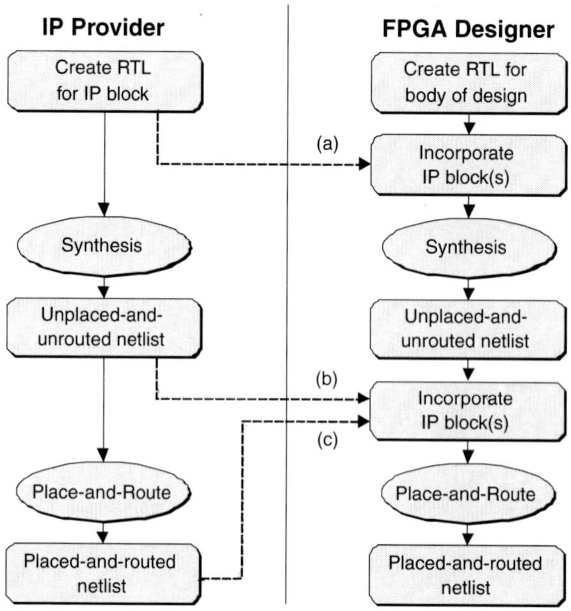

Figure 17-1. Alternative potential IP acquisition points.

already have simulated, synthesized, and verified the IP blocks before handing over the RTL source code).

Generally speaking, this is an expensive option because IP providers typically don't want anyone to see their RTL source code. Certainly, FPGA vendors are usually reluctant to provide unencrypted RTL because they don't want anyone to retarget it toward a competitor's device offering. So if you really wish to go this route, whoever is providing the IP will charge you an arm and a leg, and you'll end up signing all sorts of licensing and *nondisclosure agreements (NDAs)*.

Assuming you do manage to lay your hands on unencrypted RTL, one advantage of this approach is that you can modify the code to remove any functions you don't require in your design (or in some cases you might add new functions). Another advantage, assuming that you purchase the IP from a third party rather than from an FPGA vendor, is that you can quickly and easily retarget the IP across different device families and FPGA vendors. The big disadvantage is that the resulting implementation will typically be less efficient in

NDA is pronounced by spelling it out as "N-D-A."

terms of resource requirements and performance when compared to an optimized version delivered at the netlist level as discussed below.

IP at the encrypted RTL level

Unfortunately, at the time of this writing, there is no industry-standard encryption technique for RTL that has popular tool support. This has led companies like Altera and Xilinx to develop their own encryption schemes and tools. RTL encrypted by a particular FPGA vendor's tools can only be processed by that vendor's own synthesis tools (or sometimes by a third-party synthesis tool that has been OEM'd by the FPGA vendor).

IP at the unplaced-and-unrouted netlist level

Perhaps the most common scenario is for FPGA designers to purchase IP at the unplaced-and-unrouted LUT/CLB netlist level (Figure 17-1b). Such netlists are typically provided in encrypted form, either as encrypted EDIF or using some FPGA vendor-specific format.

In this case, the IP vendor may also provide a compiled cycle-accurate C/C++ model to be used for functional verification because such a model will simulate much faster than the LUT/CLB netlist-level model.

The main advantage of this scenario is that the IP provider has often gone to a lot of effort tuning the synthesis engine and handcrafting certain portions of the function so as to achieve an optimal implementation in term of resource utilization and performance. One disadvantage is that the FPGA designer doesn't have any ability to remove unwanted functionality. Another disadvantage is that the IP block is tied to a particular FPGA vendor and device family.

IP at the placed-and-routed netlist level

In certain cases, the FPGA designer may purchase IP at the placed-and-routed LUT/CLB netlist level (Figure 17-1c). Once again, such netlists are typically provided in encrypted

EDIF is pronounced "E-DIF;" that is, by spelling out the 'E' followed by "dif" to rhyme with "miff."

1927:
First public demonstration of long-distance television transmission (basically a Nipkow disk).

form, either as encrypted EDIF or using some FPGA vendor-specific format.

The reason for having placed-and-routed representations is to obtain the highest levels of performance. In some cases the placements will be relative, which means that the locations of all of the LUT, CLB, and other elements forming the block are fixed with respect to each other, but the block as a whole may be positioned anywhere (suitable) within the FPGA. Alternatively, in the case of IP blocks such as communications or bus protocol functions with specific I/O pin requirements, the placements of the elements forming the block may be absolute, which means that they cannot be changed in any way.

Once again, the IP vendor may also provide a compiled cycle-accurate C/C++ model to be used for functional verification because such a model will simulate much faster than the LUT/CLB netlist-level model.

IP core generators

Another very common practice is for FPGA vendors (sometimes EDA vendors, IP providers, and even small, independent design houses) to provide special tools that act as IP block/core generators. These generator applications are almost invariably parameterized, thereby allowing you to specify the widths and depths, or both of buses and functional elements.

First, you get to select from a list of different blocks/cores, and then you get to specify the parameters to be associated with each. Furthermore, in the case of some blocks/cores, the generator application may allow you to select from a list of functional elements that you wish to be included or excluded from the final representation. In the case of a communications block, for example, it might be possible to include or exclude certain error-checking logic. Or in the case of a CPU core, it might be possible to omit certain instructions or addressing modes. This allows the generator application to create the most efficient IP block/core in terms of its resource requirements and performance.

Depending on the origin of the generator application (or sometimes the licensing option you've signed up for), its output may be in the form of encrypted or unencrypted RTL source code, an unplaced-and-unrouted netlist, or a placed-and-routed netlist. In some cases, the generator may also output a cycle-accurate C/C++ model for use in simulation (Figure 17-2).

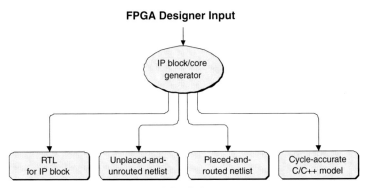

Figure 17-2. IP block/core generator.

Miscellaneous stuff

There is currently a push by the main FPGA vendors to provide special *system generator* utilities. These tools are essentially IP integrators that allow you to quickly build up very sophisticated designs using the various IP building blocks available from the respective FPGA vendor.

These system generator tools essentially spit out netlists for systems defined in some abstract form (as opposed to detailed end-user RTL coding.) These tools aim to change the FPGA design model by providing a system-level design paradigm that sits on top of the standard RTL-based design flow. This concept is of particular interest for designers who don't write RTL or who prefer to work at a higher level of abstraction (see also Chapter 12).

In addition to providing system generators, FPGA vendors are also working to simplify the use of IP by incorporating IP-based design-flow capabilities into their *independent development environments (IDEs)*.

IDE is pronounced by spelling it out as "I-D-E."

Depending on whom you are talking to, the 'D' in IDE can stand for "design" or "development."

Last, but not least, some IP that used to be "soft" is now becoming "hard." For example, the most current generation of FPGAs contains hard processor, clock manager, Ethernet, and gigabit I/O blocks, among others. These help bring high-end ASIC functionality into standard FPGAs. Over time, it is likely that additional functions of this ilk will be incorporated into the FPGA fabric.

Migrating ASIC Designs to FPGAs and Vice Versa

Alternative design scenarios

When it comes to creating an FPGA design, there are a number of possible scenarios depending on what you are trying to do (Figure 18-1).

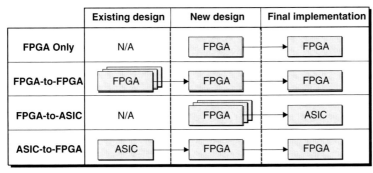

	Existing design	New design	Final implementation
FPGA Only	N/A	FPGA	FPGA
FPGA-to-FPGA	FPGA	FPGA	FPGA
FPGA-to-ASIC	N/A	FPGA	ASIC
ASIC-to-FPGA	ASIC	FPGA	FPGA

Figure 18-1. Alternative design scenarios.

FPGA only

This refers to a design that is intended for an FPGA implementation only. In this case, one might use any of the design flows and tools introduced elsewhere in this book.

FPGA-to-FPGA

This refers to taking an existing FPGA-based design and migrating it to a new FPGA technology (the new technology is often presented in the form of a new device family from the same FPGA vendor you used to implement the original design, but you may be moving to a new vendor also).

With this scenario, it is rare that you will be performing a simple one-to-one migration, which means taking the contents of an existing component and migrating them directly to a new device. It is much more common to migrate the functionality from multiple existing FPGAs to a single new FPGA. Alternatively, you might be gathering the functionality of one or more existing FPGAs, plus a load of surrounding discrete logic, and bundling it all into a new device.

In these cases, the typical route is to gather all of the RTL code describing the original devices and discrete logic into a single design. The code may be tweaked so as to take advantage of any new features available in the targeted device and then resynthesized.

FPGA-to-ASIC

This refers to using one or more FPGAs to prototype an ASIC design. One big issue here is that, unless you're working with a small to medium ASIC, it is often necessary to partition the design across multiple FPGAs. Some EDA and FPGA vendors either have (or used to have) applications that will perform this partitioning automatically,[1] but tools like this come and go with the seasons. Also, their features and capabilities, along with the quality of their results, can change on an almost weekly basis (which is my roundabout way of telling you that you'll have to evaluate the latest offerings for yourself).

Another consideration is that functions like RAMs configured to act as FIFO memories or dual-port memories have specific realizations when they are implemented using embedded RAM blocks in FPGAs. These realizations are typically different from the way in which these functions will be implemented in an ASIC, which may cause problems. One solution is to create your own RTL library of ASIC functions for such things as multipliers, comparators, memory blocks, and the

Literally as this book was heading to press, Synopsys (www.synopsys.com) made a rather interesting announcement. Using its well-known *Design Compiler*® ASIC synthesis engine as a base, they've created an FPGA-optimized version called *Design Compiler FPGA*. Among other things, DC FPGA features some innovative new Adaptive Optimization™ Technology that looks to be very interesting.

---

[1] A good example of an application that provides this sort of functionality is *Certify*® from Synplicity (www.synplicity.com).

like that will give you a one-for-one mapping with their FPGA counterparts. Unfortunately, this means instantiating these elements in the RTL code for your design, as opposed to using generic RTL and letting the synthesis engine handle everything (so it's a balancing act like everything else in engineering).

As we discussed in Chapter 7, a design intended for an FPGA implementation typically contains fewer levels of logic between register stages than would a pure ASIC design. In some cases, it's best to create the RTL code associated with the design with the final ASIC implementation in mind and just take the hit with regard to reduced performance in the FPGA prototype.

Alternatively, one might generate two flavors of the RTL—one for use with the FPGA prototype and the other to provide the final ASIC. But this is generally regarded to be a horrible way to do things because it's easy for the two representations to lose synchronization and end up going in two totally different directions.

One way around this might be to use the pure C/C++ based tools introduced in chapter 11. As you may recall, the idea here is that, as opposed to adding intelligence to the RTL source code by hand (thereby locking it into a target implementation), all of the intelligence is provided by your controlling and guiding the C/C++ synthesis engine itself (Figure 18-2).

But the main point is that DC ASIC and DC FPGA can use the same RTL source code, constraints, etc. to create both ASIC and FPGA implementations of the same design. (Each engine can be instructed to use different microarchitecture schemes such as resource sharing and the number of pipeline stages. Furthermore, DC FPGA can perform automatic transformation on any ASIC-centric clock-gating embedded in the RTL.) All of this makes the combination of DC FPGA and DC ASIC very interesting in the context of using FPGAs as prototypes for final ASIC implementations.

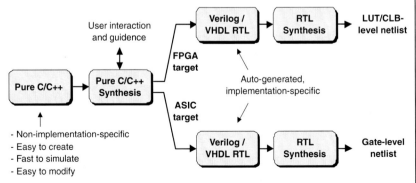

Figure 18-2. A pure C/C++–based design flow.

1927:
Harold Stephen Black
conceives the idea of
negative feedback,
which, amongst other
things makes Hi-Fi
amplifiers possible.

Once the synthesis engine has parsed the C/C++ source code, you can use it to perform microarchitecture tradeoffs and evaluate their effects in terms of size and speed. The user-defined configuration associated with each "what-if" scenario can be named, saved, and reused as required. Thus, you could first create a configuration for use as an FPGA prototype and, once this had been verified, you could create a second configuration to be used for the final ASIC implementation. The key point is that the same C/C++ source code is used to drive both flows.

Another point to ponder is that a modern ASIC design can contain an unbelievable number of clock domains and subdomains (we're talking about hundreds of domains/subdomains here). By comparison, an FPGA has a limited number of primary clock domains (on the order of 10). This means that if you're using one or more FPGAs to prototype your ASIC, you're going to have to put a lot of thought into how you handle your clocks.

Last but not least, there's an interesting European Patent numbered EP0437491 (B1), which, when you read it—and, good grief, it's soooo boring—seems to lock down the idea of using multiple programmable devices like FPGAs to temporarily realize a design intended for final implementation as an ASIC. In reality, I think this patent was probably targeted toward using FPGAs to create a logic emulator, but the way it's worded would prevent anyone from using two or more FPGAs to prototype an ASIC.

ASIC-to-FPGA

This refers to taking an existing ASIC design and migrating it to an FPGA. The reasons for doing this are wide and varied, but they often involve the desire to tweak an existing ASIC's functionality without spending vast amounts of money. Alternatively, the original ASIC technology may have become obsolete, but parts might still be required to support ongoing contracts (this is often the case with regard to military programs). One point of interest is that the latest generation of FPGAs has usually jumped so far so fast that it's possible to place an entire ASIC design from just a few years ago into a single modern FPGA (if you do have to partition the design across multiple FPGAs, then there are tools to aid you in this task, as discussed in the "FPGA-to-ASIC" section above).

First of all, you are going to have to go through your RTL code with a fine-tooth comb to remove (or at least evaluate) any asynchronous logic, combinatorial loops, delay chains, and things of this ilk (see also Chapter 7). In the case of flip-flops with both set and reset inputs, you might wish to recode these to use only one or the other (see also Chapter 7). You might also wish to look for any latches and redesign the circuit to use flip-flops instead. Also, you should keep a watchful eye open for statements like *if-then-else* without the *else* clause because, in these cases, synthesis tools will infer latches (see also Chapter 9).

In the case of clocks, you will have to ensure that your target FPGA provides enough clock domains to handle the requirements of the original ASIC design—otherwise, you'll have to redesign your clock circuitry. Furthermore, if your original ASIC design made use of clock-gating techniques, you will have to strip these out and possibly replace them with clock-enable equivalents (see also Chapter 7). Once again, some FPGA and EDA vendors provide synthesis tools that can automatically convert an ASIC design using gated clocks to an equivalent FPGA design using clocks with enables.[2]

In the case of complex functional elements such as memory blocks (e.g., FIFOs and dual-port RAMs), it will probably be necessary to tweak the RTL code to fit the design into the FPGA. In some cases, this will involve replacing generic RTL statements (that will be processed by the synthesis engine) with calls to instantiate specific subcircuits or FPGA elements.

1927: America. Philo Farnsworth assembles a complete electronic TV system.

---

[2] A good example of an application that provides this sort of functionality is *Amplify*® from Synplicity (www.synplicity.com).

1928: America.
First quartz crystal clock
is introduced.

Last, but not least, the original pipelined ASIC design probably had more levels of logic between register elements than you would like in the FPGA implementation if you wish to maintain performance. Most modern logic synthesis and physically aware tools provide retiming capability, which allows them to move logic back and forth across pipeline register boundaries to achieve better timing (the physically aware synthesis engines typically do a much better job at this; see also chapter 19).

It's also true that your modern FPGA is probably based on a later technology node (say, 130 nano) than your original ASIC design (say, 250 nano). This gives the FPGA an inherent speed advantage, which serves to offset its inherent track-delay disadvantages. At the end of the day, however, you may still end up having to hand-tweak the code to add in more pipeline stages.

Simulation, Synthesis, Verification, etc. Design Tools

Introduction

Design engineers typically need to use a tremendous variety of tools to capture, verify, synthesize, and implement their designs. Introducing all of these tools would require a book in itself,[1] so this chapter focuses on some of the more significant contenders in the context of FPGA designs (along with a couple I threw in for interest's sake):

- Simulation (cycle-based, event-driven, etc.)
- Synthesis (logic/HDL versus physically aware)
- Timing analysis (static versus dynamic)
- Verification in general
- Formal verification
- Miscellaneous

Simulation (cycle-based, event-driven, etc.)

What are event-driven logic simulators?

Logic simulation is currently one of the main verification tools in the design (or verification) engineer's arsenal. The most common form of logic simulation is known as *event driven* because, perhaps not surprisingly, these tools see the world as a series of discrete events. As an example, consider a very simple circuit comprising an OR gate driving both a BUF (buffer) gate and a brace of NOT (inverting) gates, as shown in Figure 19-1.

---

[1] I'd be more than happy to write such a book if anyone would be prepared to fund the effort!

1928:
John Logie Baird
demonsrates colr TV on
an electronic TV system.

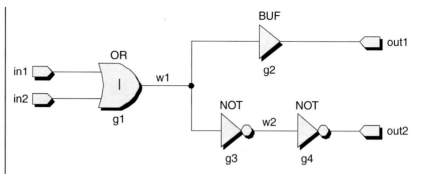

Figure 19-1. An example circuit.

Just to keep things simple, let's assume that NOT gates have a delay of 5 picoseconds (ps), BUF gates have a delay of 10 ps, and OR gates have a delay of 15 ps. On this basis, let's consider what will happen when a signal change occurs on one of the input pins (Figure 19-2).

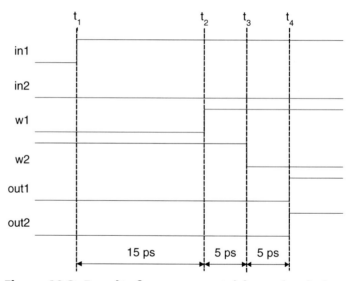

Figure 19-2. Results from an event-driven simulation.

Internally, the simulator maintains something called an *event wheel* onto which it places events that are to be "actioned" at some time in the future. When the first event occurs on input *in1* at a time we might refer to as t_1, the simulator looks to see what this input is connected to, which happens to be our OR gate. We are assuming that the OR gate has a delay of 15 ps, so the simulator schedules an event on the output of the OR gate—a rising (0 to 1) transition on wire *w1*—for 15 ps in the future at time t_2.

The simulator then checks if any further actions need to be performed at the current time (t_1), then it looks at the event wheel to see what is to occur next. In the case of our example, the next event happens to be the one we just scheduled at time t_2, which was for a rising transition on wire *w1*. At the same time as the simulator is performing this action, it looks to see what wire w1 is connected to, which is BUF gate *g2* and NOT gate *g3*.

As NOT gate *g3* has a delay of 5 ps, the simulator schedules a falling (1 to 0) transition on its output, wire *w2*, for 5 ps in the future at time t_3. Similarly, as BUF gate *g2* has a delay of 10 ps, the simulator schedules a rising (0 to 1) transition on its output, output *out1*, for 10 ps in the future at time t_4. And so it goes until all of the events triggered by the initial transition on input *in1* have been satisfied.

The advantage of this event-driven approach is that simulators based on this technique can be used to represent almost any form of design, including synchronous and asynchronous circuits, combinatorial feedback loops, and so forth. These simulators also offer extremely good visibility into the design for debugging purposes, and they can evaluate the effects of delay-related narrow pulses and glitches that are very difficult to find using other techniques (see also the discussions on delays in the next section). The big disadvantage associated with these simulators is that they are extremely compute-intensive and correspondingly slow.

A brief overview of the evolution of event-driven logic simulators

As we discussed in chapter 8, the first event-driven digital logic simulators (circa the late 1960s and early 1970s) were based on the concept of *simulation primitives*. At a minimum,

1928: John Logie Baird invents a videodisc to record television programs.

1929:
Joseph Schick invents
the electric razor.

these primitive elements would include logic gates such as
BUF, NOT, AND, NAND, OR, NOR, XOR, and XNOR,
along with a number of tri-state buffers. Some simulators also
offered a selection of registers and latches as primitive elements, while others required you to create these functions as
subcircuits formed from a collection of the more primitive
logic gates.

At that time, the functionality of the design would be captured using a standard text editor as a gate-level netlist.
Similarly, the testbench would be captured as a textual (tabular) stimulus file. The simulator would accept the netlist and
testbench along with any control files and command-line
instructions; it would use the netlist to build a model of the
circuit in the computer's memory; it would apply the stimulus
from the testbench to this model; and it would output results
in the form of a textual (tabular) file (Figure 19-3).

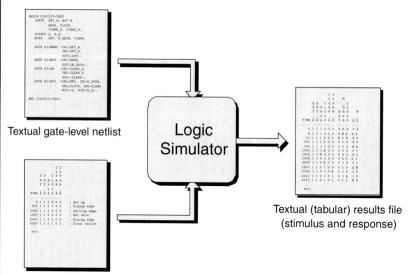

Textual gate-level netlist

Textual (tabular) stimulus

Textual (tabular) results file
(stimulus and response)

Figure 19-3. Running a logic simulator.

RTL is pronounced by
spelling it out as "R-T-L."

Over time things started to become a little more sophisticated. First, schematic capture packages were used to capture
the design and to generate the gate-level netlist. Next, special

display tools were used to read in the textual results files and to present the results as graphical waveforms. In some cases, these waveform tools were also used to capture the testbench in a graphical manner and to generate the tabular stimulus file.

Still later, the creators of digital simulators started to experiment with more sophisticated languages that could describe logical functions at higher levels of abstraction such as the register transfer level, or RTL. A good example of such a language was the *GenRad Hardware Description Language (GHDL)* used by the System HILO simulator.

Similarly, more sophisticated testbench languages started to evolve, such as the *GenRad Waveform Description Language (GWDL)*. Languages of this type could support complex constructs like loops, and they could even access the current state of the circuit and vary their tests accordingly (along the lines of," If this output is a logic 0, then jump to Test B or else jump to Test C").

In some respects, these early languages were ahead of their time. For example, GWDL had a really useful feature in that, in addition to specifying the input stimulus (e.g., "input-A = 0"), you could also specify the expected output response (e.g., "output-Y == 1"). (Note the use of one equals sign to assign a value to an input and of a pair of equal signs to indicate an expected response.) If you then used a special STROBE statement, the simulator would check to see if the actual response (from the circuit) matched the expected response (specified in the waveform) and generate a warning if there was a discrepancy between the two.

As the years passed by, industry-standard HDLs such as Verilog and VHDL started to appear. These had the advantage that the same language could be used to represent both the functionality of the circuit and the testbench.[2] (See also the discussions on special verification languages like *e* in the "Verification in general" section later in this chapter.)

VCD is pronounced by spelling it out as "V-C-D."

FSDB is pronounced by spelling it out as "F-S-D-B."

_____
[2] The chief architect of the Verilog language—Phil Moorby—was also one of the designers of the original HILO language and simulator.

SDF is pronounced
by spelling it out as
"S-D-F."

As opposed to using the
'X' character to represent
"unknown" or "don't
know," data books typi-
cally use it to represent
"don't care." By compari-
son, hardware description
languages tend to use '?'
or '-' to represent "don't
care" values.

Also, "don't care" values
cannot be assigned to
outputs as driven states.
Instead, they are used to
specify how a model's
inputs should respond to
different combinations of
signals.

Digital simulation logic
value systems (such as
the *cross-product* versus
interval-value
approaches) and various
aspects of unknown X
values are introduced in
more detail in my book
*Designus Maximus
Unleashed (Banned in Ala-*

Also, standard file formats for capturing simulation output results, such as the *value change dump (VCD)* format, started to appear on the scene. This facilitated third-party EDA companies creating sophisticated waveform display and analysis tools that could work with the outputs from multiple simulators. (A more recent entry here is the *Fast Signal Database™ (FSDB)* format from Novas Software (www.novas.com), which provides much smaller file sizes than VCD while offering extremely fast information-retrieval capabilities.)

Similarly, innovations like the *standard delay format (SDF)* specification facilitated third-party EDA companies' creating sophisticated timing analysis tools that could evaluate circuits, generate timing reports highlighting potential problems, and output SDF files that could be used to provide more accurate timing simulations (see also the discussion on alternative delay formats below).

Logic values and different logic value systems

The overwhelming majority of today's digital electronics systems are based on *binary logic* with digits called *bits*; that is, logic gates using two different voltages to represent the binary digits 0 and 1 or the Boolean logic values *True* and *False*. Some experiments have been performed on *tertiary logic*, which is based on three different logic levels and whose digits are referred to as *trits*. Thus far, however, this technology hasn't made any inroads into commercial applications (for which what's left of my brain is truly thankful).

But we digress. The minimum set of logic values required to represent the operation of binary logic gates is 0 and 1. The next step is the ability to represent unknown values, for which we typically use the character X. These unknown values may be used to represent a variety of conditions, such as the contents of an uninitialized register or the clash resulting from two gates driving the same wire with opposing logical values. And it's also nice to be able to represent high-impedance values driven by the outputs of tri-state gates, for which we typically use the character Z.

But the 0, 1, X, and Z states are only the tip of the iceberg. More advanced logic simulators have ways to associate different drive strengths with the outputs of different gates. This is combined with ways in which to resolve and represent situations where multiple gates are driving the same wire with different logic values of different strengths. Just to make life fun, of course, VHDL and Verilog handle this sort of thing in somewhat different ways.

Mixed-language simulation

The problem with having two industry-standard languages like Verilog and VHDL is that it's not long before you find yourself with different portions of a design represented in different languages. Anything you design from scratch will obviously be written in the language du jour favored by your company. However, problems can arise if you wish to reuse legacy code that is in the other language. Similarly, you may wish to purchase blocks of IP from a third party, but this IP may be available only in the language you aren't currently using yourself. And there's also the case where your company merges with, commences a joint project with, another company, where the two companies are entrenched in design flows using disparate languages.

This leads to the concept of *mixed-language simulation*, of which there have historically been several flavors. One technique used in the early days was to translate the "foreign" language (the one you weren't using) into the language you were working with. This was painful to say the least because the different languages supported different logic states and language constructs (even similar language statements had different semantics). The end result was that when you simulated the translated design, it rarely behaved the way you expected it to, so this approach is rarely used today.

Another technique was to have both a VHDL simulator and a Verilog simulator and to cosimulate the two simulation kernels. In this case the performance of the ensuing simulation was sadly lacking because each kernel was forever stopping

bama), ISBN
0-7506-9089-5

while it waited for the other to complete an action. Thus, once again, this approach is rarely used today.

The optimum solution is to have a single-kernel simulator that supports designs represented as a mixture of VHDL and Verilog blocks. All of the big boys in EDA have their own version of such a tool, and some go far beyond anything envisaged in the past because they can support multiple languages such as Verilog, SystemVerilog, VHDL, SystemC, and PSL (where PSL is introduced in more detail in the "Formal verification" section in this chapter).[3]

Alternative delay formats

How you decide to represent delays in the models you are creating for use with an event-driven simulator depends on two things: (a) the delay modeling capabilities of the simulator itself and (b) where in the flow (and with what tools) you intend to perform your timing analysis.

A very common scenario is for *static timing analysis (STA)* to be performed externally from the simulation (this is discussed in more detail later in this chapter). In this case, logic gates (and more complex statements) may be modeled with zero (0 timebase unit) delays or unit (1 timebase unit) delays, where the term *timebase unit* refers to the smallest time segment recognized by the simulator.

Alternatively, we might associate more sophisticated delays with logic gates (and more complex statements) for use in the simulation itself. The first level of complexity is to separate rising delays from falling delays at the output from the gate (or more complex statement). For historical reasons, a rising (0-to-1) delay is often referred to as LH (standing for "low-to-high"). Correspondingly, a falling (1-to-0) delay may be referred to as HL (meaning "high-to-low"). For example, consider what happens if we were to apply a 12 ps positive-

STA is pronounced by spelling it out as "S-T-A."

---

[3] A good example of this type of single-kernel solution is *ModelSim®* from Mentor Graphics (www.mentor.com).

going (0-1-0) pulse to the input of a simple buffer gate with delays of LH = 5 ps and HL = 8 ps (Figure 19-4).

1929:
British mechanical TVs roll off the production line.

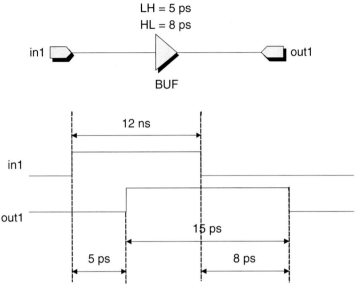

Figure 19-4. Separating LH and HL delays.

Not surprisingly, the output of the gate rises 5 ps after the rising edge is applied to the input, and it falls 8 ps after the falling edge is applied to the input. The really interesting point is that, due to the unbalanced delays, the 12 ps input pulse has been stretched to 15 ps at the output of the gate, where the additional 3 ps reflect the difference between the LH and HL values. Similarly, if a negative-going 12 ps (1-0-1) pulse were applied to the input of this gate, the corresponding pulse at the output would shrink to only 9 ps (try sketching this out on a piece of paper for yourself).

In addition to LH and HL delays, simulators also support minimum:typical:maximum (min:typ:max) values for each delay. For example, consider a positive-going pulse of 16 ps presented to the input of a buffer gate with rising and falling delays specified as 6:8:10 ps and 7:9:11 ps, respectively (Figure 19-5).

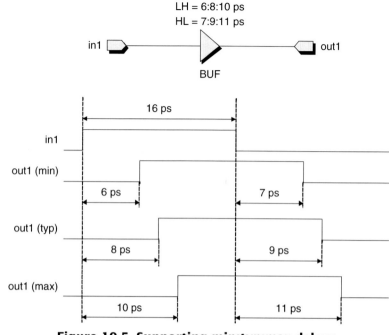

Figure 19-5. Supporting min:typ:max delays.

TTL (which is pronounced by spelling it out as "T-T-L") refers to bipolar junction transistors (BJTs) connected together in a certain fashion.

BJT is pronounced by spelling it out as "B-J-T."

This range of values is intended to accommodate variations in the operating conditions such as temperature and voltage. It also covers variations in the manufacturing process because some chips may run slightly faster or slower than others of the same type. Similarly, gates in one area of a chip (e.g., an ASIC or an FPGA) may switch faster or slower than identical gates in another area of the chip. (See also the discussions on timing analysis, particularly dynamic timing analysis, later in this chapter).

In the early days, all of the input-to-output delays associated with a multi-input gate (or more complex statement) were identical. For example, consider a 3-input AND gate with an output called y and inputs a, b, and c. In this case, any LH and HL delays would be identical for the paths a-to-y, b-to-y, and c-to-y. Initially, this didn't cause any problems because it matched the way in which delays were specified in data books. Over time, however, data books began to specify individual input-to-output delays, so simulators had to be enhanced to support this capability.

Another point to consider is what will happen when a narrow pulse is applied to the input of a gate (or more complex statement). By "narrow" we mean a pulse that is smaller than the propagation delay of the gate. The first logic simulators were largely targeted toward simple ICs implemented in *transistor-transistor logic (TTL)* being used at the circuit board level. These chips typically rejected narrow pulses, so that's what the simulators did. This became known as the *inertial delay model*. As a simple example, consider two positive-going pulses of 8 ps and 4 ps applied to a buffer gate whose min:typ:max rising and falling delays are all set to 6 ps (Figure 19-6).

ECL (which is pronounced by spelling it out as "E-C-L") refers to bipolar junction transistors connected together in a different fashion to TTL. Logic gates implemented in ECL switch faster than their TTL counterparts, but they also consume more power (and thus dissipate more heat).

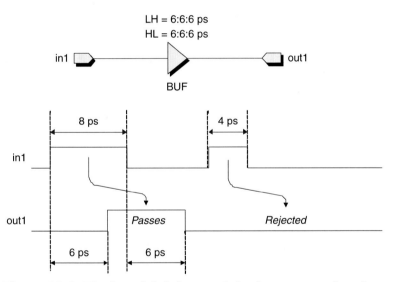

Figure 19-6. The inertial delay model rejects any pulse that is narrower than the gate's propagation delay.

By comparison, logic gates implemented in later technologies such as *emitter-coupled logic (ECL)* would pass pulses that were narrower than the propagation delay of the gate. In order to accommodate this, some simulators were equipped with a mode called the *transport delay model*. Once again, consider

1929:
Experiments begin on electronic colour television.

two positive-going pulses of 8 ps and 4 ps applied to a buffer gate whose min:typ:max rising and falling delays are all set to 6 ps (Figure 19-7).

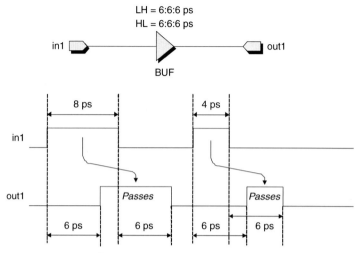

Figure 19-7. The transport delay model propagates any pulse, irrespective of its width.

The problem with both the inertial and transport delay models is that they only provide for extreme cases, so the creators of some simulators started to experiment with more sophisticated narrow-pulse handling techniques, such as the *three-band delay model*.[4] In this case, each delay may be qualified with two values called r (for "reject") and p (for "pass"), specified as percentages of the total delay. For example, assume we have a buffer gate whose min:typ:max delays have all been set to 6 ps qualified by r and p values of 33 percent and 66 percent, respectively (Figure 19-8).

Any pulses presented to the input that are greater than or equal to the p value will propagate; any pulses that are less than the r value will be completely rejected; and any pulses that fall between these two extremes will be propagated as a

[4] The *System HILO* simulator from GenRad started to employ the 3-band delay model shortly before it disappeared off the face of the planet.

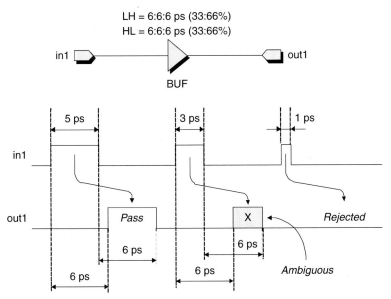

Figure 19-8. The three-band delay model.

1929:
First ship-to shore communications (passenger can call relatives at home ... at a price).

pulse with an unknown X value to indicate that they are ambiguous because we don't know whether or not they will propagate through the gate in the real world. (Setting both *r* and *p* to 100 percent equates to an inertial delay model, while setting them both to 0 percent reflects a pure transport delay model.)

Cycle-based simulators

An alternative to the event-driven approach is to use a cycle-based simulation technique. This is particularly well suited to pipelined designs in which "islands" of combinational logic are sandwiched between blocks of registers (Figure 19-9).

In this case, a cycle-based simulator will throw away any timing information associated with the gates forming the combinational logic and convert this logic into a series of Boolean operations that can be directly implemented using the CPU's logical machine code instructions.

Given an appropriate circuit with appropriate activity, cycle-based simulators may offer significant run-time advantages over their event-driven counterparts. The downside,

1929: Germany. Magnetic sound recording on plastic tape.

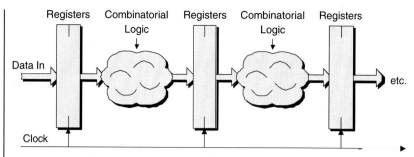

Figure 19-9. A simple pipelined design.

however, is that they typically only work with 0 and 1 logic values (no X or Z values, and no drive strength representations). Also, cycle-based simulators can't represent asynchronous logic or combinatorial feedback loops.

These days it's rare to see anyone using a pure cycle-based simulator. However, several event-driven simulators have been augmented to have hybrid capabilities. In this case, if you instruct the simulator to aim for extreme performance (as opposed to timing accuracy), it will automatically handle some portions of the circuit using an event-driven approach and other portions using cycle-based techniques.

Choosing the best logic simulator in the world!

Choosing a logic simulator is, as with anything else in engineering, a balancing act. If you are a small startup and cost is your overriding metric, for example, then bounce over to the discussions on creating an open-source-based flow in Chapter 25.

One point to consider is whether or not you require mixed-language capability. If you are a small startup, you may be planning on using only one language, but remember that any IP you decide to purchase down the road may not be available in this language. Having a solution that can work with VHDL, Verilog, and SystemVerilog would be a good start, and if it can also handle SystemC along with one or more formal verification languages, then it will probably stand you in good stead for some time to come.

Generally speaking, performance is the number-one criterion for most folks. The trick here is how to determine the performance of a simulator without being bamboozled. The only way to really do this is to have your own benchmark design and to run it on a number of simulators. Creating a good benchmark design is a nontrivial exercise, but it's way better than using a design supplied by an EDA vendor (because such a design will be tuned to favor their solution, while delivering a swift knee to the metaphorical groins of competing tools).

However, there's more to life than raw performance. You also need to look for a good interactive debugging solution such that when you detect a problem, you can stop the simulator and poke around the design. All simulators are not created equal in this department. Different tools have different levels of capability; in some cases, even if the simulator does let you do what you want, you may have to jump through hoops to get there. So the trick here is—after running your performance benchmark—bring up the same circuit with a known bug and see how easy it is (and how long it takes) to detect and isolate the little rapscallion. In reality, some simulators that give you the performance you require do such a poor job in this department that you are obliged to use third-party postsimulation analysis tools.[5]

Another thing to consider is the capacity of the simulator. The tools supplied by the big boys in EDA essentially have no capacity limitations, but simulators from smaller vendors might be based on ported 32-bit code if you were to look under the hood. Of course, if you are only going to work with smaller designs (say, equivalent to 500,000 gates or less), then you will probably be okay with the simulators supplied by the FPGA vendors (these are typically "lite" versions of the tools supplied by the big EDA vendors).

1929:
The first car radio is installed.

---

[5] Novas Software Inc. (www.novas.com) are at the top of the pile here with their *Debussy*® and *Verdi*™ tools.

1930:
America. Sliced bread is
introduced.

Of course, you will have your own criteria in addition to the topics raised above, such as the quality of the code coverage and performance analysis provided by the various tools. These used to be the province of specialist third-party tools, but most of the larger simulators now provide some level of integrated code coverage and performance analysis in the simulation environment itself. However, different simulators offer different feature sets (see also the discussions on code coverage and performance analysis in the "Miscellaneous" section later in this chapter).

Synthesis (logic/HDL versus physically aware)

Logic/HDL synthesis technology

Traditional *logic synthesis* tools appeared on the scene around the early to mid-1980s. Depending on whom you are talking to, these tools are now often referred to as *HDL synthesis technology.*

The role of the original logic/HDL synthesis tools was to take an RTL representation of an ASIC design along with a set of timing constraints and to generate a corresponding gate-level netlist. During this process, the synthesis application performed a variety of minimizations and optimizations (including optimizing for area and timing).

Around the middle of the 1990s, synthesis tools were augmented to understand the concept of FPGA architectures. These architecturally aware applications could output a LUT/CLB-level netlist, which would subsequently be passed to the FPGA vendor's place-and-route software (Figure 19-10).

In real terms, the FPGA designs generated by architecturally aware synthesis tools were 15 to 20 percent faster than their counterparts created using traditional gate-level synthesis offerings.

Physically aware synthesis technology

The problem with traditional logic/HDL synthesis is that it was developed when logic gates accounted for most of the

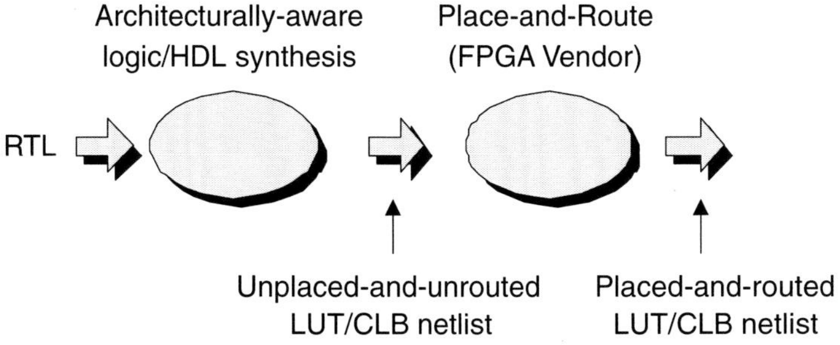

Figure 19-10. Traditional logic/HDL synthesis.

1930:
America. Vannevar Bush designs an analogue computer called a Differential Analyzer.

delays in a timing path, while track delays were relatively insignificant. This meant that the synthesis tools could use simple wire-load models to evaluate the effects of the track delays. (These models were along the lines of, One load gate on a wire equates to x pF of capacitance; two load gates on a wire equates to y pF of capacitance; etc.) The synthesis tool would then estimate the delay associated with each track as a function of its load and the strength of the gate driving the wire.

This technique was adequate for the designs of the time, which were implemented in multimicron technologies and which contained relatively few logic gates by today's standards. By comparison, modern designs can contain tens of millions of logic gates, and their deep submicron feature sizes mean that track delays can account for up to 80 percent of a delay path. When using traditional logic/HDL synthesis technology on this class of design, the timing estimations made by the synthesis tool bear so little resemblance to reality that achieving timing closure can be well-nigh impossible.

For this reason, ASIC flows started to see the use of *physically aware synthesis* somewhere around 1996, and FPGA flows began to adopt similar techniques circa 2000 or 2001. Of course there are a variety of different definition, as to exactly what the term *physically aware synthesis* implies. The core concept is to use physical information earlier in the synthesis process, but what does this actually mean? For example, some companies have added interactive floor-planning capabilities

1933:
Edwin Howard
Armstrong conceives a
new system for radio
communication:
wideband frequency's
modulation (FM).

to the front of their synthesis engines, and they class this as
being physical synthesis or physically aware synthesis. For
most folks, however, physically aware synthesis means taking
actual placement information associated with the various logical elements in the design, using this information to estimate
accurate track delays, and using these delays to fine-tune the
placement and perform other optimizations. Interestingly
enough, physically aware synthesis commences with a first-
pass run using a relatively traditional logic/HDL synthesis
engine (Figure 19-11).

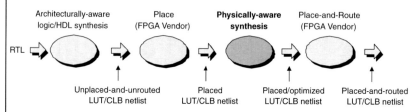

Figure 19-11. Physically aware synthesis.

Retiming, replication, and resynthesis

There are a number of terms that one tends to hear in the
context of physical synthesis, including *retiming*, *replication*,
and *resynthesis*.[6] The first, retiming, is based on the concept of
balancing out positive and negative slacks throughout the
design. In this context, *positive slack* refers to a path with some
delay available that you are not using, while *negative slack*
refers to a path that is using more delay than is available to it.

For example, let's assume a pipelined design whose clock
frequency is such that the maximum register-to-register delay
is 15 ps. Now let's assume that we have a situation as shown
in Figure 19-12a, whereby the longest timing path in the first
block of combinational logic is 10 ps (which means it has a

---

6 These concepts may also be used with traditional logic/HDL synthesis,
but they are significantly more efficacious when applied in the context of
physically aware synthesis.

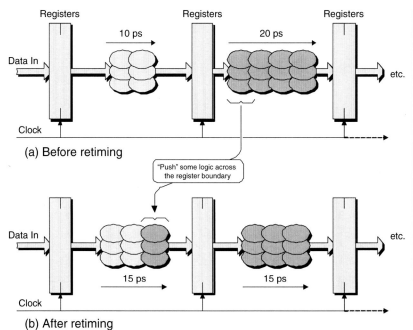

Registers   Registers   Registers

10 ps   20 ps

Data In   etc.

Clock

(a) Before retiming

"Push" some logic across
the register boundary

Data In   etc.

15 ps   15 ps

Clock

(b) After retiming

Figure 19-12. Retiming.

1934:
Half the homes in the
USA have radios.

positive slack of 5 ps), while the longest path in the next block of combinational logic is 20 ps (which means it has a *negative slack* of 5ps).

Once the initial path timing, including routing delays, has been calculated, combinational logic is moved across register boundaries (or vice versa, depending on your point of view) to steal from paths with positive slack and donate to paths with negative slack (Figure 19-12b). Retiming is very common in physically aware FPGA design flows because registers are plentiful in FPGA devices.

Replication is similar to retiming, but it focuses on breaking up long interconnect. For example, let's assume that we have a register with 4 ps of positive slack on its input. Now let's assume that this register is driving three paths, whose loads each see negative slack (Figure 19-13a).

By replicating the register and placing the copies close to each load, we can redistribute the slack so as to make all of the timing paths work (Figure 19-13b).

1935:
All-electronic VHF
television comes out of
the lab.

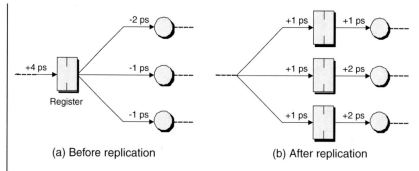

(a) Before replication   (b) After replication

Figure 19-13. Replication.

Last, but not least, the concept of resynthesis is based on the fact that there are many different ways of implementing (and placing) different functions. Resynthesis uses the physical placement information to perform local optimizations on critical paths by means of operations like logic restructuring, reclustering, substitution, and possible elimination of gates and wires.

Choosing the best synthesis tool in the world!

Come on, be serious, you didn't really expect to find the answer to this here, did you? In the real world, the capabilities of the various synthesis engines, along with associated features like autointeractive floor planning, change on an almost daily basis, and the various vendors are constantly leapfrogging each other.

There's also the fact that different engines may work better (or worse) with different FPGA vendors' architectures. One thing to look for is the ability (or lack thereof) of the engine to infer things automatically, like clocking elements and embedded functions, from your source code or constraints files without your having to define them explicitly. At the end of the day, however, you are on your own when it comes to evaluating and ranking the various offerings (but please feel free to e-mail me to let me know how you get on at max@techbites.com.

STA is pronounced by spelling it out as "S-T-A."

Timing analysis (static versus dynamic)

Static timing analysis

The most common form of timing verification in use today is classed as STA. Conceptually, this is quite simple, although in practice things are, as usual, more complex than they might at first appear.

The timing analyzer essentially sums all of the gate and track delays forming each path to give you the total input-to-output delays for each path. (In the case of pipelined designs, the analyzer calculates delays from one bank of registers to the next.)

Prior to place-and-route, the analyzer may make estimations as to track delays. Following place-and-route, the analyzer will employ extracted parasitic values (for resistance and capacitance) associated with the physical tracks to provide more accurate results. The analyzer will report any paths that fail to meet their original timing constraints, and it will also warn of potential timing problems (e.g., setup and hold violations) associated with signals being presented to the inputs of any registers or latches.

STA is particularly well suited to classical synchronous designs and pipelined architectures. The main advantages of STA are that it is relatively fast, it doesn't require a test bench, and it exhaustively tests every possible path into the ground. On the other hand, static timing analyzers are little rascals when it comes to detecting false paths that will never be exercised during the course of the design's normal operation. Also, these tools aren't at their best with designs employing latches, asynchronous circuits, and combinational feedback loops.

Statistical static timing analysis

STA is a mainstay of modern ASIC and FPGA design flows, but it's starting to run into problems with the latest process technology nodes. At the time of this writing, the 90-nano node is coming online, with the 45-nano node expected around 2007.

As previously discussed, in the case of modern silicon chips, interconnect delays dominate logic delays, especially with respect to FPGA architectures. In turn, interconnect delays are dependent on parasitic capacitance, resistance, and inductance values, which are themselves functions of the topology and cross-sectional shape of the wires.

The problem is that, in the case of the latest technology process nodes, photolithographic processes are no longer capable of producing exact shapes. Thus, as opposed to working with squares and rectangles, we are now working with circles and ellipsoids. Feature sizes like the widths of tracks are now so small that small variations in the etching process cause deviations that, although slight, are significant with relation to the main feature size. (These irregularities are made more significant by the fact that in the case of high-frequency designs, the so-called skin-effect comes into play, which refers to the fact that high-frequency signals travel only through the outer surface, or skin, of the conductor.) Furthermore, there are variations in the vertical plane of the track's cross section caused by processes like *chemical mechanical polishing (CMP)*.

As an overall result, it's becoming increasingly difficult to calculate track delays accurately. Of course, it is possible to use the traditional engineering fallback of guard-banding (using worst-case estimations), but excessively conservative design practices result in device performance significantly below the silicon's full potential, which is an extremely unattractive option in today's highly competitive marketplace. In fact, the effects of geometry variations are causing the probability distributions of delays to become so wide that worst-case numbers may actually be slower than in an earlier process technology!

One potential solution is the concept of the *statistical static timing analyzer (SSTA)*. This is based on generating a probability function for the delay associated with each signal for each segment of a track, then evaluating the total delay probability functions of signals as they propagate through entire paths. At the time of this writing, there are no commercially

---

CMP is pronounced by spelling it out as "C-M-P."

SSTA is so new at the time of writing that no one knows how to pronounce it, but my guess is that folks will say "Statistical S-T-A" (or spell it out as "S-S-T-A.")

DTA is pronounced by spelling it out as "D-T-A".

deliverable SSTA products, but a number of folks in EDA and the academic arena are looking into this technology.

Dynamic timing analysis

Another form of timing verification, known as *dynamic timing analysis (DTA)*, really isn't seen much these days, but it is mentioned here for the sake of interest. This form of verification is based on the use of an event-driven simulator, and it does require the use of a testbench. The key difference between a standard event-driven simulator and a dynamic timing analyzer is that the former only uses a single minimum (min), typical (typ), or maximum (max) delay for each path, while the latter uses a delay pair (either min:typ, typ:max, or min:max). For example, consider how the two simulators would evaluate a simple buffer gate (Figure 19-14).

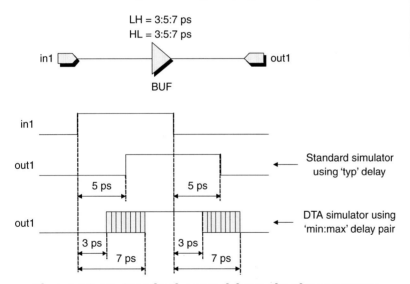

Figure 19-14. Standard event-driven simulator versus dynamic timing analyzer.

In the case of the standard simulator, a signal change at the input to the gate will cause an event to be scheduled for some specific time in the future. By comparison, in the case of the dynamic timing analyzer, assuming a min:max delay pair, the gate's output will begin to transition after the minimum delay, but it won't end its transition until it reaches the maximum delay.

The ambiguity between these two values is different from an unknown X state, because we know that a good 0-to-1 or a 1-to-0 transition is going to take place, we just don't know when. For this reason, we introduce two new states called "Gone high, but don't know when" and "Gone low, but don't know when."[7]

DTA can detect subtle, potential problems that are almost impossible to find using any other form of timing analysis. Unfortunately, these tools are so compute intensive that you don't really see them around much these days, but who knows what the future holds?

Verification in general

Verification IP

As designs increase in complexity, verifying their functionality consumes more and more time and resources. Such verification includes implementing a verification environment, creating a testbench, performing logic simulations, analyzing the results to detect and isolate problems, and so forth. In fact, verifying one of today's high-end ASIC, SoC, or FPGA designs can consume 70 percent or more of the total development effort from initial concept to final implementation.

DUT is pronounced by spelling it out as "D-U-T".

One way to alleviate this problem is to make use of *verification IP*. The idea here is that the design, which is referred to as the *device under test (DUT)* for the purposes of verification, typically communicates with the outside world using standard interfaces and protocols. Furthermore, the DUT is typically communicating with devices such as microprocessors, peripherals, arbiters, and the like.

---

[7] Dynamic timing analysis is discussed in a tad more detail in my book *Designus Maximus Unleashed (Banned in Alabama)*, ISBN 0-7506-9089-5

The most commonly used technique for performing functional verification is to use an industry-standard event-driven logic simulator. One way to test the DUT would be to create a testbench describing the precise bit-level signals to be applied to the input pins and the bit-level responses expected at the outputs. However, the protocols for the various interfaces and buses are now so complex that it is simply not possible to create a test suite in this manner.

Another technique would be to use RTL models of all of the external devices forming the rest of the system. However, many of these devices are extremely proprietary and RTL models may not be readily available. Furthermore, simulating an entire system using fully functional models of all of the processor and I/O devices would be prohibitively expensive in terms of time and computing requirements.

The solution is to use verification IP in the form of *bus functional models (BFMs)* to represent the processors and the I/O agents forming the system under test (Figure 19-15).[8]

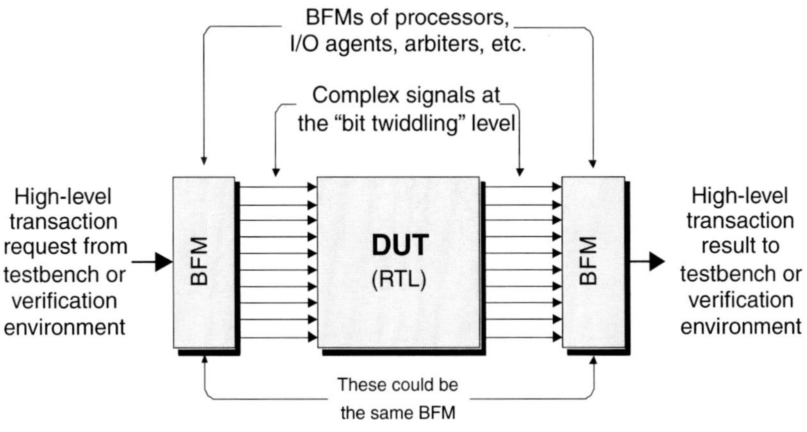

BFM is pronounced by spelling it out as "B-F-M."

Figure 19-15. Using verification IP in the form of BFMs.

---

[8] One source of very sophisticated verification IP is TransEDA PLC (www.transeda.com).

A BFM doesn't replicate the entire functionality of the device it represents; instead, it emulates the way the device works at the bus interface level by generating and accepting transactions. In this context, the term *transaction* refers to a high-level bus event such as performing a read or write cycle. The verification environment (or testbench) can instruct a BFM to perform a specific transaction like a memory write. The BFM then generates the complex low-level ("bit-twiddling") signal interactions on the bus driving the DUT's interface transparently to the user.

Similarly, when the DUT (the design) responds with a complex pattern of signals, another BFM (or maybe the original BFM) can interpret these signals and translate them back into corresponding high-level transactions. (See also the discussions on verification environments and creating testbenches below.)

It should be noted that, although they are much smaller and simpler (and hence simulate much faster) than fully functional models of the devices they represent, BFMs are by no means trivial. For example, sophisticated BFMs, which are often created as cycle-accurate, bit-accurate C/C++ models, may include internal caches (along with the ability to initialize them), internal buffers, configuration registers, write-back queues, and so forth. Also, BFMs can provide a tremendous range of parameters that provide low-level control of such things as address timing, snoop timing, data wait states for different memory devices, and the like.

Verification environments and creating testbenches

When I was a young man starting out in simulation, we created test vectors (stimulus and response) to be used with our simulations as tabular ASCII text files containing logic 0 and 1 values (or hexadecimal values if you were lucky). At that time, the designs we were trying to test were incredibly simple compared to today's monsters, so an English translation of our tests would be something along the lines of

At time 1,000 make the reset signal go into its active state.
At time 2,000 make the reset signal go into its inactive state.
At time 2,500 check to see that the 8-bit data bus is 00000000.
At time … and so it went.

Over time, designs became more complex, and the way in which they could be verified became more sophisticated with the advent of high-level languages that could be used to specify stimulus and expected response. These languages sported a variety of features such as loop constructs and the ability to vary the tests depending on the state of the outputs (e.g., "If the status bus has a value of 010, then jump to test *xyz*"). At some stage, folks started referring to these tests as *testbenches*.[9]

The current state of play is that many of today's designs are now so complex that it's well nigh impossible to create an adequate test bench by hand. This has paved the way for sophisticated verification environments and languages. Perhaps the most sophisticated of the languages, known by some as *hardware verification languages (HVLs)*, is the aspect-oriented *e* offering from Verisity Design (www.verisity.com).[10]

In case you were wondering, *e* doesn't actually stand for anything now, but originally it was intended to reflect the idea of "English-like" in that it has a natural language feel to it. You can use *e* to specify directed tests if you wish, but you would typically only wish to do this for special cases. Instead, the concept behind *e*, which you can think of as a blend of C and Verilog with a hint of Pascal, is more about declaring valid ranges and sequences of input values (along with their invalid counterparts) and high-level verification strategies. This *e* description is then used by an appropriate verification environ-

HVL is pronounced by spelling it out as "H-V-L."

---

[9] To be a tad more pedantic, the term "testbench" really refers to the infrastructure supporting test execution.

[10] By and large, the industry tends to view proprietary languages with suspicion, so Verisity are working with the IEEE to make *e* an industry-standard language. At the time of this writing, the IEEE working group P1647 has been established and the *e* language reference manual (LRM) has been published.

ment to guide the simulations.

Speaking of which, the first (and only, at the time of this writing) verification environment to make full use of the power of *e* is Verisity's *Specman Elite®*. We can think of Specman as being a cross between a compiler and an event-driven simulator that links to and controls the standard HDL event-driven simulators you are already using. Specman uses your *e* program to generate stimuli that are applied to your design (via your HDL simulator) on the fly. It also monitors the results and the functional coverage of the simulations and reacts to what it sees by dynamically retargeting subsequent stimuli to address any remaining coverage holes.

Analyzing simulation results

Almost every simulator comes equipped with a graphical waveform viewer that can be used to display results interactively (as the simulator runs) or to accept and display postsimulation results from a *value change dump* (VCD) file.

Sad to relate, however, some of these tools are not as effective as one might hope when it comes to really analyzing this information and tracking down problems. In this case, you might wish to use a tool from a third-party vendor.[11]

Formal verification

Although large computer and chip companies like IBM, Intel, and Motorola have been developing and using formal tools internally for decades (since around the mid-1980s), the whole field of *formal verification* (FV) is still relatively new to a lot of folks. This is particularly true in the FPGA arena, where the adoption of formal verification is lagging behind its use in

VCD is pronounced by spelling it out as "V-C-D."

In conversation, one almost invariably says "formal verification" (I've never heard anyone spelling it out as "F-V").

---

[11] In the context of classical waveform analysis, debugging, and display tools, one of the acknowledged industry leaders is Novas Software Inc. (www.novas.com) with its *Debussy®* offering. Another tool from Novas that is well worth looking at is *Verdi™*, which provides an extremely innovative and powerful way of extracting, visualizing, analyzing, exploring, and debugging a design's temporal behavior across multiple clock cycles.

ASIC design flows. Having said this, formal verification can be such an incredibly powerful tool that more and more folks are starting to use it in earnest.

One big problem is that formal verification is still so new to mainstream usage that there are a lot of players, all of whom are happily charging around in a bewildering variety of different directions. Also, as opposed to a lack of standards, there are now so many different offerings that the mind boggles. The confusion is only increased by the fact that almost everyone you talk to puts his or her unique spin on things (if, for example, you ask 20 EDA vendors to define and differentiate the terms *assertion* and *property*, your brains will leak out of your ears at the diametrically opposing responses).[12]

Trying to unravel this morass is a daunting task to say the least. However, there is nothing to fear but fear itself, as my dear old dad used to say, so let's take a stab at rending the veils asunder and describing formal verification in a way that we can all understand.

Different flavors of formal verification

In the not-so-distant past, the term *formal verification* was considered synonymous with *equivalency checking* for the majority of design engineers. In this context, an equivalency checker is a tool that uses formal (rigorous mathematical) techniques to compare two different representations of a design—say an RTL description with a gate-level netlist—to determine whether or not they have the same input-to-output functionality.

In fact, equivalency checking may be considered a subclass of formal verification called *model checking*, which refers to techniques used to explore the state-space of a system to test whether or not certain properties, typically specified in the form of *assertions*, are true. (Definitions of terms like *property* and *assertion* are presented a little later in this section.)

Formal tools were originally developed for internal use by large computer and chip companies. One of the first commercially available formal tools to be widely accepted was an equivalency checker called Design VERIFYer®, which was introduced in 1993 by Chrysalis Symbolic Design Inc.

Model checking tools were also first developed by large companies for internal use. The introduction of Design inSIGHT® by Chrysalis in 1996 signaled the first commercial rollout of model checking technology.

---

[12] I speak from painful experience on this point!

For the purposes of the remainder of our discussions here, we shall understand formal verification to refer to model checking. It should be noted, however, that there is another category of formal verification known as *automated reasoning*, which uses logic to prove, much like a formal mathematical proof, that an implementation meets an associated specification.

But just what is formal verification, and why is it so cool?

In order to provide a starting point for our discussions, let's assume we have a design comprising a number of subblocks and that we are currently working with one of these blocks, whose role in life is to perform some specific function. In addition to the HDL representation that defines the functionality of this block, we can also associate one or more assertions/properties with that block (these assertions/properties may be associated with signals at the interface to the block or with signals and registers internal to the block).

A very simple assertion/property might be along the lines of "Signals A and B should never be active (low) at the same time." But these statements can also extend to extremely complex transaction-level constructs, such as "When a PCI write command is received, then a memory write command of type *xxxx* must be issued within 5 to 36 clock cycles."

Thus, assertions/properties allow you to describe the behavior of a time-based system in a formal and rigorous manner that provides an unambiguous and universal representation of the design's intent (try saying that quickly). Furthermore, assertions/properties can be used to describe both expected and prohibited behavior.

ABV is pronounced by spelling it out as "A-B-V."

The fact that assertions/properties are both human- and machine-readable makes them ideal for the purposes of capturing an executable specification, but they go far beyond this. Let's return to considering a very simple assertion/property such as "Signals A and B should never be active (low) at the same time." One term you will hear a lot is *assertion-based veri-*

fication (ABV), which comes in several flavors: simulation, static formal verification, and dynamic formal verification. In the case of *static formal verification*, an appropriate tool reads in the functional description of the design (typically at the RTL level of abstraction) and then exhaustively analyzes the logic to ensure that this particular condition can never occur. By comparison, in the case of *dynamic formal verification*, an appropriately augmented logic simulator will sum up to a certain point, then pause and automatically invoke an associated formal verification tool (this is discussed in more detail below).

Of course, assertions/properties can be associated with the design at any level, from individual blocks, to the interfaces linking blocks, to the entire system. This leads to a very important point, that of *verification reuse*. Prior to formal verification, there was very little in the way of verification reuse. For example, when you purchase an IP core, it will typically come equipped with an associated testbench that focuses on the I/O signals at the core's boundary. This allows you to verify the core in isolation, but once you've integrated the core into the middle of your design, its testbench is essentially useless to you.

Now consider purchasing an IP core that comes equipped with a suite of predefined assertions/properties, like "Signal A should never exhibit a rising transition within three clocks of Signal B going active." These assertions/properties provide an excellent mechanism for communicating interface assumptions from the IP developer to downstream users. Furthermore, these assertions/properties remain true and can be evaluated by the verification environment, even when this IP core is integrated into your design.

With regard to assertions/properties associated with the system's primary inputs and outputs, the verification environment may use these to automatically create stimuli to drive the design. Furthermore, you can use assertions/properties throughout the design to augment code and functional coverage analysis (see also the "Miscellaneous" section below) so as to

1935: England. First demonstration of Radar at Daventry.

1936: America.
Efficiency expert August
Dvorak invents a new
typewriter layout called
the Dvorak Keyboard.

ensure that specific sequences of actions or conditions have been performed.

Terminology and definitions

Now that we've introduced the overall concept of the model checking aspects of formal verification, we are better equipped to wade through some terminology and definitions. To be fair, this is relatively uncharted water ("Here be dragons"); the following was gleaned from talking with lots of folks and then desperately trying to rationalize the discrepancies between the tales they told.

- *Assertions/properties:* The term *property* comes from the model checking domain and refers to a specific functional behavior of the design that you want to (formally) verify (e.g., "after a request, we expect a grant within 10 clock cycles"). By comparison, the term *assertion* stems from the simulation domain and refers to a specific functional behavior of the design that you want to monitor during simulation (and flag a violation if that assertion "fires").

  Today, with the use of formal tools and simulation tools in unified environments and methodologies, the terms *property* and *assertion* tend to be used interchangeably; that is, a property is an assertion and vice versa. In general, we understand an assertion/property to be a statement about a specific attribute associated with the design that is expected to be true. Thus, assertions/properties can be used as checkers/monitors or as targets of formal proofs, and they are usually used to identify/trap undesirable behavior.

- *Constraints:* The term *constraint* also derives from the model checking space. Formal model checkers consider all possible allowed input combinations when performing their magic and working on a proof. Thus, there is often a need to constrain the inputs to their legal behavior; otherwise, the tool would report false negatives, which are property violations that would not normally occur in the actual design.

As with properties, constraints can be simple or complex. In some cases, constraints can be interpreted as properties to be proven. For example, an input constraint associated with one module could also be an output property of the module driving this input. So, properties and constraints may be dual in nature. (The term *constraint* is also used in the "constrained random simulation" domain, in which case the constraint is typically used to specify a range of values that can be used to drive a bus.)

■ *Event:* An *event* is similar to an assertion/property, and in general events may be considered a subset of assertions/properties. However, while assertions/properties are typically used to trap undesirable behavior, events may be used to specify desirable behavior for the purposes of functional coverage analysis.

In some cases, assertions/properties may consist of a sequence of events. Also, events can be used to specify the window within which an assertion/property is to be tested (e.g., "After a, b, c, we expect d to be true, until e occurs," where a, b, c, and e are all events, and d is the behavior being verified).

Measuring the occurrence of events and assertions/properties yields quantitative data as to which corner cases and other attributes of the design have been verified. Statistics about events and assertions/properties can also be used to generate functional coverage metrics for a design.

■ *Procedural:* The term *procedural* refers to an assertion/property/event/constraint that is described within the context of an executing process or set of sequential statements, such as a VHDL process or a Verilog "always" block (thus, these are sometimes called "in-context" assertions/properties). In this case, the assertion/property is built into the logic of the design and

1936: America. Psychologist Benjamin Burack constructs the first electrical logic machine (but he doesn't publish anything about it until 1949).

1936:
First electronic speech synthesis (Vodar).

will be evaluated based on the path taken through a set of sequential statements.

- *Declarative:* The term *declarative* refers to an assertion/property/event/constraint that exists within the structural context of the design and is evaluated along with all of the other structural elements in the design (for example, a module that takes the form of a structural instantiation). Another way to view this is that a declarative assertion/property is *always* "on/active," unlike its procedural counterpart that is only "on/active" when a specific path is taken/executed through the HDL code.

- *Pragma:* The term *pragma* is an abbreviation for "pragmatic information," which refers to special pseudocomment directives that can be interpreted and used by parsers/compilers and other tools. (Note that this is a general-purpose term, and pragma-based techniques are used in a variety of tools in addition to formal verification technology.)

Alternative assertion/property specification techniques

This is where the fun really starts, because there are various ways in which assertions/properties and so forth can be implemented, as summarized below.

- *Special languages:* This refers to using a formal property/assertion language that has been specially constructed for the purpose of specifying assertions/properties with maximum efficiency. Languages of this type, of which Sugar, PSL, and OVA are good examples, are very powerful in creating sophisticated, regular, and temporal expressions, and they allow complex behavior to be specified with very little code (Sugar, PSL, and OVA are introduced in more detail later in this chapter).

  Such languages are often used to define assertions/properties in "side-files" that are maintained outside the main HDL design representation. These side-files may be accessed during parser/compile time and implemented in a declarative fashion. Alternatively, a parser/compiler/simulator may be augmented so as to allow statements in the special language to be embedded directly in the HDL as in-line code or as pragmas (see the definition of "pragma" in the previous section); in both of these cases, the statements may be implemented in a declarative and/or procedural manner (see the definitions of "declarative" and "procedural" in the previous section).

■ *Special statements in the HDL itself*: Right from the get-go, VHDL came equipped with a simple *assert* statement that checks the value of a Boolean expression and displays a user-specified text string if the expression evaluates *False*. The original Verilog did not include such a statement, but SystemVerilog has been augmented to include this capability.

The advantage of this technique is that these statements are ignored by synthesis engines, so you don't have to do anything special to prevent them from being physically implemented as logic gates in the final design. The disadvantage is that they are relatively simplistic compared to special assertion/property languages and are not well equipped to specify complex temporal sequences (although SystemVerilog is somewhat better than VHDL in this respect).

■ *Models written in the HDL and called from within the HDL*: This concept refers to having access to a library of internally or externally developed models. These models represent assertions/properties using standard HDL statements, and they may be instantiated in the design like any other blocks. However, these instantiations will be wrapped by synthesis OFF/ON pragmas to ensure that they aren't physically implemented. A good example of this approach is the *open verification library (OVL)* from the Accellera standards committee

1936:
Fluorescent lighting is introduced.

1936:
The Berlin Olympics
are televised

(www.accellera.org), as discussed in the next section.

■ *Models written in the HDL and accessed via pragmas*: This is similar in concept to the previous approach in that it involves a library of models that represent assertions/properties using standard HDL statements. However, as opposed to instantiating these models directly from the main design code, they are pointed to by pragmas. A good example of this technique is the *Checker-Ware*® library from 0-In Design Automation (www.0-In.com). For example, consider a design containing the following line of Verilog code:

```
reg [5:0] STATE_VAR; // 0in one_hot
```

The left-hand side of this statement declares a 6-bit register called STATE_VAR, which we can assume is going to be used to hold the state variables associated with an FSM. Meanwhile, the right-hand side ("0in one–hot") is a pragma. Most tools will simply treat this pragma as a comment and ignore it, but 0-In's tools will use it to call a corresponding "one-hot" assertion/property model from their CheckerWare library. Note that the 0-In implementation means that you don't need to specify the variable, the clocking, or the bit-width of the assertion; this type of information is all picked up automatically. Also, depending on a pragma's position in the code, it may be implemented in a declarative or procedural manner.

Static formal versus dynamic formal

This is a little tricky to wrap one's brain around, so let's take things step by step. First of all, you can use assertions/properties in a simulation environment. In this case, if you have an assertion/property along the lines of "Signals A and B should never be active (low) at the same time," then if this illegal case occurs during the course of a simulation, a warning flag will be raised, and the fact this happened can be logged.

Simulators can cover a lot of ground, but they require some sort of testbench or a verification environment that is dynamically generating stimulus. Another consideration is that some portions of a design are going to be difficult to verify via simulation because they are deeply buried in the design, making them difficult to control from the primary inputs. Alternatively, some areas of a design that have large amounts of complex interactions with other state machines or external agents will be difficult to control.

At the other end of the spectrum is *static formal verification*. These tools are incredibly rigorous and they examine 100 percent of the state space without having to simulate anything. Their disadvantage is that they can typically be used for small portions of the design only, because the state space increases exponentially with complex properties, and one can quickly run into a "state space explosion." By comparison, logic simulators, which can also be used to test for assertions, can cover a lot of ground, but they do require stimuli, and they don't cover every possible case.

In order to address these issues, some solutions combine both techniques. For example, they may use simulation to reach a corner condition and then automatically pause the simulator and invoke a static formal verification engine to exhaustively evaluate that corner condition. (In this context, a general definition of a "corner condition" or "corner case" is a hard-to-exercise or hard-to-reach functional condition associated with the design.) Once the corner condition has been evaluated, control will automatically be returned to the simulator, which will then proceed on its merry way. This combination of simulation and traditional static formal verification is referred to as *dynamic formal verification*.

As one simple example of where this might be applicable, consider a FIFO memory, whose "Full" and "Empty" states may be regarded as corner cases. Reaching the "Full" state will require a lot of clock cycles, which is best achieved using simu-

1937: American. George Robert Stibitz, a scientist at Bell Labs, builds a simple digital calculator machine based on relays called the Model K.

OVA is pronounced by spelling it out as "O-V-A."

With regards to OVA, the original version drew on Synopsys's strength in simulation technologies. The folks at Synopsys subsequently desired to extend OVA to support formal property verification, so they partnered with the guys and gals at Intel to build on their experience in formal verification with their internally developed For-Spec assertion language. The result was OVA 2.0, which included powerful constructs for both static and dynamic formal verification.

OVL is pronounced by spelling it out as "O-V-L."

PSL is pronounced by spelling it out as "P-S-L."

lation. But exhaustively evaluating attributes/properties associated with this corner case, such as the fact that it should not be possible to write any more data while the FIFO is full, is best achieved using static techniques.

Once again, a good example of this dynamic formal verification approach is provided by 0-In. Corner cases are explicitly defined as such in their CheckerWare library models. When a corner case is reached during simulation, the simulator is paused, and a static tool is used to analyze that corner case in more detail.

Summary of different languages, etc.

This is where things could start to get really confusing if we're not careful (so let's be careful). We'll begin with something called *Vera*®, which began life with work done at Sun Microsystems in the early 1990s. It was provided to Systems Science Corporation somewhere around the mid-1990s, which was in turn acquired by Synopsys in 1998.

Vera is essentially an entire verification environment, similar to, but perhaps not as sophisticated as, the *e* verification language/environment introduced earlier in this chapter. Vera encapsulates testbench features and assertion-based capabilities, and Synopsys promoted it as a stand-alone product (with integration into the Synopsys logic simulator). Sometime later, due to popular demand, Synopsys opened things up to for third-party use by making *OpenVera*™ and *OpenVera Assertions* (OVA) available.

Somewhere around this time, SystemVerilog was equipped with its first pass at an *assert* statement. Meanwhile, due to the increasing interest in formal verification technology, one of the Accellera standards committees started to look around for a formal verification language it could adopt as an industry standard. A number of languages were evaluated, including OVA, but in 2002, the committee eventually opted for the *Sugar* language from IBM. Just to add to the fun and frivolity, Synopsys then donated OVA to the Accellera committee in charge of SystemVerilog (this was a different committee from the one evaluating formal property languages).

Yet another Accellera committee ended up in charge of something called the *open verification library*, or OVL, which refers to a library of assertion/property models available in both VHDL and Verilog 2K1.

So now we have the assert statements in VHDL and SystemVerilog, OVL (the library of models), OVA (the assertion language), and the *property specification language (PSL)*, which is the Accellera version of IBM's Sugar language (Figure 19-16).[13] The advantage of PSL is that it has a life of its own in that it can be used independently of the languages used to represent the functionality of the design itself. The disadvantage

1937: England. Graduate student Alan Turing invents a theoretical (thought experiment) computer called the Turing Machine.

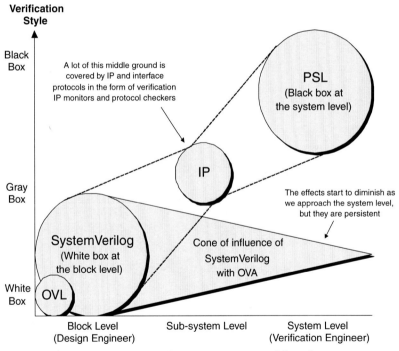

Figure 19-16. Trying to put everything into context and perspective.

---

[13] Don't make the common mistake of referring to "PSL/Sugar" as a single/combined language. There's PSL and there's Sugar and they're not the same thing. PSL is the Accellera standard, while Sugar is the language used inside IBM.

1937: England. Graduate student Alan Turing writes his groundbreaking paper *On Computable Numbers with an Application to the Entscheidungsproblem.*

is that it doesn't look like anything the hardware description languages design engineers are familiar with, such as VHDL, Verilog, C/C++, and the like. There is some talk of spawning various flavors of PSL, such as a VHDL PSL, a Verilog PSL, a SystemC PSL, and so forth; the syntax would differ among these flavors so as to match the target language, but their semantics would be identical.

It's important to note that figure 19-16 just reflects one view of the world, and not everyone will agree with it (some folks will consider this to be a brilliant summation of an incredibly confusing situation, while others will regard it as being a gross simplification at best and utter twaddle at worst).

Miscellaneous

HDL to C conversion

As we discussed in chapter 11, there is an increasing push toward capturing designs at higher levels of abstraction such as C/C++. In addition to facilitating architectural exploration, high-level (behavioral and/or algorithmic) C/C++ models can simulate hundreds or thousands of times faster than can their HDL/RTL counterparts.

Having said this, many design engineers still prefer to work in their RTL comfort zone. The problem is that when you are simulating an entire SoC with an embedded processor core, memory, peripherals, and other logic all represented in RTL, you are lucky to achieve simulation speeds of more than a couple of hertz (that is, a few cycles of the main system clock for each second in real time).

In order to address this problem, some EDA companies are starting to offer ways to translate your "Golden RTL" models into faster-simulating alternatives that can achieve kilohertz simulation speeds.[14] This is fast enough to allow you to run

---

14 One interesting solution is the *VTOC*™ (Verilog-to-C) translator from Tenison Technology Ltd. (www.tenison.com). Another is the *SPEEDCompiler*™ and *DesignPlayer*™ concept from Carbon Design Systems Inc. (www.carbondesignsystems.com).

software on your hardware representation for milliseconds of real run time. In turn, this allows you to test critical foundation software, such as drivers, diagnostics, and firmware, thereby facilitating system validation and verification to occur much faster than with traditional methods.

Code coverage, etc.

In the not-so-distant past, code coverage tools were specialist items provided by third-party EDA vendors. However, this capability is now considered important enough that all of the big boys have code coverage integrated into their verification (simulation) environments, but, of course, the feature sets vary among offerings.

By now, it may not surprise you to learn that there are a lot of different flavors of code coverage, summarized briefly in order of increasing sophistication as follows:

- *Basic code coverage:* This is just line coverage; that is, how many times each line in the source code is hit (executed).
- *Branch coverage:* This refers to conditional statements like *if-then-else*; how many times do you go down the *then* path and how many down the *else* path.
- *Condition coverage:* This refers to statements along the lines of "if (a OR b == TRUE) then." In this case, we are interested in the number of times the *then* path was taken because variable *a* was TRUE compared to the number of times variable *b* was TRUE.
- *Expression coverage:* This refers to expressions like "a = (b AND c) OR !d". In this case, we are interested in analyzing the expression to determine all of the possible combinations of input values and also which combinations triggered a change in the output and which variables were never tested.
- *State coverage:* This refers to analyzing state machines to determine which states were visited and which ones were neglected, as well as which guard conditions and

1937: Pulse-code modulation points the way towards digital radio transmission.

1938:
American Claude E.
Shannon publishes an
article (based on his
master's thesis at MIT)
showing how Boolean
Algebra can be used to
design digital circuits.

paths between states are taken, and which aren't, and so forth. You can derive this sort of information from line coverage, but you have to read between the lines (pun intended).

■ *Functional coverage:* This refers to analyzing which transaction-level events (e.g., memory-read and memory-write transactions) and which specific combinations and permutations of these events have been exercised.

■ *Assertion/property coverage:* This refers to a verification environment that can gather, organize, and make available for analysis the results from all of the different simulation-driven, static formal, and dynamic formal assertion-/property-based verification engines.

This form of coverage can actually be spilt into two camps: specification-level coverage and implementation-level coverage. In this context, specification-level coverage measures verification activity with respect to items in the high-level functional or macroarchitecture definition. This includes the I/O behaviors of the design, the types of transactions that can be processed (including the relationships of different transaction types to each other), and the data transformations that must occur. By comparison, implementation-level coverage measures verification activity with respect to microarchitectural details of the actual implementation. This refers to design decisions that are embedded in the RTL that result in implementation-specific corner cases, for example, the depth of a FIFO buffer and the corner cases for its "high-water mark" and "full" conditions. Such implementation details are rarely visible at the specification level.

Performance analysis

One final feature that's important in a modern verification environment is its ability to perform *performance analysis*. This refers to having some way of analyzing and reporting exactly where the simulator is spending its time. This allows you to focus on high-activity areas of your design, which may reap huge rewards in terms of final system performance.

1938: Argentina. Hungarian Lazro Biro invents and patterns the first ballpoint pen.

1938: Germany. Konrad Zuse finishes the construction of the first working mechanical digital computer (the ZI)

1938: John Logie Baird demonstrated live TV in colour.

1938: America. Radio drama *War of the Worlds* causes wide spread panic.

1938: Television broadcasts can be taped and edited.

1938: Walter Schottky discovers the existence of holes in the band structure of semiconductors and explains metal/ semiconductor interface rectification.

Choosing the Right Device

So many choices

Many aspects of life would be so much simpler if we were presented with fewer alternatives. For example, ordering a seemingly simple American Sunday brunch comprising eggs, bacon, hash browns (fried potatoes), and toast can take an inordinate amount of time because there are so many options to choose from.

First, your waitress is going to ask you how you want your eggs (sunny-side up, over-easy, over-medium, over-hard, scrambled, poached, hard-boiled, in an omelet, etc.). Next, you will be asked if you want American or Canadian bacon; should your hash browns be complemented by onions, tomatoes, cheese, ham, chili, or any combination thereof; would you like the bread for your toast to be white, rye, whole wheat, stone ground, sourdough ...

The frightening thing is that the complexity of ordering brunch pales in comparison to choosing an FPGA because there are so many product families from the different vendors. Product lines and families from the same vendor overlap; product lines and families from different vendors both overlap and, at the same time, sport different features and capabilities; and things are constantly changing, seemingly on a daily basis.

If only there were a tool

Before we start, it's worth noting that size isn't everything in the FPGA design world. You really need to base your FPGA selection on your design needs, such as number of I/O pins,

1939: America. George Robert Stibitz builds a digital calculator called the Complex Number Calculator.

1939: America John Vincent Atanasoff (and Clifford Berry) may or may not have constructed the first truly electronic special-purpose digital computer called the ABC (but it didn't work till 1942).

available logic resources, availability of special functional blocks, and so forth.

Another consideration is whether you already have dealings with a certain FPGA vendor and product family, or whether you are plunging into an FPGA design for the very first time. If you already have a history with a vendor and are familiar with using its components, tools, and design flows, then you will typically stay within that vendor's offerings unless there's an overriding reason for change.

For the purposes of the remainder of these discussions, however, we'll assume that we are starting from ground zero and have no particular affiliation with any vendor. In this case, choosing the optimum device for a particular design is a daunting task.

Becoming familiar with the architectures, resources, and capabilities associated with the various product families from the different FPGA vendors demands a considerable amount of time and effort. In the real world, time-to-market pressures are so intense that design engineers typically have sufficient time to make only high-level evaluations before settling on a particular vendor, family, and device. In this case, the selected FPGA is almost certainly not the optimum component for the design, but this is the way of the world.

Given a choice, it would be wonderful to have access to some sort of FPGA selection wizard application (preferably Web based). This would allow you to choose a particular vendor, a selection of vendors, or make the search open to all vendors.

For the purposes of a basic design, the wizard should then prompt you to enter estimates for such things as ASIC equivalent gates or FPGA system gates (assuming there are good definitions as to what equivalent gates and system gates are—see also chapter 4). The wizard should also prompt for details on I/O pin requirements, I/O interface technologies, acceptable packaging options, and so forth.

In the case of a more advanced design, the wizard should prompt you for any specialist options such as gigabit transceivers or embedded functions like multipliers, adders, MACs, RAMs (both distributed and block RAM), and so forth. The wizard should also allow you to specify if you need access to embedded processor cores (hard or soft) along with selections of associated peripherals.

Last, but not least, it would be nice if the wizard would prompt you as to any IP requirements (hey, since we're dreaming, let's dream on a grand scale). Finally, clicking the "Go" button would generate a report detailing the leading contenders and their capabilities (and costs).

Returning to the real world with a sickening thump, we remember that no such utility actually exists at this time[1], so we have to perform all of these evaluations by hand, but wouldn't it be nice ... Of course, creating this sort of application would be nontrivial, and maintaining it would be demanding and time-consuming, but I'm sure that system houses or design engineers would happily pay some sort of fee for such a service should anyone be brave enough to pick up the challenge and run with it.

Technology

One of your first choices is going to be deciding on the underlying FPGA technology. Your main options are as follows:

- *SRAM based:* Although very flexible, this requires an external configuration device and can take up to a few seconds to be configured when the system is first powered up. Early versions of these devices could have substantial power supply requirements due to high transient startup currents, but this problem has been addressed in the current generation of devices. One key advantage of this option is that it is based on standard CMOS technology and doesn't require any esoteric process steps.

---

[1] There used to be tools like this to aid in selecting PLDs, but that was a significantly less complex solution space.

1939:
Bell Labs begin testing high-frequency radar.

1939:
Light-emitting diodes (LEDs) are patented by Messers Bay and Szigeti.

1939: England.
Regular TV broadcasts
begin.

1940: America.
George Robert Stibitz
performs first example
of remote computing
between New York and
New Hampshire.

This means that SRAM-based FPGAs are at the forefront of the components available with the most current technology node.

- *Antifuse based:* Considered by many to offer the most security with regard to design IP, this also provides advantages like low power consumption, instant-on availability, and no requirement for any external configuration devices (which saves circuit board cost, space, and weight). Antifuse-based devices are also more radiation hardened than any of the other technologies, which makes them of particular interest for aerospace-type applications. On the downside, this technology is a pain to prototype with because it's OTP. Antifuse devices are also typically one or more generations behind the most current technology node because they require additional process steps compared to standard CMOS components.

- *FLASH based:* Although considered to be more secure than SRAM-based devices, these are slightly less secure than antifuse components with regard to design IP. FLASH-based FPGAs don't require any external configuration devices, but they can be reconfigured while resident in the system if required. In the same way as antifuse components, FLASH-based devices provide advantages like instant-on capability, but are also typically one or more generations behind the most current technology node because they require additional process steps compared to standard CMOS components. Also, these devices typically offer a much smaller logic (system) gate-count than their SRAM-based counterparts.

Basic resources and packaging

Once you've decided on the underlying technology, you need to determine which devices will satisfy your basic resource and packaging requirements. In the case of core resources, most designs are pin limited, and it's typically only in the case of designs featuring sophisticated algorithmic processing like color space conversion that you will find yourself logic limited. Regardless of the type of design, you will need to decide on the number of I/O pins you are going to require and the approximate number of fundamental logical entities (LUTs and registers).

As discussed in chapter 4, the combination of a LUT, register, and associated logic is called a *logic element (LE)* by some and a *logic cell (LC)* by others. It is typically more useful to think in these terms as opposed to higher-level structures like *slices* and *configurable logic blocks (CLBs)* or *logic array blocks (LABs)* because the definition of these more sophisticated structures can vary between device families.

Next, you need to determine which components contain a sufficient number of clock domains and associated PLLs, DLLs, or *digital clock managers (DCMs)*.

Last, but not least, if you have any particular packaging requirements in mind, it would be a really good idea to ensure that the FPGA family that has caught your eye is actually available in your desired package. (I know this seems obvious, but would you care to place a bet that no one ever slipped up on this point before?)

General-purpose I/O interfaces

The next point to ponder is which components have configurable general-purpose I/O blocks that support the signaling standard(s) and termination technologies required to interface with the other components on the circuit board.

Let's assume that way back at the beginning of the design process, the system architects selected one or more I/O standards for use on the circuit board. Ideally, you will find an FPGA that supports this standard and also provides all of the other capabilities you require. If not, you have several options:

1941:
First touch-tone phone system (too expensive for general use).

1942:
Germany between 1942 and 1945/6, Konrad Zuse develops the idea for a high-level computer programming language called Plankakul.

- If your original FPGA selection doesn't provide any must-have capabilities or functionality, you may decide to opt for another family of FPGAs (possibly from another vendor).
- If your original FPGA selection does provide some must-have capabilities or functionality, you may decide to use some external bridging devices (this is expensive and consumes board real estate). Alternatively, in conjunction with the rest of the system team, you may decide to change the circuit board architecture (this can be really expensive if the system design has progressed to any significant level).

Embedded multipliers, RAMs, etc.

At some stage you will need to estimate the amount of distributed RAM and the number of embedded block RAMs you are going to require (along with the required widths and depths of the block RAMs).

Similarly, you will need to muse over the number of special embedded functions (and their widths and capabilities) like multipliers and adders. In the case of DSP-centric designs, some FPGAs may contain embedded functions like MACs that will be particularly useful for this class of design problem and may help to steer your component selection decisions.

Embedded processor cores

If you wish to use an embedded processor core in your design, you will need to decide whether or not a soft core will suffice (such a core may be implemented across a number of device families) or if a hard core is the order of the day (see also the discussion in Chapter 13).

In the case of a soft core, you may decide to use the offering supplied by an FPGA vendor. In this case, you are going to become locked into using that vendor, so you need to evaluate the various alternatives carefully before taking the plunge. Alternatively, you may decide to use a third-party soft-core solution that can be implemented using devices from multiple vendors.[2]

If you decide on a hard core, you have little option but to become locked into a particular vendor. One consideration that may affect your decision process is your existing experience with different types of processors. Let's say that you hold a black belt in designing systems based around the PowerPC, for example. In such a case, you would want to preserve your investment in PowerPC design tools and flows (and your experience and knowledge in using such tools and flows). Thus, you would probably decide on an FPGA offering from Xilinx because they support the PowerPC. Alternatively, if you are a guru with respect to ARM or MIPS processors, then selecting devices from Altera or QuickLogic, respectively, may be the way to go.

Gigabit I/O capabilities

If your system requires the use of gigabit transceivers, then points to consider are the number of such transceivers in the device and the particular standard that's been selected by your system architects at the circuit board level (see also Chapter 21).

IP availability

Each of the FPGA vendors has an IP portfolio. In many cases there will be significant overlap between vendors, but more esoteric functions may only be available from selected vendors, which may have an impact on your component selection.

Alternatively, you may decide to purchase your IP from a third-party provider. In such a case, this IP may be available for use with multiple FPGAs from different vendors, or it may only be available for use with a subset of vendors (and a subset of device families from those vendors).

---

[2] An example of this type of solution is the *Nexar* offering from Altium Ltd. (www.altium.com), which was introduced in Chapter 13.

1943: Germany. Konrad Zuse starts work on his general-purpose relay-based computer called the Z4.

1944: America. Howard Aiken and team finish building an electromechanical .computer called the Harvard Mark I (also known as the IBM ASCC).

1945: America. Hungarian/American mathematician Johann (John) Von Neumann publishes a paper entitled *First draft on a report on the EDVAC.*

1945: Percy L Spensor invents the Microwave Oven (the first units go on sale in 1947)

One further point: We commonly think of IP in terms of hardware design functions, but some IP may come in the form of software routines.[3] For example, consider a communications function that might be realized as a hardware implementation in the FPGA fabric or as a software stack running on the embedded processor. In the latter case, you might decide to purchase the software stack routines from a third party, in which case you are essentially acquiring software IP.

Speed grades

Once you've decided on a particular FPGA component for your design, one final decision is the speed grade of this device. The FPGA vendors' traditional pricing model makes the performance (speed grade) of a device a major factor with regard to the cost of that device.

As a rule of thumb, moving up a speed grade will increase performance by 12 to 15 percent, but the cost of the device will increase by 20 to 30 percent. Conversely, if you can manipulate the architecture of your design to improve performance by 12 to 15 percent (say, by adding additional pipelining stages), then you can drop a speed grade and save 20 to 30 percent on the cost of your silicon (FPGA).

If you are only contemplating a single device for prototyping applications, then this may not be a particularly significant factor for you. On the other hand, if you are going to be purchasing hundreds or thousands of these little rascals, than you should start thinking very seriously about using the lowest speed grade you can get away with.

The problem is that modifying and reverifying RTL to perform a series of what-if evaluations of alternative implementations is difficult and time-consuming. (Such evaluations may include performing certain operations in parallel versus sequentially, pipelining portions of the design versus nonpipelining, resource sharing, etc.) This means that

---

[3] There's also *Verification IP*, as discussed in chapter 19.

the design team may be limited to the number of evaluations it can perform, which can result in a less-than-optimal implementation.

As discussed in chapter 11, one alternative is to use a pure untimed C/C++-based flow. Such a flow should feature a C/C++ analysis and synthesis engine that allows you to perform microarchitecture trade-offs and evaluate their effects in terms of size/area and speed/clock cycles. Such a flow facilitates improving the performance of a design, thereby allowing it to make use of a slower speed grade if required.

On a happier note

My friend Tom Dillon said that after scaring everyone with the complexities above, I should end on a happier note. So, on the bright side, once a design team has selected an FPGA vendor and become familiar with a product family, it tends to stick with that family for quite some time, which makes life (in the form of the device selection process) a lot easier for subsequent projects.

1945: Sci-fi author Arthur C. Clark envisions geo-synchronous communications satellites.

1946: America. John William Mauchly, J. Presper Eckert and team finish building a general-purpose electronic computer called ENIAC.

1946: Automobile radiotelephones connect to the telephone network.

1947: America. Physicists William Shockley, Walter Brattain, and John Bardeen create the first point-contact germanium Transistor on the 23rd December.

Gigabit Transceivers

Introduction

As we discussed in chapter 4, the traditional way to move large amounts of data between two (or more) devices on the same circuit board is to use a bus, which refers to a collection of signals that carry similar data and perform a common function (Figure 21-1).

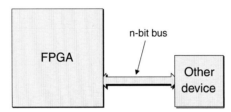

Figure 21-1. Using a bus to communicate between devices.

Early microprocessor-based systems circa 1975 used 8-bit buses to pass data around. As the need to push more data around and to move it faster grew, buses increased to 16 bits in width, then 32 bits, then 64 bits, and so forth. The problem is that this consumes a lot of pins on each device and requires a lot of tracks to connect the devices together. Routing these tracks such that they are all the same length and impedance and so forth becomes increasingly painful as boards grow in complexity. Furthermore, it becomes increasingly difficult to manage SI issues (such as susceptibility to noise and crosstalk effects) when you are dealing with large numbers of bus-based tracks.

1948: America. Airplane re-broadcasts TV signals to nine states.

For this reason, today's high-end FPGAs include special hard-wired *gigabit transceiver* blocks. These high-speed serial interfaces use one pair of differential signals to *transmit (TX)* data and another pair to *receive (RX)* data (Figure 21-2).

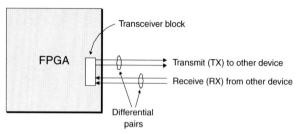

Figure 21-2. Using high-speed transceivers to communicate between devices.

Note that, unlike a traditional data bus in which you can have lots of devices hanging off the bus, these high-speed serial interfaces are point-to-point connections, which means that each transceiver can only talk to a single transceiver on one other device.

At the time of this writing, relatively few designs (probably only a few percent of total design starts) make use of these high-speed serial interfaces, but this number is expected to rise dramatically over the next few years. Using these gigabit transceivers is something of an art form, but each FPGA vendor will provide detailed user guides and application notes for its particular technology.

One problem with these interfaces is that there are so many nitty-gritty details to wrap one's brain around. For the purposes of this book, however, we shall introduce only enough of the main concepts to give the unwary sufficient information to make them dangerous!

Differential pairs

The reason for using differential pairs (which refers to a pair of tracks that always carry complementary logical levels) is that these signals are less susceptible to noise from an external source, such as radio interference or another signal switching in close proximity to these tracks. In order to illustrate this, consider the same amount of noise applied to both a single wire and a differential pair (Figure 21-3).

1948: America. Work starts on what is supposed to be the first commercial computer, UNIVAC-1.

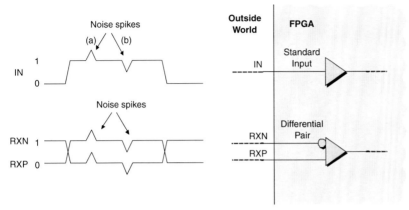

Figure 21-3. Using high-speed transceivers to communicate between devices.

In the case of the standard input, we have a pin called IN connected to a buffer gate. For the purposes of this example, we aren't particularly interested in the first noise spike (a), but the second spike (b) could cause problems. If this noise spike crosses the input switching threshold of the buffer gate, it could cause a glitch (pulse) on the output of the gate. In turn, this glitch could cause some undesired activity (such as registers loading incorrect values) inside the FPGA.

Things were somewhat easier in the not-so-distant past when the difference between logic 0 and logic 1 values was 5 volts because a noise spike of, say, 1 volt wouldn't cause any problems. But the sands of time have slipped through the hourglass as is their wont, and depending on the I/O standard you are using, the difference between a logic 0 and a logic 1 may now be only 1.8 volts, 1.5 volts, or even less. In this case, a noise spike much smaller than 1 volt could be devastating.[1]

---

[1] In the case of differential pairs, one standard has a differential voltage—the difference between a logic 0 and a logic 1—of only 0.175 volts (175 millivolts)!

1948:
First atomic clock is
constructed.

Now consider the differential pair, whose signals are generated by a special type of driving gate in the transmitting device (Figure 21-4). For the purists among us, we should note that the positive (true) halves of the differential pairs (RXP and TXP in Figures 21-3 and 21-4, respectively) are usually drawn on the top, while the negative (inverse or complementary) halves (RXN and TXN)—along with the bobbles (circles) on their buffer symbols—are usually drawn on the bottom. The reason we drew them the other way round was to make the RXP signal match up with the IN signal in Figure 21-3, thereby making this figure a little easier to follow.

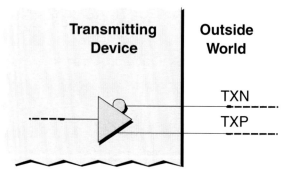

Figure 21-4. Generating a differential pair.

Remember that the two signals on a differential pair always carry complementary logical values. So when RXP in Figure 21-3 is a logic 0, RXN will be a logic 1, and vice versa. The point is that, as we see in Figure 21-3, the fact that the two tracks forming the differential pair are routed very closely together means that any noise spikes will affect both tracks identically. The receiving buffer gate is essentially interested only in the difference between the two signals, which means that differential pairs are much less susceptible to the effects of noise than are connections formed from individual wires.

The end result is that, assuming the circuit board is designed appropriately, these transceivers can operate at incredibly high speeds. Furthermore, each FPGA may contain a number of these transceiver blocks and, as we shall see, several transceivers can be "ganged together" to provide even higher data transfer rates.

1949: America. MIT's first real-time computer called Whirlwind is launched.

Multiple standards

Of course, electronics wouldn't be electronics if there weren't a variety of standards for this sort of thing. Each standard defines things from the high-level protocols all the way down to the *physical layer (PHY)*. A few of the more common standards are as follows:

- Fibre Channel
- InifiniBand®
- PCI Express (started and pushed by Intel Corporation)
- RapidIO™
- SkyRail™ (from Mindspeed Technologies™)
- 10-gigabit Ethernet

This situation is further complicated by the fact that, in the case of some of these standards, like PCI Express and Sky-Rail, device vendors might use the same underlying concepts, but rebrand things using their own names and terminology. Also, implementing some standards requires the use of multiple transceiver blocks (see also the "Ganging multiple transceiver blocks together" section later in this chapter).

Let's assume that we are building a circuit board and wish to use some form of high-speed serial interface. In this case, the system architects will determine which standard is to be used. Each of the gigabit transceiver blocks in an FPGA can generally be configured to support a number of different standards, but usually not all of them. This means that the system architects will either select a standard that is supported by the FPGAs they intend to use, or they will select FPGAs that will support the interface standard they wish to employ.

If the system under consideration includes creating one or more ASICs, we can of course implement the standard of our choice from the ground up (or, more likely, we would purchase an appropriate block of IP from a third-party vendor). Off-

1949: America.
Start of network TV.

the-shelf (ASSP-type) devices, however, will typically support only one, or a subset, of the above standards. In this case, an FPGA may be used to act as an interface between two (or more) standards (Figure 21-5).

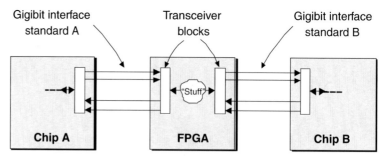

Figure 21-5. Using an FPGA to interface between multiple standards.

8-bit/10-bit encoding, etc.

One problem that rears its ugly head when you are talking about signals with data rates of gigabits per second is that the circuit board and its tracks absorb a lot of the high-frequency content of the signal, which means that the receiver only gets to see a drastically attenuated version of that signal.

Unfortunately, this is something that doesn't make much sense in words, so let's take a peek at some illustrations. First, let's consider an ideal signal that's alternating between logic 0 and logic 1 values (Figure 21-6).

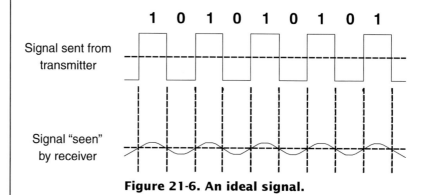

Figure 21-6. An ideal signal.

Full-blown engineers will immediately spot some errors in this diagram. For example, the signal generated by the transmitting chip is shown as being a pure digital square wave, but in the real world such a signal would actually have significant analog characteristics. In reality, the best you can say at these frequencies is that the signal is horrible coming out (from the transmitting chip), and it's even worse going in (to the receiving chip). Also, the signal seen by the receiver would be phase shifted from that shown in Figure 21-6, but we've aligned the two signals so that we can see which bits at the transmitting and receiving ends of the track are associated with each other.

As this illustration shows, the signal seen at the receiving end of the track has been severely attenuated, but it still oscillates above and below some median level, which will allow the receiver to detect it and pull useful information out of it. Now, let's consider what would happen if we were to modify the previous sequence such that it commenced by transmitting a series of three consecutive logic 1 values (Figure 21-7).

1949: England. EDSAC computer uses first assembly language called Initial Orders.

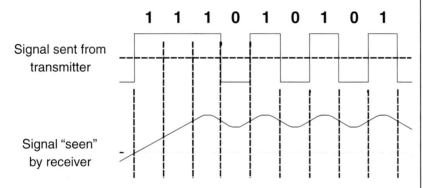

Figure 21-7. The effects of transmitting a series of identical bit values.

In this case (and remembering that this is an over-the-top, pessimistic scenario intended purely for the purposes of providing an example for us to talk about), the signal seen by the receiver continues to rise throughout the course of the first three bits. This takes the signal above the median value, which means that when the sequence eventually returns to its original 010101… sequence, the receiver will actually continue to see it as a sequence of logic 1 values.

In the context of data communications, the individual binary digits (or sometimes words formed from a collection of digits) are referred to as *symbols*. The spreading or "smearing" of symbols where the energy from one symbol affects subsequent (downstream) symbols such that the received signal might be interpreted incorrectly is referred to as *intersymbol interference (ISI)*.

ISI is pronounced by spelling it out as "I-S-I."

Another term that you often hear in conjunction with this is *consecutive identical digits (CIDs)*, which refers to occurrences such as our three logic 1 values shown in Figure 21-7. As we noted earlier, the example shown in Figure 21-7 is overly pessimistic. In reality, it is only necessary to ensure that we never send more than five identical bits in a row. Thus, our high-speed transceiver blocks have to include some form of encoding—such as the 8-bit/10-bit (abbreviated to 8b/10b or 8B/10B) standard—in which each 8-bit chunk of data is augmented by two extra bits to ensure that we never send more than five 0s or five 1s in a row. Furthermore, this standard ensures that the signal is always DC-balanced (that is, it has the same amount of energy above and below the median) over the course of 20 bits (two chunks).

There are alternative encoding schemes to the 8B/10B standard, including 64B/66B (or 64b/66b) and *SONET Scrambling*. The "scrambling" portion of the latter appellation comes from the fact that, like all of the schemes discussed here, this standard serves to randomize ("scramble") the patterns of 0s and 1s to prevent long strings of all 0s or all 1s.

One last point worth noting while we are here is that, in addition to addressing the problem presented in Figure 21-7, one of the main reasons for using these encoding schemes is to ease the task of recovering the clock signal from the data stream (see also the discussions on "Clock recovery, jitter, and eye diagrams" later in this chapter).

Delving into the transceiver blocks

Now that we've introduced the concept of 8B/10B encoding, we're in a better position to take a slightly closer look at the main elements comprising a transceiver block (Figure 21-8).

1949: England. Cambridge University. Small experimental computer called EDSAC performs its first calculation.

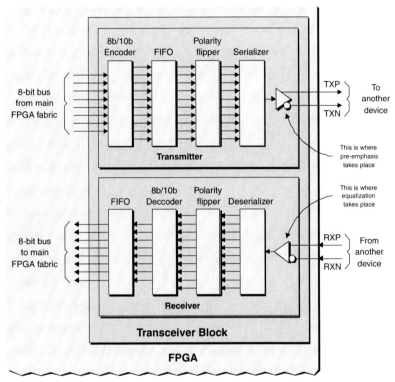

Figure 21-8. The main elements composing a transceiver block.

As usual, this is a highly simplified representation that omits a lot of bits and pieces, but it serves to cover the points of interest to us here. With regard to the annotations on "pre-emphasis" and "equalization," these topics are introduced later in this chapter.

On the transmitter side, bytes of data are presented to the transceiver from user-defined logic in the main FPGA fabric via an 8-bit bus. This is passed through an 8B/10B encoder and handed over into a FIFO buffer, which is used to store data temporarily when too many words arrive too closely together.

1950: America.
Jay Forrester at MIT
invents magnetic core
store.

The output from the FIFO passes through a polarity flipper, which may be used to pass the data through unmodified or to flip each bit from a 0 to a 1 and vice versa (polarity flipping will only be required if the device we're passing data to is expecting to see flipped data). In turn, the output from the polarity flipper is passed to a serializer, which converts the parallel input data into a serial stream of bits. This serial stream is then handed over to a special output driver/buffer that generates a differential signal pair.

Similarly, on the receiver side, a serial data stream presented as a differential signal pair is passed through a special input buffer into a deserializer, which converts the serial data into 10-bit words. These words are passed into a polarity flipper, which may be used to pass the data through unmodified or to flip each bit from a 0 to a 1 and vice versa (polarity flipping will only be required if the device we're receiving data from is sending us flipped data). The output from the polarity flipper is handed over to an 8B/10B decoder, which descrambles the data. The resulting 8-bit bytes are passed via a FIFO buffer into the main FPGA fabric, where they can be processed by whatever logic the design engineers decide to implement.

Note that, depending on the FPGA technology you are using, some transceiver blocks may support a variety of encoding standards, such as 8B/10B, 64B/66B, SONET Scrambling, and so forth. Others may support only a single standard like 8B/10B, but in this case it may be possible to switch out these blocks and implement your own encoding scheme in the main FPGA fabric if required.

Ganging multiple transceiver blocks together

The term *baud rate* refers to the number of times a signal in a communications link changes (or can change) per second. Depending on the encoding technique used, a communications link can transmit one data bit—or fewer or more bits—with each baud, or change in state.

At the time of this writing, the current state of play is that each transceiver channel can transmit and receive 8B/10B-encoded data (or data encoded using a similar scheme) at baud rates up to 3.125 gigabits per second (Gbps).[2] This translates to 2.5 Gbps of real, raw data if we ignore the overhead of the additional bits added by the 8B/10B-encoding scheme (that is, a baud rate of 3.125 Gbps divided by 10 bits and multiplied by 8 bits equals a true data rate of 2.5 Gbps).

The problem is that, by definition, standards such as 10-gigabit Ethernet have data transfer requirements of 10 Gbps. For this reason, there are additional standards like the *10-gigabit attachment unit interface (XAUI)* approach that defines how to achieve 10 Gbps of data throughput using four differential signal pairs in each direction (Figure 21-9).

XAUI is pronounced "zow-ee."

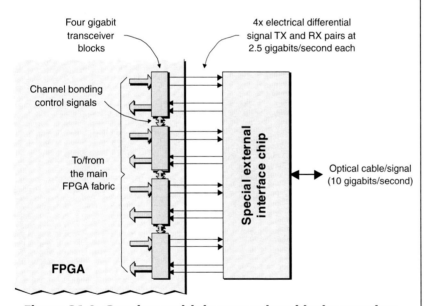

Figure 21-9. Ganging multiple transceiver blocks together.

---

[2] Once we go over baud rates of 3.175 Gbps, the overhead associated with the 8B/10B-encoding scheme becomes too high, which means we have to go to another scheme such as 64B/66B encoding.

In this case, the four transceiver blocks are linked using special channel bonding control signals so that each block knows what it is supposed to do and when it is supposed to do it.

At some stage in the future—largely dictated by the rate of adoption of high-speed serial interface technology at the circuit board level—it is likely that the functions currently embodied by the external interface chip will be incorporated into the FPGA itself, which will then have the ability to transmit and receive optical signals directly (see also the discussions in Chapter 26).

Configurable stuff

The gigabit transceiver blocks embedded in FPGAs typically have a number of configurable (programmable) features. Different vendors and device families may support different subsets of these features, a selection of the main ones being as follows.

Comma detection

The 8B/10B-encoding scheme (and other schemes) includes special *comma characters*. These are null characters that may be transmitted to keep the line "alive" or to initiate a data transfer by indicating to the receiver that things are about to start happening and it needs to wake up and prepare itself for action.

Another point is that these high-speed serial interfaces are asynchronous in nature, which means that the clock is embedded in the data signal (see also the discussions on "Clock recovery, jitter, and eye diagrams" later in this chapter). So when a transceiver block is ready to initiate a transfer, it will send a whole series of comma characters (several hundred bits) to allow the receiver at the other end of the line to synchronize itself. (Comma characters are also employed when aligning multiple bitstreams as discussed in the previous section.)

The point is that some transceiver blocks allow the comma character that will be transmitted (and received) to be configured to be any 10-bit value, thereby allowing the transceiver to support a variety of communications protocols.

Output differential swing

Different standards support different differential output swings, which refers to the peak-to-peak difference in voltage between logic 0 and logic 1 values. Thus, transceiver blocks typically allow the differential output voltage swing to be configured across a range of values so as to support compatibility with a variety of serial system voltage levels.

On-chip termination resistors

The data rates supported by high-speed serial interfaces mean that using external termination resistors can cause discontinuities in the signals, so it's typically recommended to use the on-chip termination resistors provided in the FPGA. The values of these on-chip terminating resistors are typically configurable (they can usually be set to 50 ohms or 75 ohms) so as to support a variety of different interface standards and circuit board environments.

Pre-emphasis

As was noted in the discussion associated with Figure 21-6, signals traveling across a high-speed serial interface are severely distorted (attenuated) by the time they arrive at the receiver because the circuit board and its tracks absorb a lot of the high-frequency content of the signal, leaving only the lower-frequency (more slowly changing) portions of the signal.

One technique that may be used to mitigate this effect is pre-emphasis, in which the first 0 in a string of 0s and the first 1 in a string of 1s is given a bit of a boost with a slightly higher voltage (in this context, we will consider "string" to refer to one or more bits). In a way, we can think of this as applying our own distortion in the opposite direction to the distortion coming from the circuit board (Figure 21-10).

1950:
Maurice Karnaugh invents Karnaugh Maps (circa 1950), which quickly become one of the mainstays of the logic designer's tool-chest.

1950:
Konrad Zuse's Z4 is sold
to a bank in Zurich,
Switzerland, thereby
making the Z4 the
world's first
commercially available
computer.

1951: America.
The first UNIVAC 1 is
delivered.

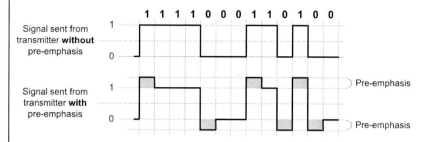

Figure 21-10. Applying pre-emphasis.

Once again, this illustration shows the signal generated by the transmitting chip as being an ideal representation (with sharp edges), but in the real world such a signal would actually have strong analog chracteristics.

The amount of pre-emphasis to be applied is typically configurable so as to accommodate different circuit board environments. The amount of pre-emphasis required for a given high-speed link is a function of the position of the FPGA in relation to other components (which equates to track lengths), a variety of board characteristics, and the high-speed standard being employed. Working out the amount of pre-emphasis to use may be determined by simulation runs or by rule of thumb.

Equalization

This is somewhat related to pre-emphasis as discussed above, except that it takes place at the receiver end of the high-speed interface (Figure 21-11).

Equalization refers to a special amplification stage that boosts higher frequencies more than lower ones. As for preemphasis, we can think of this as applying our own distortion in the opposite direction to the distortion coming from the circuit board.

The amount of equalization to be applied is typically configurable to accommodate different circuit board environments. Depending on the particular design, we might wish to use pre-emphasis, equalization, or a mixture of both.

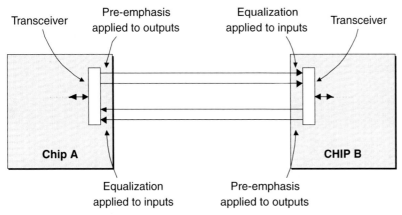

Figure 21-11. Applying equalization.

1952: America. John William Mauchly, J, Persper Eckert and team finish building a general-purpose (stored program) electronic computer called EDVAC.

1952: England. First public discussion of the concept of integrated circuits is credited to a British radar expert, G.W.A. Dummer.

One point worth noting is that, in the case of really long high-speed interface tracks on the circuit board (say, around 40 inches and above), it may be desirable to disable the internal equalization and to use an external equalizer device because the quality of equalization is typically better in a dedicated analog device than in an FPGA. Having said this, FPGAs are increasing in sophistication with regard to this sort of thing—the different vendors are constantly leapfrogging each other with regard to technology—and the quality of factors such as the quality of on-chip equalization may affect your device selection.

Clock recovery, jitter, and eye diagrams

Clock recovery

High-speed serial interfaces are asynchronous in nature, which means that the clock is embedded in the data signal. Thus, the receiver portion of the transceiver includes *clock and data recovery (CDR)* circuitry that keys off the rising and falling edges of the incoming signal and automatically derives a clock that is representative of the incoming data rate. As you can imagine, this would not be a major feat if the incoming signal were toggling back and forth between logic 0 and logic 1

1952:
Sony demonstrates the first miniature transistor radio, which is produced commercially in 1954.

1953: Americas. First TV dinner is marketed by the Swanson Company.

1954:
Launch of giant balloon called Echo 1—used to bounce telephone calls coast-to-coast in the USA.

values, in which case the clock and the data would effectively be identical (Figure 21-12a).

Things get a little trickier when the signal becomes more complex (Figure 21-12b). For example, if the incoming signal commenced with three 1s followed by three 0s, we couldn't fault the clock recovery function for making an initial guess that the clock frequency was only one third of its true value. As more data (and more transitions) arrive, however, the clock recovery function will refine its assumptions until it has derived the correct frequency.

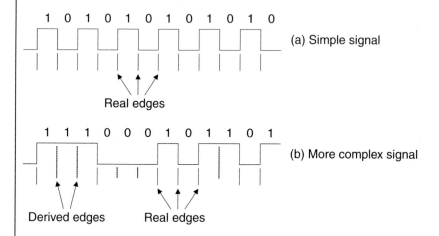

Figure 21-12. Recovering the clock signal.

Once the receiver has locked down the clock, it uses this information to sample the incoming data stream at the center point of each bit in order to determine whether that bit is a logic 0 or a logic 1 (Figure 21-13).

This is why, as we discussed earlier, a data transmission will commence with several hundred bits of comma characters to allow the receiver to lock on the clock and prepare itself for action.

The clock recovery function will continue to monitor edges and constantly tweak the clock value to accommodate

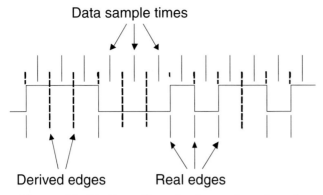

Figure 21-13. Sampling the incoming signal.

1954:
The number of radio sets in the world out-numbers newspapers sold everyday.

1954:
First silicon transistor manufactured.

1955:
Velcro is patented.

1956: America. John Backus and team at IBM introduced the first widely used high-level computer language, FORTRAN.

slight back-and-forth drifts in the clock caused by environmental conditions such as temperature and voltage variations.

Jitter and eye diagrams

The term *jitter* refers to short-term variations of signal transitions from their ideal positions in time. For example, if we were to take an incoming signal that was oscillating between logic 0 and logic 1 values (Figure 21-14a, b) and overlay the data associated with each clock cycle on top of the preceding cycles, we would start to see some fuzziness appearing (Figure 21-14c–f).

This fuzziness is caused by a variety of factors, including the clock wandering slightly in the transmitting device and also the ISI effects we noted earlier (see also the discussion associated with Figure 21-7).

In fact, we can go one step further, which is conceptually to fold each clock cycle in half, thereby overlaying the positive 0–1–0 pulses from the first half of the cycle with the negative 1–0–1 pulses from the second half of the cycle (Figure 21-14g).

Once again, the waveforms shown in Figure 21-14 are unrealistic because they feature razor-sharp edges. Real-world signals would have analog characteristics. If we were to look at a real waveform in its folded form, it would look something like that shown in Figure 21-15.

1956: America. John McCarthy develops a computer language called LISP for artificial intelligence applications.

1956: America MANIAC 1 is the first computer program to beat a human in a game (a simplified version of chess).

1956: First transatlantic telephone cable goes into operation.

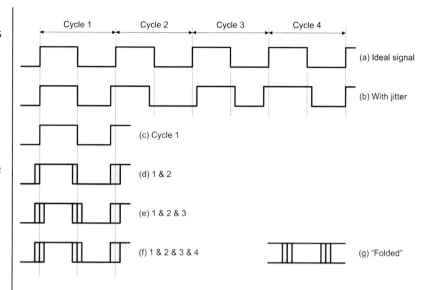

Figure 21-14. Jitter.

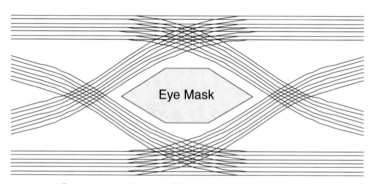

Figure 21-15. Eye diagram and eye mask.

The result is a diagram whose center looks something like a human eye, so, perhaps not surprisingly, it's referred to as an *eye diagram*. As jitter, attenuation, and other distortions increase, the center of the eye closes more and more. Thus, a lot of specifications define a geometric shape called the *eye mask*. This mask, which may be rectangular or hexagonal as shown here, represents the *data valid window*. As long as all of the curves fall outside of the eye mask, the high-speed interface will work.

The point of all of this is that if you are planning on using one of these high-speed serial communications interfaces, then you need to make sure that you have access to SI analysis tools that have been augmented to support the concept of eye diagrams.

1957: America. Gordon Gould conceives the idea of the Laser.

1957: America. IBM 610 Auto-Point computer is introduced.

1957: Russia launches the *Sputnik 1* satellite.

Reconfigurable Computing

Dynamically reconfigurable logic

The advent of SRAM-based FPGAs presented a new capability to the electronics fraternity: *dynamically reconfigurable logic*, which refers to designs that can be reconfigured on the fly while remaining resident in the system.

Just to recap, FPGAs contain a large amount of programmable logic and registers, which can be connected together in different ways to realize different functions. SRAM-based variants allow the main system to download new configuration data into the device. Although all of the logic gates, registers, and SRAM cells forming the FPGA are created on the surface of a single piece of silicon substrate, it is sometimes useful to visualize the device as comprising two distinct strata: the logic gates/registers and the programmable SRAM configuration cells (Figure 22-1).

The versatility of these devices opened the floodgates to a wealth of possibilities. For example, when the system is first powered up, the FPGAs can be configured to perform a variety of system-test (and even self-test) operations. Once the system checks out, the FPGAs can be reconfigured to perform their main function in life.

Dynamically reconfigurable interconnect

Although it's great to be able to reconfigure the function of the individual devices on a circuit board, there are occasions when design engineers would like to create board-level systems that can be reconfigured to perform a variety of radically different functions.

1957:
Russia launches the *Sputnik 1* satellite.

1958: America. Computer data is transmitted over regular telephone circuits.

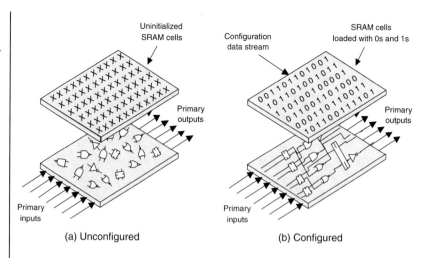

Figure 22-1. Dynamically reconfigurable logic: SRAM-based FPGAs.

The solution is to be able to configure the board-level connections between devices dynamically. A breed of devices offer just this capability: *field-programmable interconnect devices (FPIDs)*, which may also be known as *field-programmable interconnect chips (FPICs)*.[1] These devices, which are used to connect logic devices together, can be dynamically reconfigured in the same way as standard SRAM-based FPGAs. Because each FPID may have 1,000 or more pins, only a few such devices are typically required on a circuit board (Figure 22-2).

One interesting point is that the concepts discussed here are not limited to board-level implementations. Any of the technologies discussed thus far may also potentially be implemented in hybrids, *multichip modules (MCMs)*, and SoC devices.

Reconfigurable computing

As with many things in electronics, the term *reconfigurable computing (RC)* can mean different things to different people.

---

[1] FPIC is a trademark of Aptix Corporation (www.aptix.com).

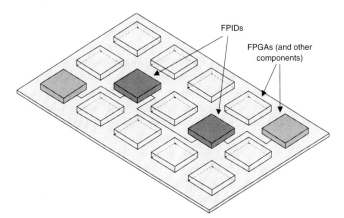

FPIDs

FPGAs (and other components)

Figure 22-2. Dynamically reconfigurable interconnect: SRAM-based FPIDs.

1958: America. Jack Kilby, working for Texas Instruments, succeeds in fabricating multiple components on a single piece of semiconductor (the first integrated circuit).

For some, it refers to special microprocessors whose instruction sets can be augmented or modified on the fly. For our purposes here, however, we understand RC to refer to a piece of general-purpose hardware—such as an FPGA (what a surprise)—that can be configured to perform a specific task, but that can subsequently be reconfigured on demand to carry out other tasks.

One limitation with the majority of SRAM-based FPGAs is the time it takes to reconfigure them. This is because they are typically programmed using a serial data stream (or a parallel stream only 8 bits wide). When we start to talk about high-end devices with tens of millions of SRAM configuration cells, it can take up to a couple of seconds to reprogram these beasts. There have been some FPGAs that address this issue by using large numbers of general-purpose I/O pins to provide a wide configuration bus (say, 256 bits) before reverting to their main I/O functionality (see also chapter 26). Also, some flavors of *field-programmable node arrays (FPNAs)* have dedicated wide programming buses (see also Chapter 23).

Another limitation with traditional FPGA architectures is that, when you wish to reconfigure any part of the device, you typically have to reprogram the entire device (some recent architectures do allow you to reconfigure them on a column-by-column basis, as discussed in chapter 14, but this offers only

1959: America. COBOL computer language is introduced for business applications.

a rather coarse level of granularity). Furthermore, it is usually necessary to halt the operation of the entire circuit board while these devices are being reconfigured. Additionally, the contents of any registers in the FPGAs are irretrievably lost during the process.

In order to address these issues, an interesting flavor of FPGA was introduced by Atmel Corporation (www.atmel.com) circa 1994. In addition to supporting the dynamic reconfiguration of selected portions of the internal logic, these devices also featured:

- No disruption to the device's inputs and outputs
- No disruption to the system-level clocking
- The continued operation of any portions of the device that are not undergoing reconfiguration
- No disruption to the contents of internal registers during reconfiguration, even in the area being reconfigured

The latter point is of particular interest because it allows one instantiation of a function to hand over data to the next function. For example, a group of registers may initially be configured to act as a binary counter. Then, at some time determined by the main system, the same registers may be reconfigured to operate as a *linear feedback shift register (LFSR)*[2] whose seed value is determined by the final contents of the counter before it is reconfigured.

Although these devices were evolutionary in terms of technology, they were revolutionary in terms of their potential. To reflect their new capabilities, appellations such as "virtual hardware" and "cache logic"[3] were quickly coined.

The term *virtual hardware* is derived from its software equivalent, *virtual memory*, and both are used to imply something that is not really there. In the case of virtual memory, a

---

[2] LFSRs are introduced in detail in Appendix C.

[3] Cache Logic is a trademark of Atmel Corporation, San Jose, CA, USA.

computer's operating system pretends that it has access to more memory than is actually available. For example, a program running on the computer may require 500 megabytes to store its data, but the computer may have only 128 megabytes of memory available. To get around this problem, whenever the program attempts to access a memory location that does not physically exist, the operating system performs a sleight of hand and exchanges some of the contents in the memory with data on the hard disk. Although this practice, known as *swapping*, tends to slow things down, it does allow the program to perform its tasks without having to wait while someone runs down to the store to buy some more memory chips.

Similarly, the term *cache logic* is derived from its similarity to the concept of *cache memory*, in which high-speed, expensive SRAM is used to store active data, while the bulk of the data resides in slower, lower-cost memory devices such as DRAM. (In this context, "active data" refers to data or instructions that a program is currently using or that the operating system believes the program will want to use in the immediate future.)

In fact, the concepts behind virtual hardware are actually quite easy to understand. Each large macrofunction in a device is usually formed by the combination of a number of smaller microfunctions, such as counters, shift registers, and multiplexers. Two things become apparent when a group of macrofunctions is divided into their respective microfunctions. First, functionality overlaps, and an element such as a counter may be used several times in different places. Second, there is a substantial amount of *functional latency*, which means that at any given time only a portion of the microfunctions are active. Put another way, relatively few micro- functions are in use during any given clock cycle. Thus, the ability to reconfigure individual portions of a virtual hardware device dynamically means that a relatively small amount of logic can be used to implement a number of different macrofunctions.

1959: America. Robert Noyce invents techniques for creating microscopic aluminum wires on silicon, which leads to the development of modern integrated circuits.

1959:
Swiss physicist Jean Hoerni invents the planar process, in which optical lithographic techniques are used to create transistors.

By tracking the occurrence and usage of each microfunction, then consolidating functionality and eliminating redundancy, virtual hardware devices can perform far more complex tasks than they would appear to have logic gates available. For example, in a complex function requiring 100,000 equivalent gates, only 10,000 gates may be active at any one time. Thus, by storing, or caching, the functions implemented by the extra 90,000 gates, a small, inexpensive 10,000-gate device can be used to replace a larger, more expensive 100,000-gate component (Figure 22-3).

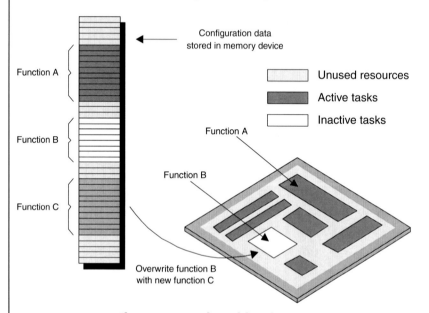

Figure 22-3. Virtual hardware.

Theoretically, it would be possible to compile new design variations in real time, which may be thought of as dynamically creating subroutines in hardware!

RC was a big buzz in the latter half of the 1990s, and there are still some who are waving the RC banner (and wearing the T-shirts). Sad to relate, however, nothing really came of this with the exception of highly specialized applications. The core problem is that traditional FPGA architectures are too fine grained, so reconfiguring them takes too long (in computer terms). In order to support true RC, one would need access to devices that could be reconfigured hundreds of thousands of times per second. The answer may be the coarser-grained architectures fielded by the FPNAs introduced in Chapter 23.

1960: America. Theodore Maimen creates the first Laser.

1960: America. The Defense Advanced Research Projects Agency (DARPA) begins work on what will become the Internet.

Field-Programmable Node Arrays

Introduction

Before we throw ourselves into this topic with wild abandon, it's probably only fair to note that the term *field-programmable node array*, or *FPNA*, was coined by the author and is not industry-standard terminology (yet).

Fine-, medium-, and coarse-grained architectures

When it comes to categorizing different IC architectures, ASICs are usually said to be *fine grained*, because design engineers can specify their functionality down to the level of individual logic gates. By comparison, the majority of today's FPGAs may be classed as *medium grained* because they consist of small blocks ("islands") of programmable logic (where each block represents a number of logic gates and registers) in a "sea" of programmable interconnect. (Even though today's FPGA offerings typically include processor cores, blocks of memory, and embedded functions like multipliers, the main underlying architecture is as described above.)

Truth to tell, many engineers would actually refer to FPGAs as being *coarse grained*, but classing them as medium grained makes much more sense when we start to bring FPNAs into the picture because these boast really coarse-grained architectures. The underlying concept behind FPNAs is that they are formed from an array of *nodes*, each of which is a sophisticated processing element (Figure 23-1).

Of course, this is a very simplified representation of an FPNA, not the least because it omits any I/O. Furthermore, we've only shown relatively few processing nodes, but such a

1960:
NASA and Bell Labs
launch the first
commercial
communications
satellite.

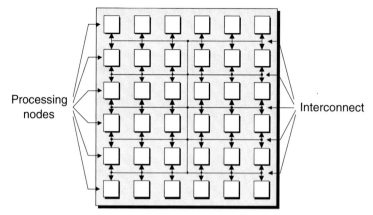

Figure 23-1. Generic representation of an FPNA.

device can potentially contain hundreds or thousands of nodes. Depending on the vendor, each node might be an *algorithmic logic unit (ALU)*, a complete microprocessor CPU, or an algorithmic processing element (this latter case is discussed in more detail later in this chapter). At the time of this writing, 30 to 50 companies are seriously experimenting with different flavors of FPNAs; a representative sample of the more interesting ones is as follows:

Company	Web site	Comment
Exilent	www.elixent.com	ALU-based nodes
IPflex	www.ipflex.com	Operation-based nodes
Motorola	www.motorola.com	Processor-based nodes
PACT XPP Technologies AG	www.pactxpp.com	ALU-based nodes
picoChip Designs	www.picochip.com	Processor-based nodes
QuickSilver Technology	www.qstech.com	Algorithmic element nodes

For the purposes of these discussions, we shall concentrate on just two of these vendors—picoChip and QuickSilver—who are conceptually at opposite ends of the spectrum: picoChip's picoArray devices are formed from arrays of processors. Their key application area is large, fixed installations such as base stations for wireless networks in which power consumption is not a major consideration. Furthermore, these chips are intended to be reconfigured now and again (for example, every hour or so as cellular phone usage profiles change throughout the day).

By comparison, QuickSilver's *adaptive computing machine (ACM)* devices are formed from clusters of algorithmic element nodes. Their key application area is small, low-power, handheld products like cameras and cell phones (although they are of interest for a wide variety of other applications). Furthermore, these chips can be reconfigured (QuickSilver prefers the term *adapted*) hundreds of thousands of times per second.

Algorithmic evaluation

FPNAs are mainly intended to execute sophisticated, compute-intensive algorithms. This means that before we go any further, we should spend a few moments ruminating on these algorithms to set the scene for what is to come.

At one end of the spectrum are word-oriented algorithms, such as the extremely compute-intensive *time division multiple access (TDMA)* algorithm used in digital wireless transmission. Any variants such as *Sirius, XM Radio, EDGE,* and so forth form a subset of this algorithmic class, so an architecture that can handle high-end TDMA should also be able to handle its less-sophisticated cousins (figure 23-2).

At the other end of the continuum, we find bit-oriented algorithms, such as *wideband code division multiple access (W-CDMA)*, and its subvariants, such as CDMA2000, IS-95A, and the like. (W-CDMA is used for the wideband digital radio communications of Internet, multimedia, video, and other capacity-demanding applications.)

And then there are algorithms that exhibit different mixes of word-oriented and bit-oriented components, such as the various flavors of MPEG, voice and music compression, and so forth.

When one evaluates these various algorithms, it soon becomes quickly apparent that conventional RC approaches tend to attack the problem at inappropriate levels (RC concepts were introduced in chapter 22). For example, some RC

1961:
Time-sharing computing is developed.

1962: America.
Steve Hofstein and
Fredric Heiman at RCA
Research Lab invent
field effect transistors
(FETS).

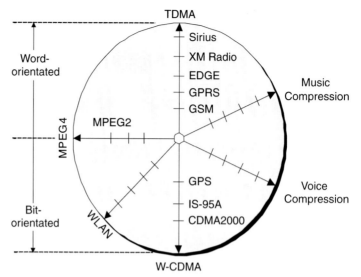

Figure 23-2. A simplified view of algorithm space.

approaches engage problems at too micro of a level, that is, at the level of individual gates or FPGA blocks. Coupled with hideously difficult application programming, this power-hungry approach results in relatively long reconfiguration times, thereby making it unsuitable for some applications. By comparison, other approaches tackle the problem at too macro of a level, that is, at the level of entire applications or algorithms, which results in inefficient use of resources.

Perhaps not surprisingly, it soon becomes apparent that algorithms are heterogeneous in nature, which means that if you take a bunch of diverse algorithms, their constituent elements are wildly different. Based on this, the obvious solution is to use heterogeneous architectures that fully address the heterogeneous nature of the algorithms they are required to implement, but what might these little scamps look like?

picoChip's picoArray technology

In order to address the processing requirements of the algorithms discussed above, picoChip came up with a device called a *picoArray*. The heterogeneous node-based architecture of the picoArray features a matrix of different flavors of *reduced instruction set computing (RISC)* processors. These 16-bit devices are optimized in a variety of different ways: for example, one processor type may have lots of memory, while another will support special algorithmic instructions that can perform operations like "spread" and "despread" from the CDMA wireless standard using a single clock cycle (as opposed to 40 cycles using a general-purpose processor).

RISC is pronounced to rhyme with "lobster bisque."

In the first incarnation of these devices, each processor node was approximately equivalent (in processing capability, not in architecture) to an ARM9 for control-style applications or a TI C54*xx* for DSP-style applications. When you take into account the fact that a single picoArray can contain hundreds of such nodes, the result is a truly ferocious amount of processing power.

As one example, when I first became aware of the picoArray technology around December 2002, one of the absolute top-of-the-line dedicated DSP chips in the world at that time was the TMS320C6415 from Texas Instruments. That bad boy could perform such a humongous number of calculations at such a breathtaking speed that it made your eyes water. However, picoChip claims that a single picoArray running at only 160 megahertz could deliver almost 20 times more processing power (measured in 16-bit ALU MOPS) than a TMS320C6415 running at 600 megahertz. Wow!

An ideal picoArray application: Wireless base stations

Cell phone companies spend billions and billions of dollars every year on wireless infrastructure, and a large portion of these funds is devoted to developing the digital baseband processing portions of wireless base stations. Depending on its location, each base station has to be capable of processing tens or hundreds of channels simultaneously.

Not surprisingly, there is a huge drive to reduce the cost of implementing each channel. The fact that a single picoArray

1962: America. Unimation introduces the first industrial robot.

can replace a number of traditional ASICs, FPGAs, and DSPs offers a way of dramatically reducing the cost of each base station channel.

In fact, one of the problems with conventional solutions is that they require at least three design environments: ASIC and/or FPGA, DSP, and RISC (where the latter refers to some microprocessor-type functionality). All of this complicates development and test and slows the base station's time-to-market, which is not considered to be a good thing (Figure 23-3a). By comparison, a major advantage of a picoArray-based solution is that it largely consolidates everything into a single design environment (Figure 23-3b).

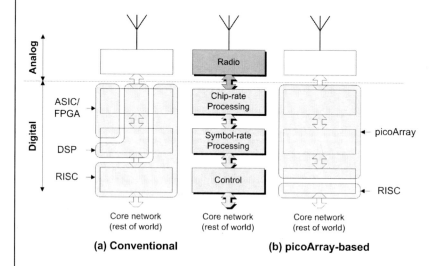

Figure 23-3. Conventional devices versus a picoArray approach.

Furthermore, in the case of conventional solutions, although ASICs can provide extremely high performance, they are very expensive to develop and they have long design cycles. Even worse, algorithms implemented in ASICs are effectively "frozen in silicon." This is a major problem because wireless standards are evolving so quickly that, by the time an ASIC design has actually been implemented, it may already be obsolete (honestly, this happens way more often than you might imagine).

By comparison, in the case of the picoArray-based approach, the fact that every processor node on the device is fully programmable means that each channel can be easily reconfigured to adapt to hourly changes in usage profiles, to weekly enhancements and bug fixes, and to monthly evolutions in wireless protocols. Thus, a base station based on picoArray technology will have a much longer life in the field, thereby reducing operating costs.

The picoArray design environment

The underlying functionality to be mapped onto the processor nodes in a picoArray is captured in pure C code or in assembly language. As we discussed in chapter 11, C is a sequential language, so we need some way to describe any parallel processing requirements. As opposed to using one of the augmented C/C++ techniques mentioned in chapter 11, the folks at picoChip have taken another approach, which is to employ a VHDL framework to capture the structure of the design, including any parallel processing requirements, and to connect design modules together at the block level. C or assembly code is then used to implement the internals of each module.

Another interesting aspect of the picoChip solution is the fact that they provide a complete library of programming/configuration modules that can be hooked together to implement a fully functioning base station (users can also tweak individual modules to implement their own algorithm variations, thereby gaining a competitive advantage). Around May 2003, picoChip announced that they had achieved a "world first" by using this library to implement a 3GPP-compliant carrier-class base station and to make a 3G call on that base station! Since that time, they have continued to progress in leaps and bounds, so you'll have to visit their Web site at www.picochip.com to apprise yourself of the current state of play.

1962:
First commercial communications satellite (*Telstar*) launched and operational.

QuickSilver's ACM technology

For several years now, the guys and gals at QuickSilver have been in "secret squirrel" mode working on their version of an FPNA (although I'm sure they are going to moan and groan about this appellation). Based on what I know (which is more than they think I know ... at least I think it is), it's fair to say that QuickSilver's technology, which they call an *adaptive computing machine*, or ACM, boasts a truly revolutionary heterogeneous node-based architecture and interconnect structure (Figure 23-4).

ACM is pronounced by spelling it out as "A-C-M."

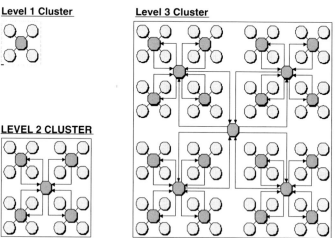

Figure 23-4. The ACM's architecture.

At the lowest level we have an *algorithmic element node*. Four of these nodes, forming a "quad," are gathered together with a *matrix interconnect network* (MIN) to form what we might call a Level 1 cluster. Four of these Level 1 clusters can be grouped to form a Level 2 cluster, and so forth.

At the time of this writing, there are a variety of different types of algorithmic element nodes (we'll talk about how these node types are mapped into the quads in a little while). We aren't going to delve into the guts of every node here, but it's important to understand that each such node performs tasks at the level of complete algorithmic elements. For example, an *arithmetic* node can be used to implement different (variable-width) linear arithmetic functions such as a FIR filter, a *discrete cosign transform (DCT)*, an FFT, and so forth. Such a node can also be used to implement (variable-width) nonlinear arithmetic functions such as ((1/sine A) (1/x)) to the 13th power.

Similarly, a *bit-manipulation* node can be used to implement different (variable-width) bit-manipulation functions, such as a *linear feedback shift register (LRSR)*, Walsh code generator, GOLD code generator, TCP/IP packet discriminator, and so forth.

Each node is surrounded by a *wrapper*, which makes all of the nodes *appear* to be identical to the outside world (that is, to the world outside the node). This wrapper is in charge of accepting packets of information (instructions, raw data, configuration data, etc.) from the outside world, unpacking this data, distributing it throughout the node, managing tasks, gathering and packing the results together, and presenting these results back to the outside world.

The concept of the wrapper isolating the node from the outside world and making all of the nodes appear to be identical is of especial interest when we come to realize that each node is "Turing complete." This means that you can present any node with any problem—say, an arithmetic node with a bit-manipulation task—and that node will solve the problem, although less efficiently than would a more appropriate type of node. Furthermore, QuickSilver also allows you to create your own types of nodes, where you define the core of the node and surround it with QuickSilver's wrapper.

Good grief! Trying to work out how best to wend our weary way through the complexities of all of this is making my brain ache. One key point is that any part of the device, from a few nodes all the way up to the full chip, can be adapted blazingly fast, in many cases within a single clock cycle. Also of interest is the fact that approximately 75 percent of each node

1962:
First commercial touch-tone phone system.

1963: America.
The LINC computer is
designed at MIT.

is in the form of local memory. This allows for a radical change in the way in which algorithms are implemented. As opposed to passing data from function to function, the data can remain resident in a node while the function of the node changes on a clock-by-clock basis. It also means that, unlike an ASIC implementation in which each algorithm requires its own dedicated silicon, the ACM's ability to be adapted tens or hundreds of thousands of times per second means that only those portions of an algorithm that are actually being executed need to be resident in the device at any one time (see also the discussions on SATS later in this chapter). This provides for tremendous reductions in silicon area and power consumption.

You define the mix of nodes

I'm not quite sure where to squeeze this topic in, so we'll give it a whirl here to see how well it flies. Just a little while ago, we noted that there are various types of algorithmic element nodes. We also noted that each cluster is formed from a quad of these nodes gathered together with a MIN. Based on this, I'm sure that you are wondering how the node types are assigned across multiple clusters.

Well, the point is that the folks at QuickSilver don't actually make and sell chips themselves (apart from proof-of-concept and evaluation devices of course). Instead, they license their ACM technology to anyone who is interested in playing with it, thereby allowing you (the end user) to determine the optimum mix of node types for your particular application and then have chips fabricated to your custom specifications. The fact that their wrappers make each node appear identical to the outside world makes it easy to exchange one type of node for another!

The system controller node, input/output nodes, etc.

In addition to the structure shown in figure 23-4, each ACM also includes a gaggle of special-purpose nodes, such as system controller, external memory controller, internal memory controller, and I/O nodes. In the case of the latter, each I/O node can be used to implement I/O tasks in such forms as a UART or bus interfaces such as PCI, USB, Firewire, and the like (as for the algorithmic element nodes, the I/O nodes can be reconfigured on a clock-by-clock basis as required). Furthermore, the I/O nodes are also used to import configuration data, which means that each ACM can have as wide a configuration bus as the total number of input pins if required.

We will consider how applications are created for, and executed on, ACMs shortly. For the nonce, it is only important to note that almost everything that makes life difficult with other implementation technologies is handled transparently to the ACM design engineer. For example, each ACM has an on-chip *operating system (OS)*, which is distributed across the system controller node and the wrappers associated with each of the algorithmic element nodes. The individual algorithmic element nodes take care of scheduling their tasks and any internode communications. This leaves the system controller node relatively unloaded because its primary responsibilities are limited to knowing which nodes are currently free and to allocating new tasks to those nodes.

From figure 23-4, it is obvious that the core ACM architecture is extremely scalable. Things start to get really clever if you have multiple ACMs on a board, their operating systems link up, and, to the rest of the system, they appear to function as a single device.

Spatial and temporal segmentation

One of the most important features of the ACM architecture is its ability to be reconfigured hundreds of thousands of times per second while consuming very little power. This allows ACMs to support the concept of *spatial and temporal segmentation (SATS)*.

In many cases, different algorithms, and even different portions of the same algorithm, can be performed at different times. SATS refers to the process of reconfiguring dynamic

1963:
PDP-8 becomes the first popular microcomputer.

OS is pronounced by spelling it out as "O-S."

SATS is pronounced to rhyme with "bats."

1963:
Philips introduces first
audio cassette.

hardware resources to rapidly perform the various portions of the algorithm in different segments of time and in different locations (nodes) on the ACM.

As a simple example, consider that some operations on a wireless phone are *modal*, which means they only need to be performed some of the time. The three main modes are *acquisition*, *idle*, and *traffic*. The acquisition mode refers to the cell phone locating the nearest base station. When in idle mode, the phone keeps track of the base station it's hooked up to and monitors the paging channel, looking for a signal that says, "Wake up because a call is being initiated." The traffic mode has two variations: *receiving* or *transmitting*. Although you may think you are talking and listening simultaneously, you actually are only doing one or the other at any particular time on a digital phone.

In the case of a wireless phone based on conventional IC technologies, each of these baseband processing functions requires its own silicon chip or some area on a common chip. This means that even when a function isn't being used, it still occupies silicon real estate, which translates into high cost and high power consumption that drains your battery faster. By comparison, a phone based on ACM technology would require only a single chip that can be adapted on the fly to perform each baseband function as required.

But this is only the beginning. In many cases, each of these major functions is composed of a suite of algorithms, which can themselves be performed at different times. For example, consider a highly simplified representation of a wireless phone receiving and processing a signal (figure 23-5).

The incoming signal consists of a series of highly compressed blocks of data, each occupying a tiny segment of time. This data proceeds through a series of algorithms, each of which performs some processing on the data and downshifts it to a lower frequency.

A key feature of this process is that each algorithmic stage occupies a different fragment of time. In traditional ASIC implementations, each function would occupy its own chip or

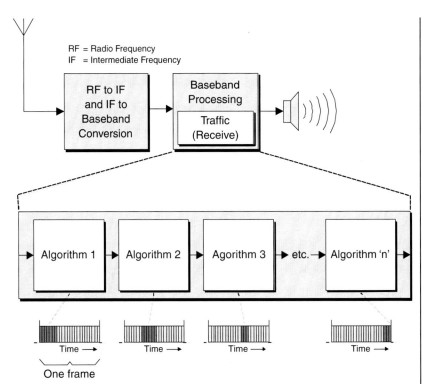

1965:
John Kemeny and Thomas Kurtz develop the BASIC computer language.

Figure 23-5. Highly simplified representation of a wireless phone receiving and processing a signal.

its own area of silicon real estate on a common device. This results in a significant waste of available resources (space and power consumption) because only a limited number of functions are actually being exercised at any particular time.

Once again, the solution is the ACM, which can be adapted on the fly to perform each algorithm as required. This concept of *on-demand hardware* results in the most efficient use of hardware in terms of cost, size (silicon real estate), performance, and power consumption (ACMs are claimed to provide 10 to 100 times or more performance increase over comparable solutions at only 1/2 to 1/20 of the power consumption).

Creating and running applications on an ACM

Of course, the next big question is how would one go about creating applications for one of these little rapscallions? Well, QuickSilver's design flow is built on a C-based system-design

1967: America. Fairchild introduce an integrated circuit called the Micromosaic (the forerunner of the modern ASIC).

language called *SilverC* (this language is similar in concept to the augmented C/C++ languages introduced in chapter 11).

SilverC preserves traditional C syntax and control structures, which makes it easy for C programmers and DSP designers to use and simplifies legacy C code conversion. SilverC also includes special *module*, *pipe*, and *process* keywords/extensions that facilitate dataflow representations and support parallel programming. Furthermore, SilverC provides special extensions for DSP programming, such as circular pointers for efficient use of DAG resources, fixed-width integer and fixed-point data types, support for saturated and nonsaturated types, and so forth.

SilverC representations can be captured and simulated much faster than the equivalent HDL representations (Verilog and VHDL) used in traditional ASIC and FPGA design flows. Once a SilverC representation has been simulated and verified, it is compiled into an executable (binary) *Silverware* application. The ACM's on-chip operating system only loads whatever portions of a Silverware application are required at any particular time, and multiple Silverware applications can be running concurrently on an ACM at any particular time.

It's important to note that when a Silverware application is created, it doesn't need to know which type of ACM chip is being used (including the mix of node types, etc.) or, indeed, how many ACM chips are available on the board. The on-chip ACM operating system takes care of handling any pesky details of this sort.

But wait, there's more

In our discussions on DSP-based design flows in chapter 12, we introduced the concept of system-level design and simulation environments such as *Simulink* from The MathWorks (www.mathworks.com). This tool, which has a wide base of users, encourages dataflow-oriented design and provides an excellent mapping to the ACM architecture.

Well, the lads and lasses at QuickSilver have been working furiously on integrating SilverC with Simulink. At the simplest level, you can use Simulink to describe the various blocks and the dataflow connections between them, and then automatically output a top-level framework of the design containing the module instantiations and the pipes connecting them together. In this case, you would then go into the framework to code the SilverC processes by hand.

Alternatively, QuickSilver has developed a library of SilverC modules that map onto existing Simulink blocks. This library includes widely used DSP components, filters, encoders, decoders, and bit and word manipulators. These SilverC modules can be used for functional and cycle-accurate simulation, and, on compilation into a Silverware executable, they can be mapped directly onto the ACM's dynamic hardware resources.

It's silicon, Jim, but not as we know it!

As you have probably surmised, I'm quite excited about the possibilities of FPNAs in general and QuickSilver's offering in particular. So, does this mean the end of ASICs and FPGAs as we know them? Of course not!

FPNAs are particularly well suited to a variety of application areas, but there is no such thing as an "all-singing, all-dancing, one-size-fits-all" chip architecture that can do everything well (and makes your teeth whiter as a by-product). In the real world, FPNAs are just one more weapon in the system architect's arsenal.

On the other hand, based on everything that has gone before, it wouldn't surprise me to see both ASICs and FPGAs with embedded FPNA cores appearing on the scene at some time in the not-so-distant future. Alternatively, as was noted earlier, QuickSilver allows you to create your own types of nodes, where you define the core of the node and surround it with QuickSilver's wrapper. So, another alternative is to use the main ACM fabric as supplied by QuickSilver, but to include some nodes implemented as FPGA fabric.

And if and when any of this comes to pass, you can bet your little cotton socks that I'll be there, gesticulating furiously and shouting, "I told you so!"

1967:
Dolby eliminates audio hiss.

Independent Design Tools

Introduction

When it comes to design tools such as logic simulators, synthesis technology, and so forth, we mostly look to the big, full-line EDA companies, to smaller EDA companies who are focused on a particular aspect of the design flow, or to the FPGA vendors themselves.

However, we shouldn't forget the guys and gals working in the open-source arena (see also Chapter 25). Furthermore, small FPGA design consultancy firms often spend some considerable time and effort creating niche tools to help with their internal development projects. Occasionally, these tools are so useful that they end up being productized and become available to the outside world. In this chapter, we briefly introduce a brace of such tools.

ParaCore Architect

Dillon Engineering (www.dilloneng.com) offers a variety of custom design services, with particular emphasis on FPGA-based DSP algorithms and high-bandwidth, real-time digital signal and image processing applications.

Toward the end of the 1990s, their engineers became conscious that they were constantly reinventing and reimplementing things like floating-point libraries, convolution kernels, and FFT processors. Thus, in order to make their lives easier, they developed a tool called ParaCore Architect™, which facilitates the design of IP cores.

The process begins by creating a source file containing a highly parameterized description of the design at an extremely high level of abstraction using a Python-based language (Python is introduced in more detail in chapter 25). ParaCore Architect takes this description, combines it with parameter values specified by the user, and generates an equivalent HDL representation, a cycle-accurate C/C++ model to speed up verification in the form of simulation, and an associated testbench (Figure 24-1).

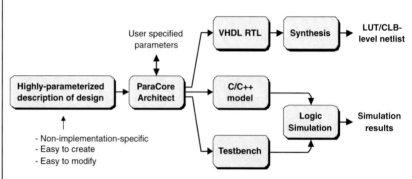

Figure 24-1. ParaCore Architect generates RTL, C/C++, and an associated testbench.

The ensuing HDL is guaranteed suitable for use with any simulation and synthesis environment, so it isn't necessary to run any form of HDL rule-checking program. The beauty of this type of highly parameterized representation is that it's extremely easy to target it toward a new application or an alternative device.

Generating floating-point processing functions

As one simple example of the use of ParaCore Architect, a number of FPGA vendors now supply devices containing embedded microprocessor cores. Sad to relate, these typically do not come equipped with an associated *floating-point unit* *(FPU)*. This means that, should the designers wish to perform floating-point operations on floating-point representations, they either have to do this in software (which is horrendously time-consuming) or they have to do it in hardware. In the latter case, this will take a lot of effort that could be better spent creating the fun part of the design.

FPU is pronounced by spelling it out as "F-P-U."

For this reason, one of the ParaCore Architect design descriptions can be used to generate corresponding floating-point cores. Different parameters can be used to define whatever exponent and mantissa precisions are required, how many pipeline stages to use, whether or not to handle IEEE floating-point special cases like infinity (some applications don't require these special cases), the type of microprocessor core being used (so as to create an appropriate interface block), and so forth.

Generating FFT functions

A good example of the power of ParaCore Architect is demonstrated by the design description used to generate FFT cores. The smallest computational element used to generate an FFT is called a *butterfly* which consists of a complex multiplication, a complex addition, and a complex subtraction (Figure 24-2).

FFT is pronounced by spelling it out as "F-F-T,"

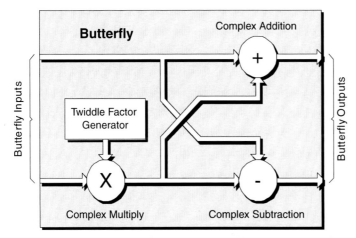

Figure 24-2. The butterfly is the smallest computational element in an FFT.

In turn, the complex multiplication requires four simple multiplications and two simple additions, while the complex addition and complex subtraction each require two simple additions. Thus, each butterfly requires a total of four simple multiplications and six simple additions.

1967: America.
First handheld
electronic calculator
invented by Jack Kilby
of Texas Instruments.

One real-world image-processing application for this core involved generating a two-dimensional 2k × 2k–point FFT that could handle 120 frames-per-second (fps). Processing a single 2,048 (2k) pixel row requires a total of 11,256 butterflies organized in eleven ranks, where the outputs from the butterflies forming the first rank are used to drive the butterflies forming the second rank, and so forth. Thus, processing a single row requires 45,025 simple multiplications and 67,536 simple additions. In order to generate the FFT for an entire 2k × 2k frame, this process has to be repeated for each of the 2,048 (2k) rows forming the frame. This means that in order to achieve a frame rate of 120 fps, the processing associated with each row must be completed within 4 microseconds. (This leads to a time budget of 90 picoseconds per simple multiplication and 60 picoseconds per simple addition.)

Let's consider the 11,256 butterfly operations required to implement a 2k-point FFT. If execution time were not a major factor, it would be necessary to use only a relatively small FPGA device—such as a Xilinx Virtex-II XC2V40—with four multiplier blocks, to create a single butterfly structure (four simple multipliers and six simple adders), and to cycle all of the butterfly operations through this function. The resulting structure would take 90 microseconds to generate each 2k-point FFT. Although this is extremely respectable, it falls well short of the 4-microsecond time budget required by the image-processing application discussed above.

The easiest way to increase the speed of this algorithm is to increase the number of butterfly structures instantiated in hardware and to perform more of the processing in parallel. In the case of Xilinx XC2V6000 devices with six million system gates, 144 × 18-bit multipliers, and 144 × 18-kilobit RAM blocks, it's possible to perform an entire 2k × 2k–point FFT fast enough to achieve a system that can process 120 fps.

The point is that targeting these different devices requires setting only a single ParaCore Architect parameter to specify the number of butterfly structures required to be instantiated in hardware.

As another example, if one were to decide to change the length of the FFT from 2K to 1K points, setting a single parameter takes care of all of the details, including resizing the RAMS used to store any internal results. Similarly, another parameter can be used to select between fixed-point and floating-point math formats (in the latter case, two further parameters are used to specify the size of the exponent and the mantissa).

In early 2002, the folks at Dillon Engineering used Para-Core Architect to create what was possibly the world's fastest FFT processor at that time. This processor subsequently found use in a variety of environments, such as the SETI project, where it is used to process huge amounts of data from radio telescopes in the search for extraterrestrial intelligence!

A Web-based interface

What is really cool is that Dillon Engineering has made ParaCore Architect available for its clients to use over the Internet. When you're creating something like an FFT, you often want to experiment with different trade-offs, such as how many bits to store for each point. Now Dillon Engineering clients can visit the www.dilloneng.com Web site, select the type of core they're interested in, specify a set of parameters, and press the "Go" button to generate the equivalent HDL, C/C++ model, and testbench.

The Confluence system design language

Like most design engineers, I quake when faced with yet another software programming or hardware design language, but Launchbird Design Systems (www.launchbird.com) has come up with a system design language called *Confluence*— along with an associated *Confluence Compiler*—that is well worth looking at.

It's hard to wrap your brain around the many facets to Confluence, but we'll give it a try. First of all, Confluence is an incredibly compact language that can be used to create representations of both hardware and embedded software. In the

1969:
First radio signal transmitted by "man on the moon."

1970: America.
Ethernet developed at
Palo Alto Research
center by Bob Metcalf
and David Boggs.

case of hardware, the Confluence Compiler then takes these descriptions and generates the corresponding RTL in VHDL or Verilog (Figure 24-3).

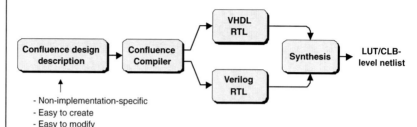

- Non-implementation-specific
- Easy to create
- Easy to modify

Figure 24-3. A highly simplified representation of the outputs from the Confluence Compiler.

One way to think about this is that you use an HDL (like VHDL or Verilog) to describe a specific circuit, but you use Confluence to describe an algorithm that can generate an entire class of circuits. The point is that you can express more in Confluence using far fewer lines of code (you can reduce your source code by 3 to 10 times, which makes designs quicker to produce, easier to manage, and faster to verify). Also, the result is "guaranteed clean" RTL, which prevents common errors and bad design practices.

In programming terms, Confluence offers recursion, high-order data types, lexical scoping, and referential transparency (more than enough to make any system designer's toes curl up in excitement).

A simple example

As a simple hardware example, consider a Confluence component that cascades any single-input-single-output element for any number of stages:

```
component Cascade +Stages +SisoComp +Input -Output
  is
    if Stages <= 0
      Output <- Input
    else
      Output <- {Cascade (Stages - 1) SisoComp
        {SisoComp Input $} $}
    end
  end
```

Although nonprogrammers may initially regard the above as being a tad scary, it's really not all that bad. The first line declares a new component we've decided to call *Cascade*, which has four parameters associated with it: *Stages* (the number of stages you require), *SisoComp* (the name of some subcomponent you wish to cascade), *Input* (the name of the input signal, or signals in the case of a bus), and *Output* (the name of the output signal, or signals in the case of a bus).

Note that the only language keywords in this line are *"component"* and *"is"*; by comparison, *Stages*, *SisoComp*, *Input*, and *Output* are all user-defined variable names. (The "+" and "-" characters in this line indicate whether the associated user-defined variables are to be regarded as input or output ports, respectively.)

Furthermore, when we said that this component cascades any single-input–single-output element, both the Input and the *Output* variables could actually be multibit buses. In fact. these signals don't even have to be bit vectors; they could be lists of bit vectors or lists of lists of bit vectors (or any data type for that matter).

As a simple example of the use of our new *Cascade* component, let's assume that for some wild reason we wish to string 1,024 NOT gates together (don't ask me why) such that the output from the first drives the input to the second, the output from the second drives the input to the third, and so forth. In this case, we could do this with a single line that calls our Cascade component and passes in the appropriate parameters:

```
{Cascade 1024 ('~') Input Output}
```

In this case, the Confluence Compiler understands "~" to be a primitive logical inversion (NOT) function.

As a slightly more interesting example, let's assume that we wish to cascade sixteen 8-bit registers such that the outputs from the first register drive the inputs to the second, the outputs from the second drive the inputs to the third, and so forth. In this case, we would first need to declare a component

1970: America. Fairchild introduced the first 256-bit static RAM called the 4100.

1970: America.
Intel announced the first
1024-bit dynamic RAM
called the 1103.

called something like *Reg8* to represent the 8-bit register, and
then use our *Cascade* component to replicate this 16 times:

```
component Reg8 +A -X is
   {VectorReg 8 A X}
end

{Cascade 16 Reg8 Input Output}
```

Pretty cool, huh? But it gets better! How about squaring a
signal's values four times with a pipeline register between each
stage? We can quickly and easily represent this as follows:

```
component RegisteredPowerOfTwo +A -X is
   {Delay 1 (A '*' A) X}
end

{Cascade 4 RegisteredPowerOfTwo Input Output}
```

As we see, our *Cascade* component provides a perfect illustration of recursion and the use of higher-order datatypes, the
two main characteristics of functional programming that provide higher levels of abstraction and increased design reuse.

And things get better and better because there's no restriction that our subcomponent variable *SisoComp* is obliged to
have input and output ports of the same width. In fact, this
variable can be associated with any user-defined function; it
can even input a component and then output a component,
or it could input a system (an instantiated component) and
then output another system. Similarly, there is no restriction
that *SisoComp* can operate only on bit vectors; it can just as
well operate on integers, floats, lists, components, systems, or
any other Confluence datatype.

As one final example, *SisoComp* could be used to concate-
nate a bit vector onto itself, thereby doubling the number of
bits. In order to illustrate this, let's assume that we create a
new component called *SelfConcat*:

```
component SelfConcat +A -X is
   X = A '++' A
end
```

where "++" is the concatenation operator. When *SelfConcat* is used in conjunction with *Cascade*, the bit vector grows by a factor of two at each stage. For example, assume that we start with a 2-bit vector set to 01 and pass *SelfConcat* into *Cascade*:

```
{Cascade 4 SelfConcat '01' Output}
```

In this case, the output will be a 32-bit vector with a value of 01010101010101010101010101010101.

Of course, VHDL has always had a generate statement, and Verilog was augmented with this capability in the 2K1 release, but Confluence blows these statements away.

But wait, there's more

As I said earlier, it's hard to wrap your brain around the many facets of Confluence. Perhaps the best way to summarize things is by means of an illustration (Figure 24-4).

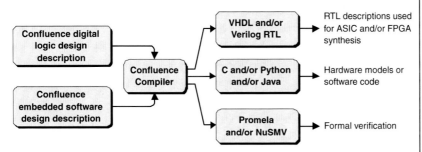

Figure 24-4. A more accurate representation of the outputs from the Confluence Compiler.

On the input side, you can use the Confluence language to create a representation of a piece of hardware or a chunk of embedded software. In the case of a hardware description, you can instruct the Confluence Compiler to generate VHDL or Verilog RTL for use with simulation and synthesis tools.

You can also use the Confluence Compiler to output ANSI C or Python or Java representations (again, the Python language is introduced in more detail in Chapter 25). If your input source represented hardware, then these outputs may be

1970:
First floppy disk (8.5 inch) is used for storing computer data.

1970:
Researchers at Corning Glass develop first commercial/feasible optical fiber.

considered to be cycle-accurate and bit-accurate high-performance simulation models, which can be linked into your custom verification environment. Alternatively, if your input source represented software, then these outputs may be considered to be executable code for use in your hardware/software coverification environment.

Last, but not least, the Confluence Compiler can be instructed to generate representations in the PROMELA or NuSMV languages for formal verification purposes using the open-source SPIN model checker and NuSMV symbolic model checker, respectively (formal verification is discussed in chapter 19, while PROMELA, SPIN, and NuSMV are introduced in more detail in Chapter 25).

Free evaluation copy

If you visit the Launchbird Web site at www.launchbird.com, you'll find a lot of Confluence source code examples. One really "cool beans" idea is that anyone can download and use a single unlimited license for free. Subsequent licenses will cost you for commercial purposes (academic usage is free), but prices are always subject to change, so you'll have to get the latest info from Launchbird on this.

What is really cool is that you own everything you develop with your free license (that is, any Confluence source code models and any ensuing VHDL, Verilog, C, etc. representations), and you can do with them what you wish, including sell them, which has to be a good deal, whichever way you look at it!

Do you have a tool?

Should you run into a useful tool from a small design house on your travels, or if you have created a tool of this type, please feel free to contact me at max@techbites.com for possible inclusion in the next edition of this tome or maybe an article in my bimonthly "Max Bytes" column at www.eedesign.com.

Creating an Open-Source-Based Design Flow

How to start an FPGA design shop for next to nothing

Something you don't really see a lot of are small two-guys-in-a-garage-type design houses focused on developing ASICs. This isn't particularly surprising because the design tools required to develop this class of device tend to be horrendously expensive at $100,000 and up on a good day. (Of course, the fact that it costs millions of dollars to actually have a chip fabricated is also a bit of a showstopper.)

By comparison, the combination of modern FPGAs and recent developments in open-source EDA and IP technology have brought the cost of starting an FPGA design outfit down to practically zero. This has paved the way for folks ranging from college graduates to full-blown professionals setting up shop in their basements.

In addition to actually knowing what you are doing with regard to creating digital logic designs, starting a successful FPGA design house requires a few fundamental pieces:

- A development platform
- A verification environment
- Formal verification (optional)
- Access to common IP components
- Synthesis and implementation tools
- FPGA development boards (optional)

The development platform: Linux

Created by the Swedish engineer Linus Torvalds (and friends) starting around 1990, Linux is quickly becoming the

Note that it is not my purpose to recommend the use of less well-supported tools. Low-cost FPGA vendor-supplied tools are preferred for cost-sensitive setups, while more powerful tools from the larger and/or specialist EDA vendors are preferred as designs increase in size and complexity.

However, if you are trying to create an FPGA design "shop" at home on a limited (or non-existent) budget, the open-source tools presented here may well be of interest.

predominant platform for ASIC and FPGA development. Even though the majority of FPGA synthesis and implementation tools originated on Microsoft Windows®, most are starting to be, or already have been, ported to Linux.

Linux and GNU provide many invaluable tools for hardware and software development. Some common Linux tools (in no particular order, excepting one that pleased the author) include the following:

- *gcc*: C remains the fastest modeling language around for simulation and verification. If your designs are so large that they choke your HDL (Verilog or VHDL) simulation capability, you might consider creating a cycle-accurate C model and compiling it using the open-source *GNU C compiler (gcc)*.
- *make*: The *make* utility is used to automate your build process. In the context of hardware, "build" can refer to anything from simulation, HDL-code generation, and logic synthesis to place-and-route. In order to tell *make* which files you wish to process and which files depend on other files, you have to define these files and their relationships in a file called a *makefile*.
- *gvim*: Derived from "visual interface," VI is the classic UNIX text editor. The *vim* utility is an enhanced version of VI, and *gvim* is a *graphical user interface (GUI)* version of *vim*. The *gvim* utility extends VI with syntax highlighting features and all sorts of other cool macros. With built-in support for both Verilog and VHDL, *gvim* is an ultrafast, never-take-your-fingers-off-the-keyword design-entry tool.
- EMACS: Considered by many hackers to be the ultimate editor, EMACS (from "Editing MACroS") is a programmable text editor with an entire LISP interpreter system inside it. More powerful and more complex than VI, EMACS now has modules available for use in developing Verilog and VHDL-based representations.

Linux is either pronounced "lee-nuks" ("lee" to rhyme with "see") or "li-nuks" ("li" to rhyme with the "li" in "lit", but NOT the "li" in "light").

GNU is pronounced "G-noo" by taking the guttural 'g' sound from "great" and following it with "noo" to rhyme with "boo" or "pooh."

LISP offocially stands for List Processor (although it's detractors say it really means "Lots of Irritating, Superfluous Parentheses.")

- *cvs*: The *Concurrent Versions System (CVS)* is the dominant open-source, network-transparent, version-control system and is applicable to everyone from individual developers to large, distributed teams. CVS supports branching, multiple users, and remote collaboration. It maintains a history of all changes made to the directory (folder) tree and files it is instructed to manage. Using this history, CVS can recreate past states of the tree and show you when, why, and by whom a given change was made. So, if you accidentally mess up your RTL code or decide you want to resynthesize a version of your design from three months ago, no problem; CVS will help you deal with this type of thing.

- *PERL*: Scripting languages are often used for one-off programming jobs and for prototyping. In the context of electronic designs, they are also used to tie a number of tools in the flow together by controlling the ways in which the tools work and by organizing how data is passed between them. The *Practical Extraction and Report Language (PERL)* is historically one of the more widely used scripting languages. Developed by Larry Wall, PERL has jokingly been described as "The Swiss Army chainsaw" of UNIX (and Linux) programming, and many hardware design flows are still glued together using PERL scripts.

- *Python*: Arguably more powerful than PERL, the Python language is an "all-singing-all-dancing" scripting language that has evolved into a full-fledged programming language. Invented by Guido Van Rossum in 1990 and named after Monty Python due to Guido's love of the *Flying Circus*, Python can be used for anything from gluing together the design flow, to high-level modeling and verification, to creating custom EDA tools (see also the additional discussions on Python later in this chapter).

- *diff*: A relatively simple, but incredibly useful, utility, *diff* is used to quickly compare source files and detect and report differences between them.

1971: America. The Datapoint 2200 computer is introduced by CTC.

1971: America.
Ted Hoff designs (and
Intel releases) the first
computer-on-a-chip, the
4004 microprocessor.

- *grep*: Standing for *globally search for a regular expression and print the lines containing matches to it* (phew!), *grep* is used to quickly search a file or group of files to locate and report on instances of a particular text string or pattern.

- *OpenSSL*: Whether you are a large or small company, it pays to ensure the security of your IP. One aspect of this comes when you wish to transmit your IP over a network or over the Internet to your collaborators or customers. In this case, you really should consider encrypting the IP before waving it a fond farewell. One solution is the open-source OpenSSL project, which features a commercial-grade, full-featured toolkit implementing the *Secure Sockets Layer (SSL)* and *Transport Layer Security (TLS)* protocols, as well as an industrial-strength general-purpose cryptography library.

- *OpenSSH*: Is your design team spread across the planet? The *Secure SHell (ssh)* utility is a program for logging into a remote machine and for executing commands on a remote machine while providing secure encrypted communications between two untrusted hosts over an insecure network. An open-source version of the *ssh* suite, OpenSSH encrypts all traffic (including passwords) to effectively eliminate eavesdropping, connection hijacking, and other network-level attacks. OpenSSH also provides a variety of secure tunneling capabilities and authentication methods.

- *tar, gzip, bzip2*: These are different utilities that can be used to compress and archive your work.

Obtaining Linux

Until recently, the leading distributors of Linux have been Red Hat (www.redhat.com) and MandrakeSoft (www.mandrakesoft.com). However, Gentoo Linux™ (www.gentoo.org) is rapidly becoming a favorite among developers. Gentoo has a unique package distribution system that automatically downloads, compiles, and installs packages to your Linux machines. Want Icarus Verilog? Just type

```
$ emerge iverilog
```

and in a few minutes you'll find that Icarus has been installed on your system and is ready to rock and roll!

The verification environment

You can argue about this back and forth, but many would say that the verification environment is the most critical part of the design flow. Anyone can bang away on the keyboard and produce HDL, but it's the verification tools that provide designers with feedback to steer the design toward a correct implementation.

Icarus Verilog

The predominant open-source verification tool is a Verilog compiler known as *Icarus* (http://icarus.com/eda/verilog). In its basic form, Icarus compiles a Verilog design into an executable that can be run as a simulation. Truth to tell, Icarus is primarily used as an event-based simulator, but it can also handle basic logic synthesis for Xilinx FPGAs.

Verilog is a complex language, and Icarus's author, Stephen Williams, has done an excellent job with his Verilog implementation. In fact, Icarus Verilog's language coverage and performance exceeds that of some commercial simulators.

Dinotrace and GTKWave

Icarus Verilog, discussed above, is strictly a command-line tool. (Command-line tools are preferred in UNIX and Linux environments because they are easy to glue together with makefiles.)

Icarus does not provide a GUI to display simulation results. Rather, it can produce industry-standard *value change dump* (*VCD*) files that can be used downstream in the design flow by stand-alone waveform viewing applications.

1971:
CTC's Kenbak-1 computer is introduced.

Enter Dinotrace and GTKWave, which are GUI utilities that can be used to display simulation results in VCD format. Both of these waveform viewers can scroll through a simulation, add trace lines, and search for patterns. Dinotrace (www.veripool.com/dinotrace) is a solid tool, but with limited functionality. By comparison, GTKWave (www.cs.man.ac.uk/apt/tools/gtkwave) started out a little rough around the edges, but has seen modest development in recent months.

Covered code coverage

When verifying a design, access to functional coverage metrics is important to ensure that your test vectors are hitting the corner cases in your design.

Covered (http://covered.sourceforge.net) is a Verilog code-coverage utility that produces the code-coverage metrics associated with a simulation. More specifically, Covered analyzes Verilog source and the VCD data produced from an Icarus Verilog simulation to determine the level of functional coverage.

Covered currently handles four types of coverage metrics: line coverage, toggle coverage, combinational coverage, and finite-state-machine coverage.

Verilator

The hot design issue these days is how to handle SoC designs, which require the integration of hardware and embedded software on a single chip. Many FPGAs host embedded hard processor cores or have access to soft processor cores (see also chapter 13).

Another useful tool is *VTOC* from Tenison EDA (www.tenison.com). This tool generates C++ or SystemC models from RTL source code.

The real trick in an SoC design involves verifying the hardware and software integration. Enter Verilator (www.veripool.com/verilator.html), which converts Verilog into cycle-accurate C++ models. The ability to autogenerate C/C++ models from RTL source code is a powerful verification tool. This allows the software to integrate directly with the C/C++ version of the RTL for simulation purposes.

In addition to hardware-software coverification, Verilator can also be used for general-purpose Verilog simulation because simulating with cycle-accurate C gives much faster run times than can be obtained with an event-based HDL simulator. All you have to do is compile the output C code using *gcc* (see "The development platform: Linux" section above) and run.[1]

Python

Python (www.python.org) is a very useful high-level scripting and programming language becoming world renowned for its rapid implementation capabilities. Not surprisingly, Python is shaping up as a power tool for digital design and verification engineers, particularly for tasks such as system modeling, testbench construction, and general design management.

In fact, many design firms are starting to discover that it's easier and faster to begin by creating Python models rather than Verilog or VHDL representations. Once these Python models have been verified via simulation, the design team can undertake the RTL coding process constantly referencing their "golden" Python models.

MyHDL (www.jandecaluwe.com/Tools/MyHDL/Overview.html) is a Python framework for high-level system modeling. It uses recent feature additions to the Python language (generators) to mimic concurrent operations. MyHDL also has the ability to connect to Icarus Verilog for mixed Python/Verilog simulation.

Formal verification

As the Dutch mathematician and computer pioneer Edsger Wybe Dijkstra once said, "Program testing can be used to show the presence of bugs, but never to show their absence."

Although hardware simulation remains the predominant means for system testing, one can only ensure a system is correct by means of *formal verification* (see also Chapter 19).

---

[1] Note that Icarus also has C code-generation capabilities.

1971:
First direct telephone dialing between the USA and Europe.

1971:
Niklaus Wirth develops
the PASCAL computer
language (named after
Blaise Pascal).

Unlike simulation, formal verification mathematically proves that a system's implementation meets some form of specification.

The two main types of formal verification are *model checking* and *automated reasoning*. Model checking is a technique that explores the state space of a system to ensure that certain properties, typically specified as "assertions," are true. A subdiscipline called *equivalence checking*, which compares two representations of a system (for example, RTL and a gate-level netlist) to determine whether or not they have the same input-to-output functionality, is a form of model checking. By comparison, automated reasoning uses logic to prove (much like a formal mathematical proof) that an implementation meets an associated specification.

Open-source model checking

The predominant open-source model checker is SPIN (http://spinroot.com), which has been under development for almost 20 years by Dr. Gerard J. Holzmann at Bell Labs. A rather cunning beast, SPIN recently received the Software and System Award by the *Association for Computing Machinery (ACM)*. This is no small honor as previous award recipients have been UNIX, SmallTalk, TCP/IP, and the World Wide Web.

SPIN accepts an input specification with an integrated system model using a language called PROMELA. By means of this language, users can create complex assertions in the form of *never-claims*, which define a series of events that should never occur in the system. Given a model and a specification, SPIN exhaustively searches the state-space for violations.

The main drawback with SPIN is that it's primarily intended for asynchronous software verification and, thus, employs a technique called *explicit verification*. Although explicit verification is ideal for verifying software protocols, the technique tends to be inefficient for large hardware-based designs.

For moderately sized hardware designs, a *symbolic model checker* is the way to go. Unlike explicit verification, symbolic model checking uses *binary decision diagrams (BDDs)* and *propositional satisfiability algorithms (SATs)*[2] to contain the problem and, if possible, avoid state-space explosion. Fortunately, there is a high-quality open-source symbolic model checker called NuSMV (http://nusmv.irst.itc.it).

Open-source automated reasoning

The advantage of the model-checking approach discussed in the previous section is that it's an automated process: click the button, then wait for the result. The drawback is that you may have to wait for a very long time.

Even though the symbolic representation used by NuSMV provides a leg up on explicit model checkers, state-space explosion is still an imminent threat. It doesn't take long before a system's size grows beyond the practical limitations of a model checker. Another problem associated with model checking is that it's limited in expression to the extent that some complex assertions simply can't be specified in a model-checking environment. Enter *automated reasoning*, otherwise known as *automated theorem proving*.

Automated reasoning does not share the limitations of model checking. For example, system size is not as relevant because automated reasoning does not search the state-space. More importantly, automated reasoning supports a much higher level of expression for accurately modeling complex and intricate specifications.

Unfortunately, what is gained in some areas is lost in others. Despite its name, automated reasoning is not a fully automatic process. In the real world, the verification engineer conducts the proofing process with the assistance of the tools. Furthermore, in order to use the tools effectively, the verifica-

1972: America. Intel introduce the 8008 microprocessor.

---

[2] The abbreviation "SAT" comes from the first three letters of "satisfiability."

1973: America. Scelbi Computer Consulting Company introduces the Scelbi-8H microcomputer-based do-it-yourself computer kit.

tion engineer needs to be well versed in proof strategies, mathematical logic, and the tools themselves. This is a non-trivial learning curve, but if you're willing to invest the time and effort, automated reasoning is arguably the most powerful form of verification.

Unlike model checking, where open-source tools struggle to compete with commercial applications, the open-source tools for automated reasoning are at the world's leading edge. Three of the most popular are HOL (http://hol.source-forge.net), TPS (http://gtps.math.cmu.edu/tps.html), and MetaPRL (http://cvs.metaprl.org:12000/metaprl/default.html).

What actually is the problem?

Like any tool, formal verification is only as good as the engineers using it. Even on a good day, formal verification can only answer the question, Does my implementation meet the specification? But the critical question remains: Is my specification correct?

Evaluations of real-world designs show that most system failures are not due to a faulty implementation per se. Even without the use of formal verification, designs tend to implement the requirements correctly more often than not. The root causes of most failures are usually the requirements themselves.

Open communication and collaboration are the best ways to ensure a correct specification, and, at the time of this writing, the only known tool that can tackle this problem is the cerebral cortex.

Access to common IP components

A useful rule of thumb if you are a small design house (or even a large design house) is to avoid reinventing the wheel. Over time, every design firm acquires a library of frequently used components that it can pull from to speed up the design process. In fact, a design firm's capabilities are sometimes judged by its IP portfolio.

OpenCores

Fortunately for aspiring designers, they already have access to a vast IP library in the form of OpenCores (www.open-cores.org). As the industry's premier open-source hardware IP repository, OpenCores collects projects with cores ranging across arithmetic units, communication controllers, coprocessors, cryptography, DSP, forward error correction coding, and embedded microprocessors. Furthermore, OpenCores also stewards Wishbone, which is a standardized bus protocol for use in SoC projects.

OVL

Designers can spend as much as 70 percent of a design's total development time in the verification portion of the flow. This has created the need for access to libraries of verification IP. For this reason, Accellera (www.accellera.org) started the Open Verification Library, or OVL, to address the need for common IP verification components.

Synthesis and implementation tools

Synthesis (both logic synthesis and physically aware synthesis) is one major step in the FPGA design flow not completely addressed by open-source technology. Unfortunately, this situation is unlikely to change in the immediate future due to the complexity of the FPGA synthesis problem.

At the time of this writing, Icarus (see "The verification environment" section above) is the only open-source tool known to synthesize HDL to FPGA primitives. The only other low-cost options are the synthesis and implementation tools from the FPGA vendors themselves (these should be the primary choice for a low-cost setup).

When a design approaches the capacity of a top-of-the-line device, however, even FPGA-vendor-provided synthesis tools start to become inadequate for the task. This means that in the case of large, bleeding-edge designs, you may have no choice but to fork out the cash for a high-end synthesis tool.

1973: America. Xerox Alto computer is introduced.

FPGA development boards

If a design firm decides to get involved with physical hardware, FPGA development boards are a must.

OpenCores (see the "Access to common IP components" section above) does offer a few FPGA development board projects, but most designers would be better served by purchasing professional development boards.

On the bright side, money spent on boards can be saved in other areas. For example, a clever engineer can turn a small FPGA evaluation board into a highly capable logic analyzer (hmmm, this sounds like a potential OpenCores project!).

Miscellaneous stuff

Some other odds and sods that might be of interest are as follows:

- *www.easics.be*   Click the "WebTools" link to find a CRC utility that allows you to select standard or custom polynomials and generate associated Verilog or VHDL modules
- *www.linuxeda.com*   EDA tools for Linux
- *http://geda.weul.org*   A collection of open-source EDA tools
- *www.veripool.com*   A collection of Verilog-based tools (this is the home of Dinotrace and Verilator)
- *http://ghdl.free.fr*   An open-source VHDL front end to *gcc*
- *http://asics.ws*   Some more open-source IP cores

While surfing the Web, one can meander into a lot of other open-source projects related to EDA and FPGAs. Unfortunately, most are dormant or have been abandoned without achieving a useful level of functionality. Having said this, should you run into something useful, or if you have created something useful, please feel free to contact me at max@techbites.com for possible inclusion in the next edition of this tome.

Future FPGA Developments

Be afraid, be very afraid

This is the scary bit, because past experience has shown that whatever I thought was coming down the pike was but a pale imitation of what actually ended up sneaking up behind me and leaping out with gusto and abandon when I was least expecting it.

You have to remember that when I started my career designing CPUs for mainframe computers back in 1980 (which really isn't all that long ago when you come to think about it), we didn't have access to any of the technologies and tools that are around today. We didn't have schematic capture packages, so we used a pencil and paper to draw gate-level circuit diagrams. We didn't have logic simulators (early versions were available, but we didn't have one), so we verified our designs by peer review, which boils down to other engineers looking at your schematics and saying, "That looks OK to me."

Sophisticated HDLs like Verilog and VHDL were a long way off in the future, and the possibility that tools like logic synthesis might one day exist simply never occurred to us. When it came to logic optimization and minimization, we had a Chinese engineer on our team who was incredible at this sort of thing; we gave him our designs and he returned optimized versions a day or so later. In the case of timing analysis, once again we were back to pencil and paper, calculating delay paths by hand (no one I knew could afford even the most rudimentary of electronic calculators).

In those days, we were working with multimicron ASIC technologies containing only a few thousand logic gates

(FPGAs had not yet been invented). If you had told me that by 2003 we'd be designing ASICs and SoCs at the 90-nanometer technology node containing tens to hundreds of millions of logic gates and that we'd have reconfigurable devices like today's SRAM-based FPGAs, I would have laughed my socks off. Similarly, if you'd told me that I'd one day have a personal computer on my desktop with hundreds of megabytes of RAM, a clock running at 2 or more gigahertz, and a hard disk with a capacity of 60 gigabytes and that I'd have access to the EDA tools that are around today, I'd have calmly smiled while furtively looking for the nearest exit.[1]

The point is that electronics is going so fast that any predictions we might make are probably going to be of interest only for the purposes of saying, "Well, we didn't see that coming, did we?" But what the heck, I'm game for a laugh, so let's throw the dice and see how well we do.

Next-generation architectures and technologies

Billion-transistor devices

One thing I feel very confident in predicting is that the next generation of FPGAs will contain a billion or more transistors (the reason I'm so self-assured on this point is that Xilinx recently announced devices of this ilk). These chips will be fabricated at the 90-nanometer technology node in late 2003 or early 2004, followed by even larger devices created at the 65- to 70-nanometer node in 2004 or 2005.

Super-fast I/O

When it comes to the gigabit transceivers discussed in chapter 21, today's high-end FPGA chips typically sport one or more of these transceiver blocks, each of which has multiple channels. Each channel can carry 2.5 Gbps of real data; so four channels have to be combined to achieve 10 Gbps. Furthermore, an external device has to be employed to convert an incoming optical signal into the four channels of electrical

In Britain, the term "billion" traditionally used to mean "a million million" (10^{12}). For reasons unknown, however, the Americans decided that "billion" should mean "a thousand million" (10^9). In order to avoid the confusion that would otherwise ensue, most countries in the world (including Britain) have decided to go along with the Americans on this one.

---

[1] The first IBM PC wouldn't see the light of day until 1981.

data that are passed to the FPGA. Conversely, this device will accept four channels of electrical data from the FPGA and convert them into a single outgoing optical signal. At the time of this writing, some FPGAs are coming online that can accept and generate these 10 Gbps optical signals internally.

Another technology that may come our way at some stage in the future is FPGA-to-FPGA and FPGA-to-ASIC wireless or wireless-like interchip communications. With regard to my use of the term *wireless-like*, I'm referring to techniques such as the experimental work currently being performed by Sun Microsystems on interchip communication based on extremely fast, low-powered capacitive coupling. This requires the affected chips to be mounted very (VERY) close to each other on the circuit board, but should offer interchip signal speeds 60 times higher than the fastest board-level interconnect technologies available today.

Super-fast configuration

The vast majority of today's FPGAs are configured using a serial bit-stream or a parallel stream only 8 bits wide. This severely limits the way in which these devices can be used in reconfigurable computing-type applications. Quite some time ago (somewhere around the mid-1990s), a team at *Pilkington Microelectronics (PMEL)* in the United Kingdom came up with a novel FPGA architecture in which the device's primary I/O pins were also used to load the configuration data. This provided a superwide bus (256 or more pins/bits) that could program the device in a jiffy.[2]

---

2 The official definition of "jiffy" is "a short space of time," "a moment," or "an instant." Engineers may use "jiffy" to refer to the duration of one tick of a computer's system clock. This is often based on one cycle of the mains power supply, which is 1/60 of a second in the U.S. and Canada and 1/50 of a second in England and most other places. More recently, equating a jiffy to 1/100 of a second has started to become common. Just to add to the fun, physicists sometimes use "jiffy" to refer to the time required for light to travel one foot in a vacuum (this is close to one nanosecond).

Founded in 1826, Pilkington is one of the world's largest manufacturers of glass products.

Pilkington is widely recognized as the world's technological leader in glass. For example, in 1952, Sir Alastair Pilkington invented the float process in which molten glass, at approximately 1000°C, is poured continuously from a furnace onto one end of a shallow bath of molten tin. The glass floats on the tin, which gives it an incredibly smooth surface. The glass cools and solidifies as it progresses across the bath, and is pulled off the far end in a continuous sheet.

Having said all of this, I have no idea why Pilkington became involved in microelectronics.

1973:
June, the term
microcomputer first
appears in print in
reference to the
8008-based Micral
microcomputer.

As an example of where this sort of architecture might be applicable, consider the fact that there are a wide variety of *compressor/decompressor* (CODEC) algorithms that can be used to compress and decompress audio and video data. If you have a system that needs to decompress different files that were compressed using different algorithms, then you are going to need to support a variety of different CODECs.

Assuming that you wished to perform this decompression in hardware using an FPGA, then with traditional devices you would either have to implement each CODEC in its own device or as a separate area in a larger device. You wouldn't wish to reprogram the FPGA to perform the different algorithms on the fly because this would take from 1 to 2.5 seconds with a large component, which is too long for an end user to wait (we demand instant gratification these days). By comparison, in the case of the PMEL architecture, the reconfiguration data could be appended to the front of the file to be processed (Figure 26-1).

Files containing configuration data
for different CODEC algorithms

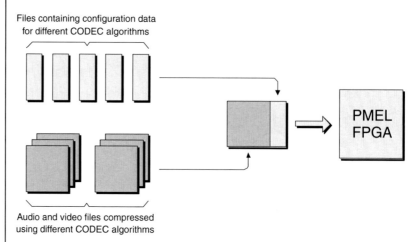

PMEL
FPGA

Audio and video files compressed
using different CODEC algorithms

Figure 26-1. A wide configuration bus.

The idea was that the configuration data would flood through the wide bus, program the device in a fraction of a second, and be immediately followed by the main audio or video data file to be decompressed. If the next file to be processed required a different CODEC, then the appropriate configuration file could be used to reprogram the device.

This concept was applicable to a wide variety of applications. Unfortunately, the original incarnation of this technology fell by the wayside, but it's not beyond the bounds of possibility that something like this could reappear in the not-so-distant future.[3]

More hard IP

In the case of technology nodes of 90 nanometers and below, it's possible to squeeze so many transistors onto a chip that we are almost certainly going to see an increased amount of hard IP blocks for such things as communications functions, special-purpose processing functions, microprocessor peripherals, and the like.

Analog and mixed-signal devices

Traditional digital FPGA vendors have a burning desire to grab as many of the functions on a circuit board as possible and to suck these functions into their devices. In the short term, this might mean that FPGAs start to include hard IP blocks with analog content such as *analog-to-digital (A/D)* and *digital-to-analog (D/A)* converters. Such blocks would be programmable with regard to such things as the number of quanta (width) and the dynamic range of the analog signals they support. They might also include amplification and some filtering and signal conditioning functions.

Furthermore, over the years a number of companies have promoted different flavors of *field-programmable analog arrays (FPAAs)*.[4] Thus, there is more than a chance that predominantly digital FPGAs will start to include areas of truly programmable analog functionality similar to that provided in pure FPAA devices.

> **1974:** America. Intel introduces the 8080 microprocessor, the first true general-purpose device.

---

3 A wide-bus configuration scheme is used by some of the *field programmable node array (FPNA)* devices introduced in chapter 23.

4 For example, Anadigm (www.anadigm.com) have some interesting devices.

ASMBL and other architectures

Just as I started penning the words for this chapter, Xilinx formally announced their forthcoming *Application Specific Modular BLock (ASMBL™)* architecture. The idea here is that you have an underlying column-based architecture, where the folks at Xilinx have put a lot of effort into designing different flavors of columns for such things as

ASMBL is pronounced like the word "assemble."

- General-purpose programmable logic
- Memory
- DSP-centric functions
- Processing functions
- High-speed I/O functions
- Hard IP functions
- Mixed-signal functions

Xilinx will provide a selection of off-the-shelf devices, each with different mixes of column types targeted toward different application domains (Figure 26-2).

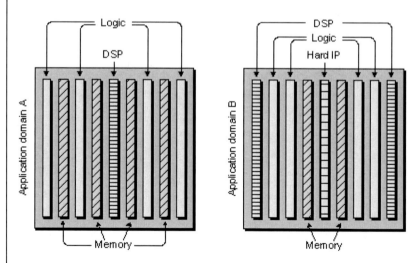

Figure 26-2. Using the underlying ASMBL architecture to create a variety of off-the-shelf devices with domain-specific functionality.

Of course, the other FPGA vendors are doubtless working on their own next-generation offerings, and we can expect to see a flurry of new architectures over the coming years.

Different granularity

As we discussed in chapter 4, FPGA vendors and university students have spent a lot of time researching the relative merits of 3-, 4-, 5-, and even 6-input LUTs.

In the past, some devices were created using a mixture of different LUT sizes, such as 3-input and 4-input LUTs, because this offered the promise of optimal device utilization. For a variety of reasons, the vast majority of today's FPGAs contain only 4-input LUTs, but it's not beyond the range of possibility that future offerings will sport a mixture of different LUT sizes.

Embedding FPGA cores in ASIC fabric

The cost of developing a modern ASIC at the 90-nanometer technology node is horrendous. This problem is compounded by the fact that once you've completed a design and built the chip, your algorithms and functions are effectively "frozen in silicon." This means that if you have to make any changes in the future, you're going to have to regenerate the design, create a new set of photo-masks (costing around $1 million), and build a completely new chip.

In order to address these issues, some users are interested in creating ASICs with FPGA cores embedded into the fabric. Apart from anything else, this means that you can use the same design for multiple end applications without having to create new mask sets. At the time of this writing, the latest incarnation of this technology is the *XBlue* architecture announced by IBM and Xilinx. Created using the 90-nanometer technology node, these devices are expected to start shipping in 2004.

I also think that we are going to see increased deployment of *structured ASICs* and that these will lend themselves to sporting embedded FPGA cores because their design styles and tools will exhibit a lot of commonality.

1974: America. Motorola introduces the 6800 microcomputer.

1974: America. *Radio Electronic* Magazine publishes an article by Jonathon (Jon) Titus on building an 8008-based microcomputer called the Mark-8.

Embedding FPNA cores in ASIC and FPGA fabric and vice versa

In Chapter 23, we discussed the concept of embedding FPNA cores in FPGA and ASIC fabric or embedding FPGA-based nodes in FPNA fabric. Should this come to pass, it's not beyond the bounds of possibility that one day we'll be designing an ASIC with an embedded FPGA core, which itself has an embedded FPNA core, which, in turn, contains FPGA-based nodes. The mind boggles!

MRAM-based devices

In Chapter 2, we introduced the concept of MRAM. MRAM cells have the potential to combine the high speed of SRAM, the storage capacity of DRAM, and the nonvolatility of FLASH, all while consuming a miniscule amount of power.

MRAM-based memory chips are predicted to become available circa 2005. Once these memory chips do reach the market, other devices, such as MRAM-based FPGAs, will probably start to appear shortly thereafter.

Don't forget the design tools

As we discussed above, the next generation of FPGAs will contain 1 billion transistors or more. Existing HDL-based design flows in which designs are captured at the RTL-level of abstraction are already starting to falter with the current generation of devices, and it won't be long before they essentially grind to a halt.

One useful step up the ladder will be increasing the level of design abstraction by using the pure C/C++-based flows introduced in Chapter 11. Really required, however, are true system-level design environments that help users explore the design space at an extremely high level of abstraction. In addition to algorithmic modeling and verification, these environments will aid in partitioning the design into its hardware and software components.

These system-level environments will also need to provide performance analysis capabilities to aid users in evaluating which blocks are too slow when realized in software and, thus, need to be implemented in hardware, and which blocks realized in hardware should really be implemented in software so as to optimize the use of the chip's resources.

People have been talking about this sort of thing for ages, and various available environments and tools go some way toward addressing these issues. In reality, however, such applications have a long way to go with regard to their capabilities and ease of use.

Expect the unexpected

That's it, the end of this chapter and the end of this book. Phew! But before closing, I'd just like to reiterate that anything you or I might guess at for the future is likely to be a shallow reflection of what actually comes to pass. There are device technologies and design tools that have yet to be conceived, and when they eventually appear on the stage (and based on past experience, this will be sooner than we think), we are all going to say, "WOW! What a cool idea!" and "Why didn't I think of that?" Good grief, I LOVE electronics!

1975: America. Microcomputer in kit form reaches U.S. home market.

Signal Integrity 101

Before we start

Before leaping into this topic, it's important to note that *signal integrity (SI)* is an incredibly complicated and convoluted subject that can quickly make your brain ache and your eyes water if you're not careful. For this reason, the discussions in this appendix are intended only to introduce some of the more significant SI concepts. If you are interested in learning more, you could do a lot worse than reading *Signal Integrity—Simplified* by SI expert Dr. Eric Bogatin, ISBN: 0130669466, and *High Speed Signal Propagation: Advanced Black Magic* by Howard W. Johnson, ISBN: 013084408X.

SI encompasses a wide range of different aspects, including the way in which the "shape" of a signal degrades as it passes through a wire, and also the way signals can effectively "bounce back" off the end of a wire that is incorrectly terminated (like a ball thrown down a corridor bouncing off the wall at the end). For our purposes here, however, we shall concentrate on those SI effects that are gathered together under the umbrella appellation of *crosstalk*.

Crosstalk-induced noise (glitches) and delays are dominated by different issues inside silicon chips from those seen at the circuit board level. For this reason, we shall commence by introducing the root causes of these effects and then consider their chip-level and board-level manifestations independently.

SI is pronounced by spelling it out as "S-I."

The amount by which a material impedes the flow of electric current is referred to as *resistance (R)*, which is measured in units of *ohms*.

The term "ohm" (represented by the Greek letter omega "Ω") is named after the German physicist Georg Simon Ohm, who defined the relationship between voltage, current, and resistance in 1827.

Capacitive and inductive coupling (crosstalk)

Consider two signal wires called *Wire1* and *Wire2*, each of which is driven by a single gate and drives a single load. In an ideal—and somewhat simplified—world, both wires would be perfectly straight with no awkward bends or discontinuities, and each could be represented by a single series resistance, series inductance, and capacitance (Figure A-1).

The property of an electric conductor that characterizes its ability to store an electric charge is referred to as *capacitance (C)*, which is measured in units of *Farads (F)*.

The term "Farad" is named after the British scientist Michael Faraday, who constructed the first electric motor in 1821.

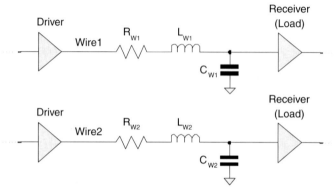

Figure A-1. Two signal wires in an ideal (simplified) world.

For the purposes of this minimalist example, the capacitances C_{W1} and C_{W2} are considered with respect to a ground plane. In its simplest form, a capacitor consists of two metal plates separated by an insulating layer called the *dielectric*. This means that if our two signal wires run in close proximity to each other, then from the perspective of an outside observer they would actually appear to form a rudimentary capacitor. This may be represented by adding a symbol C_M to reflect this mutual capacitance into our circuit diagram (Figure A-2).

When one of the signal wires is in the process of transitioning between logic values, the coupling capacitance between the wires causes a transfer of charge into the other wire, which may result in noise (glitch) and delay effects, as discussed in the following sections.

As was previously noted, each wire also has some amount of inductance associated with it. In its simplest terms, induc-

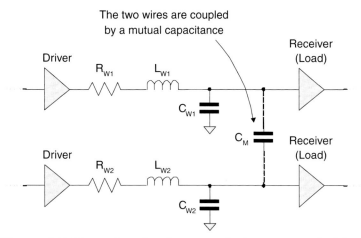

The two wires are coupled
by a mutual capacitance

Figure A-2. Two wires in close proximity are coupled by a mutual capacitance.

tance is the property associated with conductors by which changes in the current flowing through a conductor creates a magnetic field surrounding that conductor. Correspondingly, any changes in the magnetic field surrounding a conductor induce a response in that conductor.

This means that when one of our signal wires is in the process of transitioning between logic values, the change in current flowing through the wire combined with the inductance associated with that wire causes a magnetic field to build up around the wire. As it expands, this field interacts with the inductances associated with any wires in close proximity, which, once again, may result in noise and delay effects as discussed in the following sections. This mutual inductance is indicated by adding a dot to each of the inductor symbols (Figure A-3).

Chip-level effects

Chip-level effects are RC (resistance-capacitance) dominated

Early ICs had tracks that were formed from aluminum (chemical symbol Al), which has a relatively high resistance.

In 1831, the British scientist Michael Faraday discovered that a changing electromagnetic field induced a current in a nearby conductor. This effect subsequently became known as *inductance (L)*.

The symbol for inductance is the capital letter L in honor of the Russian physicist Heinrich Lenz, who discovered the relationships between the forces, voltages, and currents associated with electromagnetic induction in 1833.

Inductance is measured in units of *Henries (H)*.

The term "Henry" is named after the American scientist Joseph Henry, who independently discovered inductance around the same time as Faraday.

Pronounced "al-oo-mi-num" in America, aluminum is spelled (and pronounced) "al-u-min-ium in the UK.

Aluminium was also the accepted spelling in America until 1925. At that time, the American Chemical Society officially decided to use the name aluminum in their publications.

Dating back more than 10,000 years, copper is the oldest metal worked by man.

Most creatures on earth have blood, whose red color is caused by the iron-based pigment hemoglobin. However, some primitive creatures have green, copper-based blood, whose pigment is called cuproglobin.

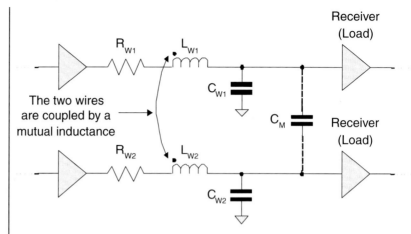

Figure A-3. Two wires in close proximity are coupled by a mutual inductance.

As device feature sizes continued to shrink with each new technology node, the resistance associated with the aluminum tracks started to increase to unacceptable levels.

IC manufacturers had long wanted to use copper tracks (chemical symbol Cu) because copper is one of the best conductors known to man, especially for high-frequency applications. However, copper also has some awkward properties, not the least of which is that it can easily diffuse into the silicon chip, thereby rendering the device useless. It was not until the late 1990s that IBM solved this problem by the inclusion of special barrier layers.

Even though copper has a much lower resistance than aluminum, signal tracks on ICs are so fine that their resistance is still extremely significant. The result is that, thus far, delay effects associated with signals propagating through IC tracks have tended to be dominated by their resistive and capacitive (RC) characteristics.

RC is pronounced by spelling it out as "R-C."

RLC is pronounced by spelling it out as "R-L-C."

At this time, inductive (L) effects are typically ignored in signal tracks and are only considered with respect to the power grid. This grid employs wider tracks with correspondingly lower resistance, such that resistance, inductance, and capacitance (RLC) characteristics all need to be accounted for.

Increased sidewall capacitive coupling

In the case of early IC implementation technologies, the aspect ratio of tracks was such that their width was significantly greater than their height (figure A-4a). As feature sizes continue to shrink, however, the processes used to create these devices result in track aspect ratios in which height predominates over width (Figure A-4b).

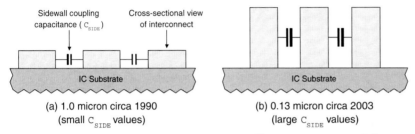

Figure A-4. Sidewall capacitance effects increase with shrinking feature sizes (not to scale—illustrates relative aspect ratios only).

The result is a dramatic increase in coupling capacitance (C_{SIDE}) between the sidewalls of adjacent tracks relative to the substrate capacitances C_{AREA} (track base to substrate) and C_{FRINGE} (sidewall to substrate). Furthermore, the high integration densities associated with today's devices, which can support eight or more metalization layers, result in significant capacitive coupling between adjacent layers. This is represented by $C_{CROSSOVER}$ (Figure A-5).

The combination of these factors leads to a tremendous increase in the complexity of crosstalk noise and timing effects, as discussed below.

Crosstalk-induced glitches

When signals in neighboring wires transition between logic values, the coupling capacitance between the wires causes a transfer of charge. Depending on the slew of the signals (the speed of switching in terms of rise and fall times) and the amount of mutual crosstalk capacitance (C_M), there can be significant crosstalk-induced glitches (Figure A-6).

The term "glitch" possibly comes from the Yiddish word *glitsh*, meaning "a slip or lapse."

1975: America.
MOS Technology
introduces the
6502-based KIM-1
microcomputer.

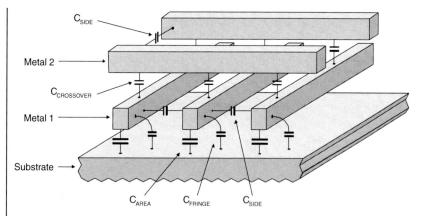

Figure A-5. Capacitance effects associated with the interconnect.

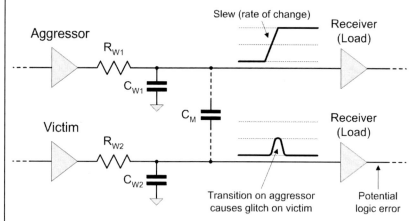

Figure A-6. A crosstalk-induced glitch.

In this example, a transition on the fast *aggressor* net causes a glitch to be presented to the input of the receiver (load) of an adjacent *victim* net. Of course, this illustration presents a very simplistic view. In reality, each track may be formed from multiple segments occupying multiple levels of metalization. Thus, the resistances (R_{W1} and R_{W2}) and capacitances (C_{W1} and C_{W2}) will each consist of multiple elements associated with the different segments. Similarly, the mutual coupling crosstalk capacitance (C_M) may consist of multiple elements.

The example glitch illustrated in figure A-6 represents only one of four generic possibilities based on the fact that a rising or falling transition on the aggressor net may be coupled with a logic 0 or logic 1 on the victim net (Figure A-7).

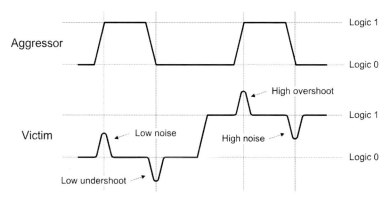

Figure A-7. Types of crosstalk-induced glitches.

If the ensuing low-noise or high-noise glitches on the victim net cross the input switching threshold of its receiver, a functional (logic) error may occur. In some cases this error may manifest itself as an incorrect data value that is subsequently loaded into a register or latch. In other cases, the error may cause a latch to perform an unintended load, set, or reset. The low-undershoot and high-overshoot glitches on the victim net pose a different problem because they can cause undesirable charge carriers to be trapped in the transistors forming the logic gates, which can degrade circuit performance. Although these effects, commonly known as *hot electron effects*, are not a major threat in the context of current IC implementation technologies, they will become increasingly significant as device geometries progress furter into the *deep-submicron (DSM)* and *ultra-deep-submicron (UDSM)* realms.

Crosstalk-induced delay effects

The situation becomes even more complex when simultaneous switching occurs on both the aggressor and victim nets.

Any IC implementation technology below 0.5 μm is referred to as being *deep submicron (DSM)*.

DSM is pronounced by spelling it out as "D-S-M."

At some point that isn't particularly well defined (or is defined differently depending on whom you are talking to), we move into the UDSM realm.

UDSM is pronounced by spelling it out as "U-D-S-M" or by saying "ultra-D-S-M."

1975: America.
Sphere Corporation
introduces the
6800-based Sphere
1 microcomputer.

For example, in the case of opposing transitions, the signal on the victim net may be slowed down (Figure A-8).

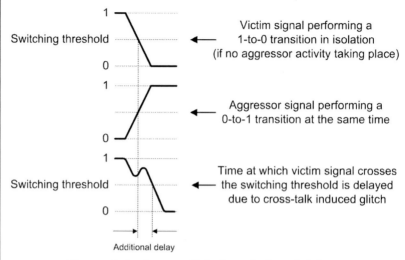

Figure A-8. Crosstalk-induced signal delay.

If the signal on the victim net were transitioning in isolation, it would take a certain amount of time to cross its receiver's switching threshold (which, for the purposes of these discussions, may be assumed to be 50 percent of the value between a logic 0 and a logic 1). However, the glitch caused by a simultaneous transition on the aggressor net holds the victim's signal above the receiver's switching threshold for an additional amount of time. This can result in a downstream setup violation.

An alternative scenario occurs when a transition on the victim is complemented by a simultaneous transition on the aggressor in the same direction, in which case the signal on the victim may speed up (Figure A-9).

In this case, the glitch caused by a simultaneous transition on the aggressor net causes the victim's signal to cross the load/receiver's switching threshold earlier than expected. This can result in a downstream hold violation.

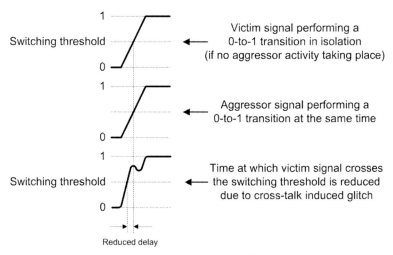

Figure A-9. Crosstalk-induced signal speed up.

1975: America. Bill Gates and Paul Allen found Microsoft.

Multiaggressor scenarios

In reality, the examples shown above are extremely simplistic. In the case of real-world designs, each victim net may be affected by multiple aggressors (Figure A-10).

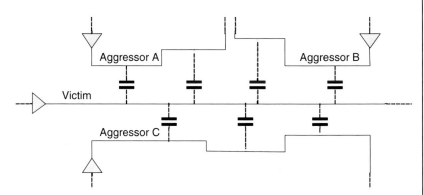

Figure A-10. Multiaggressor scenario.

Accurate analysis of today's designs requires that each aggressor's contribution be individually analyzed and accounted for.

And let's not forget the Miller effect

The Miller effect, which is of particular significance at the chip level, states that the simultaneous switching of both terminals of a capacitor will modify the effective capacitance between the terminals.

What this means in real terms becomes apparent when we consider one of the signals in the middle of a bus, for example. If one or more of the surrounding signals in the bus are switching with the same polarity (in the same direction) as the signal of interest, then the capacitance associated with this signal will appear to be reduced, and its propagation delay will decrease (this is in addition to the crosstalk-induced delay effects introduced earlier).

By comparison, if one or more of the surrounding signals in the bus are switching with the opposite polarity to the signal of interest, then the capacitance associated with this signal will appear to be larger and its propagation delay will increase.

The reason we commenced with the chip-level effects introduced above is that these provide a familiar starting point for IC design engineers. In reality, however, on-chip SI effects (excluding packaging considerations) are of little interest to engineers using FPGAs because these effects are handled behind the scenes by the device vendor. By comparison, board-level SI effects are extremely pertinent when it comes to integrating FPGAs into a circuit board environment.

Board-level effects

Board-level effects are LC (inductance-capacitance) dominated

When it comes to PCBs, the resistance of their copper tracks is almost negligible in the context of coupling effects. This is because at around 125 microns wide and 18 microns thick, board-level tracks have a huge cross-sectional area compared to their chip-level counterparts (the larger the cross section of a conductor, the lower its resistance). By compari-

In the context of an electronic circuit, the term "bus" (sometimes "buss") refers to a set of signals performing a common function and carrying similar data.

LC is pronounced by spelling it out as "L-C."

PCB is pronounced by spelling it out as "P-C-B."

son, both inductive and capacitive coupling effects are significant, so circuit board signal tracks are predominantly considered to be LC-coupled.

A *different way of thinking about things*

In the case of today's high-speed, high-performance PCBs, the tracks almost invariably act like transmission lines. This means we have to visualize a signal edge as a moving wave propagating down the wire through time (Figure A-11).

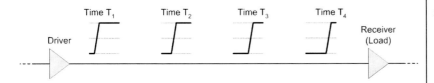

Figure A-11. A signal edge moving through time.

With regard to this transmission line view, in which the delay down the wire is long in comparison to the signal's transition times, the only place any capacitive or inductive coupling occurs is at the current location of the moving edge. This means that we have to consider the track in terms of a series of small RLC segments (which are not shown in these figures for reasons of simplicity).

Capacitive and inductive coupling effects

Things really start to get interesting when we consider two of these board-level tracks in close proximity to each other. Let's assume that we are looking at a moving edge that is in the process of propagating down an aggressor track that is inductively and capacitively coupled to a neighboring victim track (Figure A-12).

In the case of the capacitive coupling effect, the moving edge on the aggressor net induces positive-going current pulses on the victim net in both the forward and reverse directions. By comparison, in the case of the inductive coupling effect,

An alternative name for PCB is *printed wire board (PWB)*, which is pronounced by spelling it out as "P-W-B."

The most commonly used board material is FR4, which is pronounced by spelling it out as "F-R-4" (the "FR" stands for "flame retardant").

1975: England.
First liquid crystal
displays (LCDs) are used
for pocket calculators
and digital clocks.

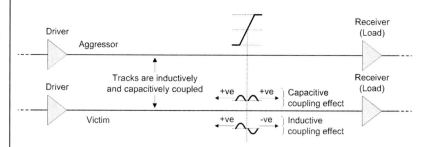

Figure A-12. Capacitive and inductive coupling effects.

the moving edge on the aggressor net induces a *negative*-going current pulse on the victim net in the forward direction and a *positive*-going current pulse on the victim net in the reverse direction.

This means that the capacitive and inductive coupling effects tend to augment each other when it comes to *near-end noise* (noise as seen at the driver end of the track). However, they tend to cancel each other out when it comes to *far-end noise* (noise as seen at the receiver end of the track). This means that the best-case scenario one can ever hope for is when the capacitive and inductive coupling effects are of comparable magnitudes, because they will cancel each other out at the far end. Unfortunately, this will only ever happen if the dielectric (insulating) layer around the signals is homogeneous, such as with a stripline stackup. In the real world, the dielectric around signals is typically inhomogeneous, such as surface traces, which have air above and FR4 below. In this case, the inductive coupling does not change, but the capacitive coupling decreases. This increases the relative amount of inductive coupling and gives rise to the generation of far-end noise at the receiver. In a typical circuit board environment, the inductive noise can be as much as two to four times the capacitive noise.

If anything occurs to degrade the return path, the inductive coupling can increase dramatically to as much as ten to thirty times the capacitive. In such a regime, where the crosstalk is dominated by inductive coupling, we call the ensuing noise *switching noise*. In the case where multiple signal paths share the same return path, the switching noise we get across the return (ground) connection is called *ground bounce*.

The anti-Miller effect

In our chip-level discussions, we introduced the concept of the Miller effect, which says that if one or more signals are switching with the same polarity (in the same direction) in close proximity to a signal of interest, then the capacitance associated with this signal will appear to be reduced, and its propagation delay will decrease.

As was previously noted, however, the propagation delays of chip-level signals are predominantly RC dependent, while the propagation delays of board-level signals are predominantly LC dependent. This means that if one or more board-level signals are switching with the *same* polarity in close proximity to a signal of interest, then the inductance associated with this signal will appear to be larger. In an inhomogeneous dielectric stackup, the relative inductive coupling is larger than the capacitive coupling, and the increased inductance of the signal trace causes the propagation delay to increase.

By comparison, if one or more board-level signals are switching with the *opposite* polarity to the signal of interest, then the inductance associated with this signal will appear to be reduced and its propagation delay will decrease.

Transmission line effects

In addition to the effects presented above, there are, of course, classical transmission line effects with associated termination considerations such as using series termination on outputs and parallel termination on inputs, but this sort of thing is beaten into the ground in standard textbooks, so we will skip over it here.

1975: America. Ed Roberts and his MIT's company introduce the 8800-based Altair 8800 microcomputer.

Things you can do to make life easier

Unfortunately, 70 to 80 percent of the SI problems associated with connecting an FPGA to a circuit board are not related to the board per se, but rather to the FPGA's package.

Ideally, the package should have as large a number of power-ground pad pairs as possible, and these pad pairs should be uniformly distributed across the base of the package so as to provide the I/O pads with plenty of adjacent return paths. In reality, the power and ground pads tend to be clustered together leaving groups of I/O pads to do the best they can with the return paths available to them.

I/O is pronounced by spelling it out as "I-O."

You can make life easier by making it a rule, if you have the option to use differential output pairs for your I/O, especially in the case of buses and high-speed interconnections, to do so. Of course, this doubles the number of pins you use for the affected I/Os, but it's well worth your time if you can afford the overhead in pins.

Another point to consider relates to the internal, programmable termination resistors provided in some FPGAs. The use of these is optional in that you can either use discrete components at the board level or enable these internal equivalents as required. These internal terminations are predominantly considered in the context of easing routing congestion at the board level, but they also have SI implications. The rule of thumb is that for any signals with rise/fall times of 500 picoseconds or less, external termination resistors cause discontinuities in the signal, so you should always use their on-chip counterparts.

Deep-Submicron Delay Effects 101

Introduction

When one is designing ASICs and ASSPs, the timing effects one needs to account for are extremely complex. As each new technology process node comes online, these effects become ever-more horrendous. At some point—which isn't particularly well defined (or which is defined differently by different people), but which we will take to be somewhere around the 0.5-micron (500 nanometer) node—we start to move into an area rife with what are known as *deep-submicron (DSM)* delay effects.

The great thing about working with FPGAs, of course, is that the folks who create these devices handle the bulk of the problems associated with DSM delay effects, leaving them largely transparent to the end users (design engineers). On this basis, it's fair to say that we really don't need to discuss DSM timing issues here. On the other hand, this is the sort of thing you tend to hear about all the time, but I've never run across an introduction to these effects that is comprehensible to anyone sporting anything less than a size-16 brain with go-faster stripes! It is for this reason that the following overview is presented for your delectation and delight.

The evolution of delay specifications

Way back in the mists of time, sometime after the Jurassic period when dinosaurs ruled the earth—say, around the late 1970s and early 1980s—the lives of ASIC design engineers were somewhat simpler than they are today. Delay specifications for the early (multimicron) technologies were

The contents of this appendix are abstracted from my book *Designus Maximus Unleashed* (ISBN 0-7506-9089-5) with the kind permission of the publisher.

1975: America.
MOS Technology
introduces the 6502
microprocessor.

rudimentary at best. Consider the case of a simple 2-input
AND gate, for which input-to-output databook delays were
originally specified as being identical for all of the inputs
and for both rising and falling transitions at the output
(Figure B-1i).

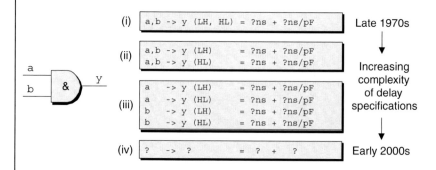

**Figure B-1. Delay specifications become increasingly complex
over time.**

As device geometries shrank, however, delay specifications became increasingly complex. The next step was to
differentiate delays for rising and falling output transitions
(Figure B-1ii), and this was followed by associating different
delays with each input-to-output path (Figure B-1iii).

All of these early delays were typically specified in the
form ?ns + ?ns/pF. The first portion (?ns) indicates a fixed
delay specified in nanoseconds[1] associated with the gate itself.
This is combined with some additional delay specified as
nanoseconds per picofarad (?ns/pF) caused by capacitive loading.[2] As we will see, this form of specification simply cannot
handle the delay effects characteristic of DSM technologies,

---

[1] Today's devices are much faster, so their delays would be measured in
picoseconds.

[2] The basic unit of capacitance is the *Farad*. This was named after the
British scientist, Michael Faraday, who constructed the first electric
motor in 1821.

not the least in the area of RLC interconnect delays, as discussed below.

1975: America. Microsoft releases BASIC 2.0 for the Altair 8800 microcomputer.

A potpourri of definitions

Before plunging headfirst into the mire of DSM delays, it is first necessary to introduce a number of definitions as follows.

Signal slopes

The *slope* (or *slew*) of a signal is its rate of change when transitioning from a *logic 0* value to a *logic 1*, or vice versa. An instantaneous transition, which cannot be achieved in the real world, would be considered to represent the maximum possible slope value (Figure B-2).

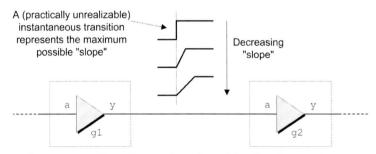

Figure B-2. The slope of a signal is the time taken to transition between logic values.

The slope of the signal is a function of the output characteristics of the driving gate combined with the characteristics of the interconnect (track) and the input characteristics of any load gate(s).

Input switching thresholds

An *input switching threshold* is the point at which an input to a load gate first sees a transition as occurring; that is, the point at which the signal presented to the input crosses some threshold value, at which point the downstream gate deigns to notice that something is happening. Input switching thresh-

1976: America.
Zilog introduces the Z80
microprocessor.

olds are usually specified as a percentage of the value (voltage differential) between a *logic 0* and a *logic 1*, and each input may have different switching thresholds for rising and falling transitions (Figure B-3).

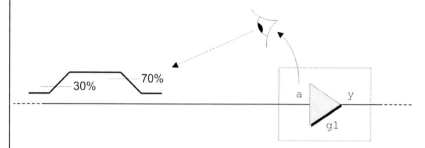

Figure B-3: Input switching thresholds may differ for rising and falling transitions.

Intrinsic versus extrinsic delays

The term *intrinsic* refers to any delay effects that are internal to a logic function, while the term *extrinsic* refers to any delay effects that are associated with the interconnect (Figure B-4).

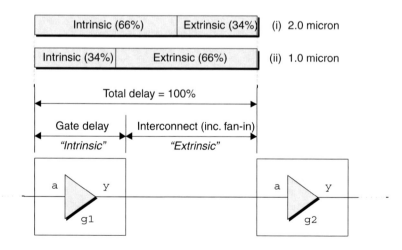

Figure B-4. Intrinsic versus extrinsic delays.

In the early multimicron technologies, intrinsic delays dominated over their extrinsic counterparts. In the case of devices with 2.0-micron geometries, for example, the intrinsic delay typically accounted for approximately two-thirds of the total delay (Figure B-4a). But extrinsic delays became increasingly important with shrinking geometries. By the time that devices with 1.0-micron geometries became available, the relative domination of the intrinsic and extrinsic delays had effectively reversed (Figure B-4b).

This trend is destined to continue because the geometry of the interconnect is not shrinking at the same rate as the transistors and logic gates. In the case of today's DSM technologies, extrinsic delays can account for 80 percent or more of the total path delays.

Pn-Pn and Pt-Pt delays

To a large extent, *pin-to-pin (Pn-Pn)* and *point-to-point (Pt-Pt)* delays are more modern terms for intrinsic and extrinsic delays, respectively. A Pn-Pn delay is measured between a transition occurring at the input to a gate and a corresponding transition occurring at the output from that gate, while a Pt-Pt delay is measured between the output from a driving gate to the input of a load gate (Figure B-5).[3]

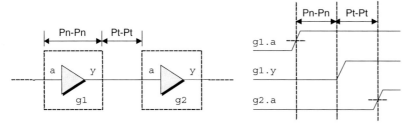

Figure B-5. Pn-Pn versus Pt-Pt delays.

---

[3] It should be noted that circuit board layout designers don't tend to worry too much about what happens inside devices, which they usually consider to be "black boxes." The reason for mentioning this is that the board designers may use the term "pin-to-pin" to refer to track delays at the board level.

1976: America. Steve Wozniak and Steve Jobs introduce the 6502-based Apple 1 microcomputer.

1976: America. Steve Wozniak and Steve Jobs form the Apple Computer Company (on April 1ˢᵗ).

To be more precise, a Pn-Pn delay is the time between a signal on a gate's input reaching that input's switching threshold to a corresponding response *beginning* at its output, while a Pt-Pt delay is the time from the output of a driving gate *beginning* its transition to a corresponding load gate perceiving that transition as crossing its input switching threshold. Good Grief!

There are a number of reasons why we're emphasizing the fact that we consider the time when the output *begins* to respond as marking the end of the Pn-Pn delay and the start of the Pt-Pt delay. In the past, these delays were measured from the time when the output reached 50 percent of the value between a *logic 0* and a *logic 1*. This was considered to be acceptable, because load gates were all assumed to have input switching thresholds of 50 percent. But consider a rising transition on the output and assume that the load gate's input switching threshold for a rising transition is 30 percent. If we're assuming that delays are measured from the time the output crosses its 50 percent value, then it's entirely possible that the load gate will see the transition before we consider the output to have changed. Also, when we come to consider mixed-signal (analog and digital) simulation, then the only meaningful time to pass an event from a gate's output transitioning in the digital realm into the analog domain is the point at which the gate's output begins its transition.

State and slope dependency

Any attribute associated with an input to a gate (including a Pn-Pn delay) that is a function of the logic values on other inputs to that gate is said to be *state dependent*. Similarly, any attribute associated with an input to a gate (including a Pn-Pn delay) that is a function of the slope of the signal presented to that input is said to be *slope dependent*. These state- and slope-dependency definitions might not appear to make much sense at the moment, but they'll come to the fore in the not-so-distant future as we progress through this chapter.

Alternative interconnect models

As the geometries of structures on the silicon shrink and the number of gates in a device increase, interconnect delays assume a greater significance, and increasingly sophisticated algorithms are required to accurately represent the effects associated with the interconnect as follows.

The lumped-load model

As was noted earlier, the Pn-Pn gate delays in early multimicron technologies dominated over Pt-Pt interconnect delays. Additionally, the rise and fall times associated with signals were typically greater than the time taken for the signals to propagate through the interconnect. In these cases, a representation of the interconnect known as the *lumped-load model* was usually sufficient (Figure B-6).

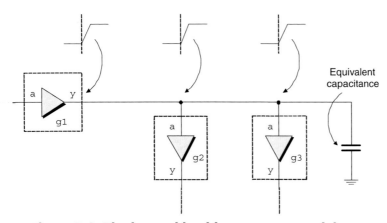

Figure B-6. The lumped-load interconnect model.

The idea here is that all of the capacitances associated with the track and with the inputs to the load gates are added together to give a single, *equivalent capacitance*. This capacitance is then multiplied by the drive capability of the driving gate (specified in terms of nanoseconds per picofarad, or equivalent) to give a resulting Pt-Pt delay. The lumped-load model is characterized by the fact that all of the nodes on the track are considered to commence transitioning at the same

1977: America. Apple introduces the Apple II microcomputer.

1977: America. Commodore Business Machines present their 6502-based Commodore PET microcomputer.

time and with the same slope. This model may also be referred to as a *pure RC model*.

The distributed RC model

The shrinking device geometries of the mid-1980s began to mandate a more accurate representation of the interconnect than was provided by the lumped-load model. Thus, the *distributed RC model* was born (where R and C represent resistance and capacitance, respectively) (Figure B-7).

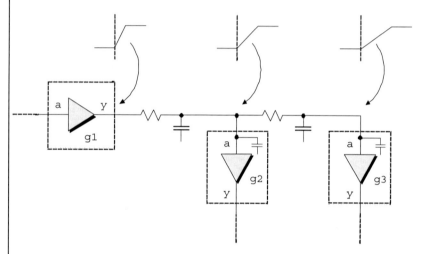

Figure B-7. The distributed RC interconnect model.

In the distributed RC model, each segment of the track is treated as an RC network. The distributed RC model is characterized by the fact that all of the nodes on the track are considered to commence transitioning at the same time but with different slopes. Another way to view this is that the signal's edge is collapsing (or deteriorating) as it propagates down the track.

The pure LC model

At the circuit board level, high-speed interconnects start to take on the characteristics of transmission lines. This *pure LC model* (where L and C represent inductance and capacitance, respectively) can be represented as a sharp transition propagating down the track as a wavefront (Figure B-8).

1977: America. Tandy/Radio Shack announce their Z80-based TRS-80 microcomputer.

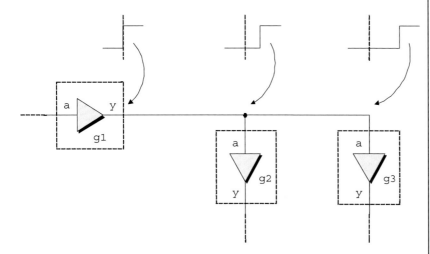

Figure B-8. The pure LC interconnect model.

Pure transmission line effects do not occur inside silicon chips, but large submicron devices do begin to exhibit certain aspects of these delay effects, as discussed below.

The RLC model

In the case of large devices with DSM geometries, the speed of the signals coupled with relatively long traces results in the interconnect exhibiting some transmission-line-type effects. However, the resistive nature of on-chip interconnect does not support pure LC effects; instead, these traces may be described as exhibiting RLC effects (Figure B-9).

The RLC model is characterized as a combination of a discrete wavefront, supplied by the interconnect's LC constituents, and a collapsing (or deteriorating) signal edge caused by the interconnect's RC components.

1977:
First implementation
of optical light-waves
in operating telephone
company.

DSM delay effects

Path-specific Pn-Pn delays

Each input-to-output path typically has its own Pn-Pn delay. In the case of a 2-input OR gate, for example, a change on input *a* causing a transition on output *y* (Figure B-10a) would have a different delay from that of a change on input *b* causing a similar transition on output *y* (Figure B-10b).

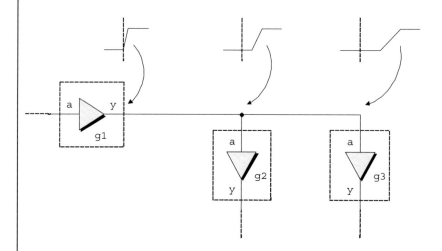

Figure B-9. The RLC interconnect model.

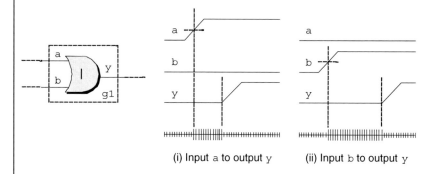

(i) Input *a* to output *y*   (ii) Input *b* to output *y*

Figure B-10. Path-specific Pn-Pn delays.

Similarly, each rising and falling transition at the output typically has its own Pn-Pn delay. In the case of our OR gate, for example, a change on input *a* causing a rising transition on output *y* would have a different delay from that of a change on input *a* causing a falling transition on output *y*.

Note that this example assumes input switching thresholds of 50 percent, and remember that Pn-Pn delays are measured from the time when a signal presented to an input crosses that input's switching threshold to the time when the output first begins to respond.

Path- and transition-specific Pn-Pn delays are not limited to DSM technologies, and they should come as no surprise, but they are presented here to prepare the stage for the horrors that are to come.

Threshold-dependent Pn-Pn delays

Pn-Pn delays depend on the switching thresholds associated with inputs, at least to the extent that the delay through the gate doesn't actually commence until the signal presented to the input crosses the threshold. For example, if the input switching threshold for a rising transition on input *a* were 30 percent of the value between the logic 0 and logic 1 levels (Figure B-11a), then the input would see the transition earlier than it would if its input switching threshold were 70 percent (Figure B-11b).

Additionally, the slope of a signal being presented to an input affects the time that signal crosses the input switching threshold. For the purposes of presenting a simple example, let's assume that input *a* has a switching threshold of 50 percent. If a signal with a steep slope is presented to input *a* (Figure B-12a), then the input will see the signal as occurring earlier than it would if the slope of the signal were decreased (Figure B-12b).

Although this affects the time at which the Pn-Pn delay commences, it is NOT the same as the slope-dependent Pn-Pn delays presented in the next section.

1978: America. Apple introduces the first hard disk drive for use with personal computers.

1979:
ADA programming
language is named after
Augusta Ada Lovelace
(now credited as being
the first computer
programmer).

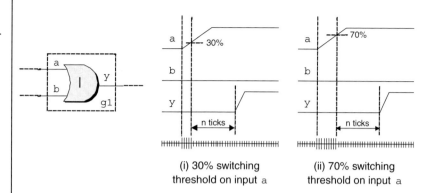

(i) 30% switching
threshold on input a

(ii) 70% switching
threshold on input a

Figure B-11. Threshold-dependent Pn-Pn delays.

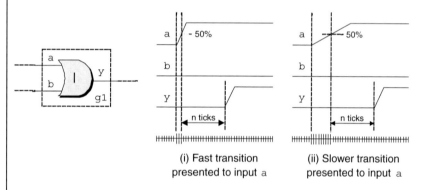

(i) Fast transition
presented to input a

(ii) Slower transition
presented to input a

**Figure B-12. The slope of an incoming signal affects the
time at which the input sees that signal.**

Slope-dependent Pn-Pn delays

Speaking of which … the previous example was somewhat
simplistic in that it showed two Pn-Pn delays as being identical, irrespective of the slope of the incoming signal. Some
vendors of computer-aided design tools refer to the previous
case as "slope dependency," but this is not a correct usage of
the term. As it happens, a variety of delay effects in DSM
technologies may be truly *slope dependent*, which means that
they may be directly modified by the slope of an incoming
signal.

Let's consider what happens from the point at which the signal presented to an input crosses that input's switching threshold. The Pn-Pn delay from this point may be a function of the rate of change of the incoming signal. For example, a fast slope presented to the input may result in a short Pn-Pn delay (Figure B-13a), while a slower slope may result in a longer delay (Figure B-13b).

1979: America. The first true commercial microcomputer program, the VisiCalc spreadsheet, is made available for the Apple II.

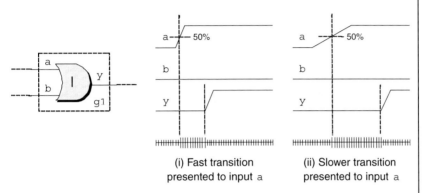

(i) Fast transition presented to input a

(ii) Slower transition presented to input a

Figure B-13. Slope-dependent Pn-Pn delays.

Actually, the effect illustrated in Figure B-13, in which a *decreasing* slope causes an *increasing* Pn-Pn delay, is only one possible scenario. This particular case applies to gates or technologies where the predominant effect is that the switching speeds of the transistors forming the gate are directly related to the rate of change of charge applied to their inputs. By comparison, in the case of certain technologies, a decreasing slope actually results in faster Pn-Pn delays (as measured from the switching threshold of the input). This latter case results from the fact that a sufficiently long slope permits internal transistors to become precharged almost to the point of switching. Thus, when the input signal actually crosses the input's switching threshold, the gate is poised at the starting blocks and appears to switch faster than it would if a sharp edge had been applied to the input.

To further increase your pleasure and double your fun, both effects may be present simultaneously. In this case, applying a

1980:
Cordless and cell
phones are developed.

sharp edge to the input may result in a certain Pn-Pn delay, and gradually decreasing the slope of the applied signal could cause a gradual increase in the Pn-Pn delay. At some point, however, further decreasing the slope of the applied input will cause a reduction in the Pn-Pn delay, possibly to the point where it becomes smaller than the Pn-Pn delay associated with our original sharp edge![4]

State-dependent Pn-Pn delays

In addition to being slope-dependent, Pn-Pn delays are often *state dependent*, which means that they depend on the logic values of other inputs (Figure B-14).

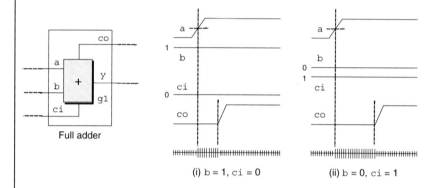

(i) b = 1, ci = 0   (ii) b = 0, ci = 1

Figure B-14. State-dependent Pn-Pn delays.

This example illustrates two cases in which a signal presented to the *a* input causes an identical response (in terms of logic values) at the *co* output. However, even assuming that the slopes of the signals presented to *a* and the switching thresholds on *a* are identical in both cases, the Pn-Pn delays may be different due to the logic values present on inputs *b* and *ci*.

---

[4] And there are those who would say that electronics is dull and boring—go figure!

Path-dependent drive capability

This is where life really starts to get interesting (trust me, have I ever lied to you before?).[5] Up to this point, we have only considered effects that impact Pn-Pn delays through a gate, but many of these effects also influence the gate's ability to drive signal at its output(s). For example, the driving capability of a gate may be path dependent (Figure B-15).

1980:
Development of the World Wide Web begins.

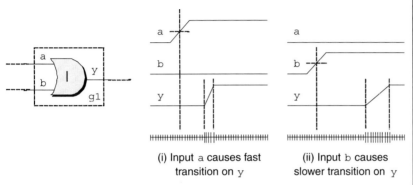

(i) Input a causes fast transition on y

(ii) Input b causes slower transition on y

Figure B-15. Path-dependent drive capability.

In this case, in addition to the fact that inputs *a* and *b* have different Pn-Pn delays, the driving capability of the gate (and hence the slope of the output signal) is dependent on which input caused the output transition to occur. This phenomenon was originally associated only with MOS technologies and was not generally linked to bipolar technologies such as TTL. As the plunge into DSM continues, however, many of the more esoteric delay effects are beginning to manifest themselves across technologies with little regard for traditional boundaries.

Slope-dependent drive capability

In addition to being dependent on which input causes an output transition to occur (as discussed in the previous point), the driving capability of the gate (and hence the slope of the

[5] Don't answer that!

1980:
Faxes can be sent over
regular phone lines.

output signal) may also be dependent on the slope of the signal presented to the input. For example, a fast transition on input *a* may cause a fast slope at the output (Figure B-16a), while a slower transition on the same input may impact the gate's driving capability and cause the slope of the output signal to decrease (Figure B-16b). Are we having fun yet?

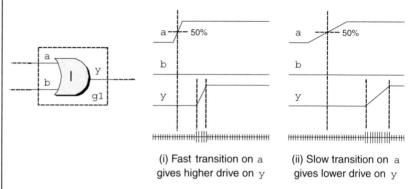

Figure B-16. Slope-dependent drive capability.

State-dependent drive capability

Yet another factor that can influence the drive capability of an output is the logic values present on inputs other than the one actually causing the output transition to occur. This effect is known as *state-dependent drive capability* (Figure B-17).

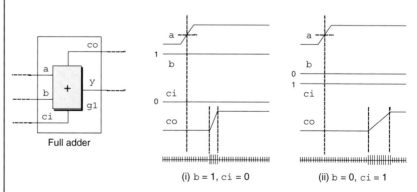

Figure B-17. State-dependent drive capability.

Figure B-17 illustrates two cases in which a signal presented to the *a* input causes an identical response (in terms of logic values) at the *co* output. However, even assuming that the slopes of the signals presented to *a* and the switching thresholds on *a* are identical in both cases, the driving capability of the gate (and hence the slope of the output signal) may be different due to the logic values present on inputs *b* and *ci*.

State-dependent switching thresholds

As you doubtless observed, the previous point on state-dependent drive capability included the phrase "assuming that the input switching thresholds on input *a* are identical in both cases." If this caused a few alarm bells to start ringing in your mind, then, if nothing else, at least these discussions are serving to hone your abilities to survive the dire and dismal depths of the DSM domain.

The point is that by some strange quirk of fate, an input's switching threshold may be state dependent; that is, it may depend on the logic values present on other inputs (Figure B-18).

1981:
America. First IBM PC is launched.

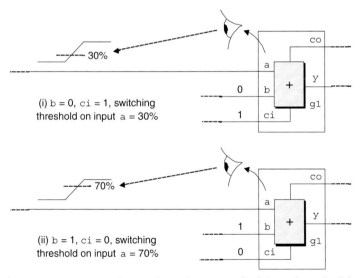

Figure B-18. State-dependent input switching thresholds.

1981: America.
First mouse pointing
device is created.

In this example, the switching threshold of input *a* (the point at which this input sees a transition as occurring) depends on the logic values presented to inputs *b* and *ci*.

State-dependent terminal parasitics

In addition to an input's switching threshold being state dependent, further characteristics associated with that input (such as its parasitic values) may also depend on the logic values presented to other inputs. For example, consider a 2-input OR gate (Figure B-19).

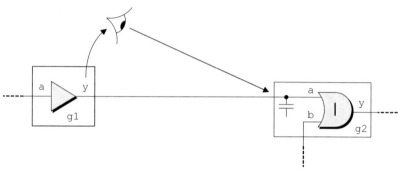

Figure B-19. State-dependent terminal parasitics.

The terminal capacitance of input *g2.a* (as seen by the driving output *g1.y*) may depend on the logic value presented to input *g2.b*. If input *g2.b* is a logic 0, a transition on input *g2.a* will cause the output of the OR gate to switch. In this case, *g1.y* (the output of the gate driving *g2.a*) will see a relatively high capacitance. However, if input *g2.b* is a logic 1, a transition on input *g2.a* will not cause the output of the OR gate to switch. In this case, *g1.y* will see a relatively small capacitance.

At this point you may be asking, "In the case where the OR gate isn't going to switch, do we really care if the parasitic capacitance on input *a* is different? Can't we just set the value of the capacitance to be that for when the OR gate will switch?" In fact, this would be okay if the output *g1.y* were only driving input *g2.a*, but problems obviously arise if we modify the circuit such that *g1.y* starts to drive two or more load gates.

This particular effect first manifested itself in ECL technologies. In fact, as far back as the late 1980s, I was made aware of one ECL gate-array technology in which the terminal capacitance of a load gate (as perceived by the driving gate) varied by close to 100 percent due to this form of state dependency. But this effect is no longer confined to ECL; once again, delay effects are beginning to manifest themselves across technologies with scant regard for traditional boundaries as we sink further into the DSM domain.

The effect of multi-input transitions on Pn-Pn delays

Prior to this point, we have only considered cases in which a signal presented to a single input causes an output response. Not surprisingly, the picture becomes more complex when multi-input transitions are considered. For example, take the case of a 2-input OR gate (Figure B-20).

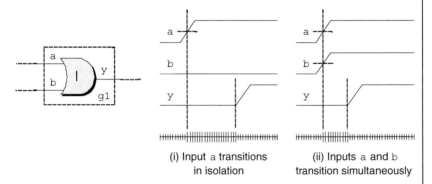

(i) Input a transitions
in isolation

(ii) Inputs a and b
transition simultaneously

Figure B-20. The effect of multi-input transitions on Pn-Pn delays.

For the sake of simplicity, we will assume that both the *a* and *b* inputs are fully symmetrical; that is, both have identical input switching thresholds and Pn-Pn delays.

1981:
First laptop computer
is introduced.

1983:
Apple's Lisa is the first personal computer to use a mouse and pull-down menus.

First, consider the case where a transition applied to a single input (for example, input *a*) causes a response at the output (Figure B-20a). The resulting Pn-Pn delay is the one that is usually specified in the databook for this cell. However, if both inputs transition simultaneously (Figure B-20b), the resulting Pn-Pn delay may be reduced to close to 50 percent of the value specified in the databook.

These two cases (a single input transition occurring in isolation versus multi-input transitions occurring simultaneously) provide us with worst-case endpoints. However, it is also necessary to consider those cases where the inputs don't transition simultaneously, but do transition close together. For example, take the OR gate shown in figure B-20 and assume that both inputs are initially at logic 0. Now assume that input *a* is presented with a rising transition, which initiates the standard databook Pn-Pn delay, but before this delay has fully completed, input *b* is also presented with a rising transition. The result is that the actual Pn-Pn delay could occur anywhere between the two worst-case endpoints.

The effect of multi-input transitions on drive capability

In addition to modifying Pn-Pn delays, multi-input transitions may also affect the driving capability of the gate, and hence the slope of the output signal (Figure B-21).

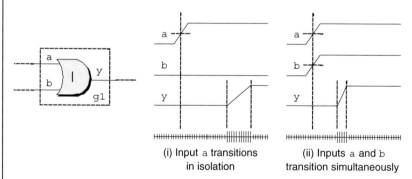

(i) Input a transitions in isolation

(ii) Inputs a and b transition simultaneously

Figure B-21. The effect of multi-input transitions on drive capability.

All of these multi-input transition effects can be estimated with simple linear approximations. Unfortunately, today's verification tools—such as STA and digital logic simulation—are not well equipped to perform on-the-fly calculations of this type.

Reflected parasitics

With the technologies of yesteryear, it was fairly safe to assume that parasitic effects had limited scope and were generally only visible to logic gates in their immediate vicinity. For example, consider the three gates shown in Figure B-22.

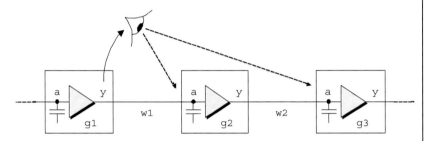

Figure B-22: Reflected parasitics.

Traditionally, it was safe to assume that gate *g2* would buffer the output of *g1* from wire *w2* and gate *g3*. Thus, the output *g1.y* would only see any parasitics such as the capacitances associated with wire *w1* and gate terminal *g2.a*.

These assumptions become less valid in the DSM domain. Returning to the three gates shown in figure B-22, it is now possible for some proportion of the parasitics associated with wire *w2* and gate terminal *g3.a* to be reflected back through gate *g2* and made visible to output *g1.y*. Additionally, if gate *g2* were a multi-input gate such as a 2-input XOR, then the proportion of these parasitics reflected back through *g2* might well be state dependent; that is, they might vary depending on the logic value presented to the other input of *g2*.

At the time of this writing, reflected parasitics remain relatively low-order effects in the grander scheme of things. If

1983:
Time magazine names the computer as *Man of the year.*

history has taught us anything, however, it is to be afraid (very afraid), because it's not beyond the bounds of possibility that these effects will assume a much greater significance as we continue to meander our way through new technology nodes.

Summary

The majority of the delay effects introduced in this chapter have always been present, even in the case of multimicron technologies, but many of these effects have traditionally been fourth or third order and were therefore considered to be relatively insignificant. As device geometries plunged through the 0.5-micron barrier to 0.35 microns, some of these effects began to assume second- and even first-order status, and their significance continues to increase with new technology nodes operating at lower voltage levels.

Unfortunately, many design verification tools are not keeping pace with silicon technology. Unless these tools are enhanced to account fully for DSM effects, designers will be forced to use restrictive design rules to ensure that their designs actually function. Thus, design engineers may find it impossible to fully realize the potential of the new and exciting technology developments that are becoming available.

Linear Feedback Shift Registers 101

The Ouroboras

The Ouroboros, a symbol of a serpent or dragon devouring its own tail and thereby forming a circle, has been employed by a variety of ancient cultures around the world to depict eternity or renewal.[1] The equivalent of the Ouroboros in the electronics world would be the *linear feedback shift register (LFSR)*, in which outputs from a standard shift register are cunningly manipulated and fed back into its input in such a way as to cause the function to cycle endlessly through a sequence of patterns.

Many-to-one implementations

LFSRs are simple to construct and are useful for a wide variety of applications. One of the more common forms of LFSR is formed from a simple shift register with feedback from two or more points, called *taps*, in the register chain (Figure C-1).

The taps in this example are at bit 0 and bit 2, and an easy way to represent this is to use the notation [0,2]. All of the register elements share a common clock input, which is omitted from the symbol for reasons of clarity. The data input to the LFSR is generated by XOR-ing or XNOR-ing the tap bits, while the remaining bits function as a standard shift register.

The contents of this appendix are abstracted from my book *Bebop to the Boolean Boogie (An Unconventional Guild to Electronics, Edition 2* (ISBN 0-7506-7543-8) with the kind permission of the publisher.

LFSR is pronounced by spelling it out as "L-F-S-R."

---

[1] Not to be confused with the Amphisbaena, a serpent in classical mythology having a head at each end and being capable of moving in either direction.

1985:
CD-ROMs are used to
store computer data for
the first time.

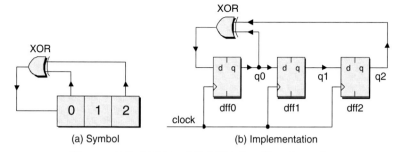

Figure C-1. LFSR with XOR feedback path.

The sequence of values generated by an LFSR is determined by its feedback function (XOR versus XNOR) and tap selection. For example, consider two 3-bit LFSRs using an XOR feedback function, the first with taps at [0,2] and the second with taps at [1,2] (Figure C-2).

Both LFSRs start with the same initial value, but due to the different taps, their sequences rapidly diverge as clock pulses are applied. In some cases, an LFSR will end up cycling

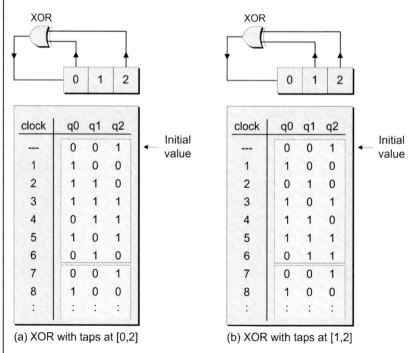

Figure C-2. Comparison of alternative tap selections.

round a loop comprising a limited number of values. However, both of the LFSRs shown in figure C-2 are said to be of *maximal length* because they sequence through every possible value (excluding all of the bits being 0) before returning to their initial values.

A binary field with n bits can assume 2^n unique values, but a maximal-length LFSR with n register bits will only sequence through $(2^n - 1)$ values. For example, a 3-bit field can support $2^3 = 8$ values, but the 3-bit LFSRs in figure C-2 sequence through only $(2^3 - 1) = 7$ values. This is because LFSRs with XOR feedback paths will not sequence through the "forbidden" value where all the bits are 0, while their XNOR equivalents will not sequence through the value where all the bits are 1 (Figure C-3).[2]

1989:
Pacific fiber-optic link/cable opens (supports 40,000 simultaneous conversation).

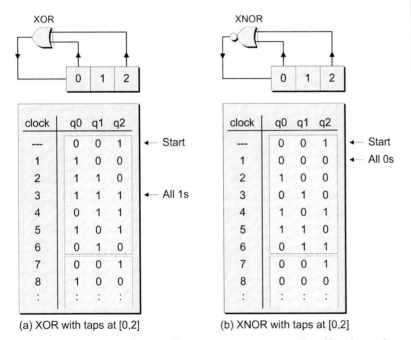

(a) XOR with taps at [0,2]

(b) XNOR with taps at [0,2]

Figure C-3. Comparison of XOR versus XNOR feedback paths.

---

[2] If an LFSR somehow finds itself containing its "forbidden value," it will lock-up in this value until some external event occurs to extract it from its predicament.

1990: Switzerland. British physicist Tim Berners-Lee sets up the world's first World Wide Web server.

More taps than you know what to do with

Each LFSR supports a number of tap combinations that will generate maximal-length sequences. The problem is weeding out the ones that do from the ones that don't, because badly chosen taps can result in the register entering a loop comprising only a limited number of states.

Purely for my own amusement, I created a simple C program to determine the taps for maximal-length LFSRs with 2 to 32 bits. These values are presented for your delectation and delight in Figure C-4 (the * annotation indicates a sequence whose length is a prime number).

The taps are identical for both XOR-based and XNOR-based LFSRs, although the resulting sequences will, of course, differ. As was previously noted, alternative tap combinations

# Bits	Loop Length	Taps
2	3 *	[0,1]
3	7 *	[0,2]
4	15	[0,3]
5	31 *	[1,4]
6	63	[0,5]
7	127 *	[0,6]
8	255	[1,2,3,7]
9	511	[3,8]
10	1,023	[2,9]
11	2,047	[1,10]
12	4,095	[0,3,5,11]
13	8,191 *	[0,2,3,12]
14	16,383	[0,2,4,13]
15	32,767	[0,14]
16	65,535	[1,2,4,15]
17	131,071 *	[2,16]
18	262,143	[6,17]
19	524,287 *	[0,1,4,18]
20	1,048,575	[2,19]
21	2,097,151	[1,20]
22	4,194,303	[0,21]
23	8,388,607	[4,22]
24	16,777,215	[0,2,3,23]
25	33,554,431	[2,24]
26	67,108,863	[0,1,5,25]
27	134,217,727	[0,1,4,26]
28	268,435,455	[2,27]
29	536,870,911	[1,28]
30	1,073,741,823	[0,3,5,29]
31	2,147,483,647 *	[2,30]
32	4,294,967,295	[1,5,6,31]

Figure C-4. Taps for maximal length LFSRs with 2 to 32 bits.

may also yield maximum-length LFSRs, although once again the resulting sequences will vary. For example, in the case of a 10-bit LFSR, there are two 2-tap combinations that result in a maximal-length sequence: [2,9] and [6,9]. There are also twenty 4-tap combinations, twenty-eight 6-tap combinations, and ten 8-tap combinations that satisfy the maximal-length criteria.[3]

VIP! It's important to note that the taps shown in figure C-4 may not be the best ones for the task you have in mind with regard to attributes such as being primitive polynomials and having their sequences evenly distributed in "random" space; they just happened to be the ones I chose out of the results I generated. If you are using LFSRs for real-world tasks, one of the best sources for determining optimum tap points is the book *Error-Correcting Codes* by W. Wesley Peterson and E. J. Weldon Jr. (published by MIT Press). Also, the CRC utility referenced under the "Miscellaneous Stuff" section at the end of chapter 25 might be of some interest.

One-to-many implementations

Consider the case of an 8-bit LFSR, for which the minimum number of taps that will generate a maximal-length sequence is four. In the real world, XOR gates only have two inputs, so a 4-input XOR function has to be created using three XOR gates arranged as two levels of logic. Even in those cases where an LFSR does support a minimum of two taps, there may be special reasons for you to use a greater number such as eight (which would result in three levels of XOR logic).

However, increasing the levels of logic in the combinational feedback path can negatively impact the maximum clocking frequency of the function. One solution is to transpose the *many-to-one implementations* discussed above into their *one-to-many counterparts* (Figure C-5).

---

[3] A much longer table (covering LFSRs with up to 168 bits) is presented in application note XAPP052 from Xilinx.

1993:
The MOSAIC web browser becomes available.

1999:
First 1 GHz
microprocessor created
by Intel.

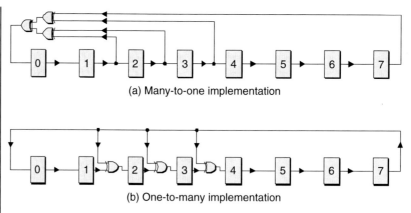

(a) Many-to-one implementation

(b) One-to-many implementation

Figure C-5: Many-to-one versus one-to-many implementations.

The traditional many-to-one implementation for the eight-bit LFSR has taps at [1,2,3,7]. To convert this into its one-to-many counterpart, the most significant tap, which is always the most significant bit (bit 7 in this case), is fed back directly into the least significant bit. This bit is also individually XOR'd with the other original taps (bits [1,2,3] in this example).

Although both of these approaches result in maximal-length LFSRs, the actual sequences of values will differ between them. But the main point is that using the one-to-many technique means that there is never more than one level of combinational logic in the feedback path, irrespective of the number of taps being employed.

Of course, FPGAs have the additional consideration that a 4-input LUT will have the same delay for 2-, 3-, and 4-input XOR trees. In this case, the many-to-one approach only starts to offer advantages when you are dealing with an LFSR that requires more than four taps.

Seeding an LFSR

One quirk with XOR-based LFSRs is that, if one happens to find itself in the all-0s value, it will happily continue to shift all 0s indefinitely (similarly for XNOR-based LFSRs and the all-1s value). This is of particular concern when power is first applied to the circuit. Each register bit can randomly initialize containing either a logic 0 or a logic 1, and the LFSR can therefore "wake up" containing its "forbidden" value. For this reason, it is necessary to initialize LFSRs with a *seed* value.

An interesting aspect of an LFSR based on an XNOR feedback path is that it does allow an all-0s value. This means that a common clear signal to all of the LFSR's registers can be used to provide an XNOR LFSR with a seed value of all 0s.

One method for loading a specific seed value is to use registers with *reset* or *set* inputs. A single control signal can be connected to the *reset* inputs on some of the registers and the *set* inputs on others. When this control signal is placed in its active state, the LFSR will load with a hard-wired seed value. With regard to certain applications, however, it is desirable to be able to vary the seed value. One technique for achieving this is to include a multiplexer at the input to the LFSR (Figure C-6).

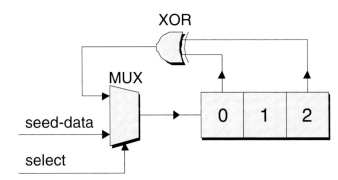

Figure C-6. Circuit for loading alternative seed values.

When the multiplexer's seed-data input is selected, the device functions as a standard shift register, and any desired seed value can be loaded. After loading the seed value, the feedback path is selected and the device returns to its LFSR mode of operation.

FIFO applications

The fact that an LFSR generates an unusual sequence of values is irrelevant in many applications. For example, let's consider a 4-bit-wide, 16-word-deep FIFO memory function (Figure C-7).

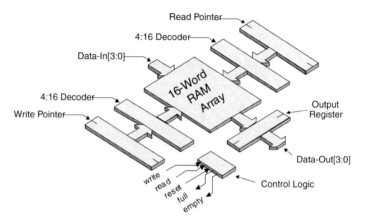

Figure C-7. A 16-word FIFO function.

In addition to some control logic and an output register, the FIFO contains a *read pointer* and a *write pointer*. These pointers are 4-bit registers whose outputs are processed by 4:16 decoders to select one of the 16 words in the memory array.

The read and write pointers chase each other around the memory array in an endless loop. An active edge on the *write* input causes any data on the input bus to be written into the word pointed to by the write pointer; the write pointer is then incremented to point to the next empty word. Similarly, an active edge on the *read* input causes the data in the word pointed to by the read pointer to be copied into the output register; the read pointer is then incremented to point to the next word containing data.[4] (There would also be some logic

---

[4] These discussions assume *write-and-increment* and *read-and-increment* techniques however, some FIFOs employ an *increment-and-write* and *increment-and-read* approach.

to detect when the FIFO is full or empty, but this is irrelevant to our discussions here.)

The write and read pointers for a 16-word FIFO are often implemented using 4-bit binary counters. However, a moment's reflection reveals that there is no intrinsic advantage to a binary sequence for this particular application, and the sequence generated by a 4-bit LFSR will serve equally well. In fact, the two functions operate in a very similar manner as is illustrated by their block diagrams (Figure C-8).

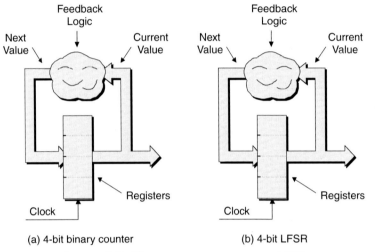

(a) 4-bit binary counter   (b) 4-bit LFSR

Figure C-8. Binary counter versus LFSR.

It doesn't take more than a few seconds before we realize that the only difference between these two diagrams is their names. The point is that the combinational feedback logic for the 4-bit binary counter requires a number of AND and OR gates, while the feedback logic for the 4-bit LFSR consists of a single, 2-input XOR gate. This means that the LFSR requires fewer tracks and is more efficient in terms of silicon real estate.

Additionally, the LFSR's feedback only passes through a single level of logic, while the binary counter's feedback passes through multiple levels of logic. This means that the new data value is available sooner for the LFSR, which can therefore be clocked at a higher frequency. These differentiations become even more pronounced for FIFOs with more words requiring pointers with more bits. Thus, LFSR's provide an interesting option for the discerning designer of FIFOs.[5]

Modifying LFSRs to sequence 2^n values

The sole downside to using 4-bit LFSRs in the FIFO scenario above is that they will sequence through only 15 values ($2^4 - 1$), as compared to the binary counter's sequence of 16 values (2^4). Depending on the application, the design engineers may not regard this to be a major problem, especially in the case of larger FIFOs. However, if it is required for an LFSR to sequence through every possible value, then there is a simple solution (Figure C-9).

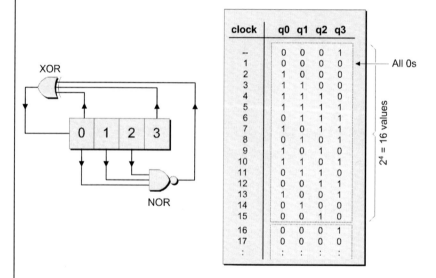

Figure C-9. LFSR modified to sequence 2^n values.

For the value where all of the bits are 0 to appear, the preceding value must have comprised a logic 1 in the

---

[5] So do *Gray Counters*, but that will have to be a topic for another time.

most significant bit (MSB)[6] and logic 0s in the remaining bit positions. In an *unmodified* LFSR, the next clock would result in a logic 1 in the *least significant bit (LSB)* and logic 0s in the remaining bit positions. However, in the *modified* LFSR shown in figure C-9, the output from the NOR is a logic 0 for every case but two: the value *preceding* the one where all the bits are 0 and the value where all the bits are 0. These two values force the NOR's output to a logic 1, which inverts the usual output from the XOR. This in turn causes the sequence first to enter the all-0s value and then to resume its normal course. (In the case of LFSRs with XNOR feedback paths, the NOR can be replaced with an AND, which causes the sequence to cycle through the value where all of the bits are 1.)

MSB and LSB are pronounced by spelling them out as "M-S-B" and "L-S-B", respectively.

Accessing the previous value

In some applications, it is required to make use of a register's previous value. For example, in certain FIFO implementations, the "full" condition is detected when the write pointer is pointing to the location preceding the location pointed to by the read pointer.[7] This implies that a comparator must be used to compare the *current* value in the write pointer with the *previous* value in the read pointer. Similarly, the "empty" condition may be detected when the read pointer is pointing to the location preceding the location pointed to by the write pointer. This implies that a second comparator must be used to compare the *current* value in the read pointer with the *previous* value in the write pointer.

In the case of binary counters (assuming that, for some reason, we decided to use them for a FIFO application), there are two techniques by which the previous value in the sequence may be accessed. The first requires the provision of an addi-

---

[6] As is often the case with any form of shift register, the MSB in these examples is taken to be on the right-hand side of the register and the LSB is taken to be on the left-hand side (this is opposite to the way we usually do things).

[7] Try saying that quickly!

tional set of *shadow registers*. Every time the counter is incremented, its current contents are first copied into the shadow registers. Alternatively, a block of combinational logic can be used to decode the previous value from the current value. Unfortunately, both of these techniques involve a substantial overhead in terms of additional logic. By comparison, LFSRs inherently remember their previous value. All that is required is the addition of a single register bit appended to the MSB (Figure C-10).

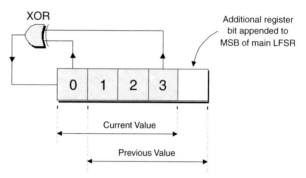

Figure C-10. Accessing an LFSR's previous value.

Encryption and decryption applications

The unusual sequence of values generated by an LFSR can be gainfully employed in the encryption (scrambling) and decryption (unscrambling) of data. A stream of data bits can be encrypted by XOR-ing them with the output from an LFSR (Figure C-11).

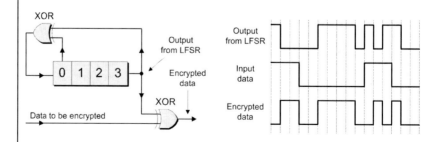

Figure C-11. Data encryption using an LFSR.

The stream of encrypted data bits seen by a receiver can be decrypted by XOR-ing them with the output of an identical LFSR. This is obviously a very trivial form of encryption that isn't very secure, but it's cheap and cheerful, and it may be useful in certain applications.

Cyclic redundancy check applications

A traditional application for LFSRs is in *cyclic redundancy check (CRC)* calculations, which can be used to detect errors in data communications. The stream of data bits being transmitted is used to modify the values being fed back into an LFSR (Figure C-12).

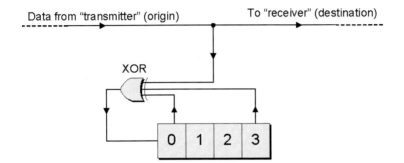

Figure C-12. CRC calculations.

CRC is pronounced by spelling is out as "C-R-C."

The final CRC value stored in the LFSR, known as a *checksum*, is dependent on every bit in the data stream. After all of the data bits have been transmitted, the transmitter sends its checksum value to the receiver. The receiver contains an identical CRC calculator and generates its own checksum value from the incoming data. Once all of the data bits have arrived, the receiver compares its internally generated checksum value with the checksum sent by the transmitter to determine whether any corruption occurred during the course of the transmission.

This form of error detection is very efficient in terms of the small number of bits that have to be transmitted in addition to the data. However, the downside is that you don't know if there was an error until the end of the transmission (and if there was an error, you have to repeat the entire transmission).

In the real world, a 4-bit CRC calculator would not be considered to provide sufficient confidence in the integrity of the transmitted data because it can only represent $(2^4 - 1) =$ 15 unique values. This leads to a problem called *aliasing*, in which the final CRC value is the same as was expected, but this value was actually caused by multiple errors canceling each other out. As the number of bits in a CRC calculator increases, however, the probability that multiple errors will cause identical checksum values approaches zero. For this reason, CRC calculators typically use 16 bits (which can accommodate 65,535 unique values) or more.

There are a variety of standard communications protocols, each of which specifies the number of bits employed in their CRC calculations and the taps to be used. The taps are selected such that an error in a single data bit will cause the maximum possible disruption to the resulting checksum value. Thus, in addition to being referred to as *maximal length*, these LFSRs may also be qualified as *maximal displacement*.

In addition to checking data integrity in communications systems, CRCs find a wide variety of other uses, for example, the detection of computer viruses. For the purposes of this discussion, a computer virus may be defined as a self-replicating program released into a computer system for a number of purposes. These purposes range from the simply mischievous, such as displaying humorous or annoying messages, to the downright nefarious, such as corrupting data or destroying (or subverting) the operating system.

One mechanism by which a computer virus may both hide and propagate itself is to attach itself to an existing program. Whenever that program is executed, it first triggers the virus to replicate itself, yet a cursory check of the system shows only the expected files to be present. In order to combat this form of attack, a unique checksum can be generated for each program on the system, where the value of each checksum is based on the binary instructions forming the program with which it is associated. At some later date, an antivirus program can be used to recalculate the checksum values for each program and to compare them to the original values. A difference in the two values associated with a program may indicate that a virus has attached itself to that program.[8]

Data compression applications

The CRC calculators discussed above can also be used in a data compression role. One such application is found in the circuit board test strategy known as *functional test*. The board, which may contain thousands of components and tracks, is plugged into a functional tester by means of its edge connector, which may contain hundreds of pins.

The tester applies a pattern of signals to the board's inputs, allows sufficient time for any effects to propagate around the board, and then compares the actual values seen on the outputs with a set of expected values stored in the system. This process is repeated for a series of input patterns, which may number in the tens or hundreds of thousands.

If the board fails the preliminary tests, a more sophisticated form of analysis known as *guided probe* may be employed to identify the cause of the failure. In this case, the tester instructs the operator to place the probe at a particular location on the board, and then the entire sequence of test patterns is rerun. The tester compares the actual sequence of values seen by the probe with a sequence of expected values that are stored in the system. This process (placing the probe and running the tests) is repeated until the tester has isolated the faulty component or track.

---

[8] Unfortunately, the creators of computer viruses are quite sophisticated, and some viruses are armed with the ability to perform their own CRC calculations. When a virus of this type attaches itself to a program, it can pad itself with dummy binary values, which are selected so as to cause an antivirus program to return a checksum value identical to the original.

A major consideration when supporting a guided probe strategy is the amount of expected data that must be stored. Consider a test sequence comprising 10,000 patterns driving a board containing 10,000 tracks. If the data were not compressed, the system would have to store 10,000 bits of expected data per track, which amounts to 100 million bits of data for the board. Additionally, for each application of the guided probe, the tester would have to compare the 10,000 data bits observed by the probe with the 10,000 bits of expected data stored in the system. Thus, using data in an uncompressed form is an expensive option in terms of storage and processing requirements.

One solution to these problems is to employ LFSR-based CRC calculators. The sequence of expected values for each track can be passed through a 16-bit CRC calculator implemented in software. Similarly, the sequence of actual values seen by the guided probe can be passed through an identical CRC calculator implemented in hardware. In this case, the calculated checksum values are also known as *signatures*, and a guided probe process based on this technique is known as *signature analysis*. Irrespective of the number of test patterns used, the system has to store only two bytes of data for each track. Additionally, for each application of the guided probe, the tester has to compare only the two bytes of data gathered by the probe with two bytes of expected data stored in the system. Thus, compressing the data results in storage requirements that are orders of magnitude smaller and comparison times that are orders of magnitude faster than the uncompressed data approach.

Built-in self-test applications

One test strategy that may be employed in complex ICs is that of *built-in self-test (BIST)*. Devices using BIST contain special test-generation and result-gathering circuits, both of which may be implemented using LFSRs (Figure C-13).

The LFSR forming the test generator is used to create a sequence of test patterns, while the LFSR forming the results

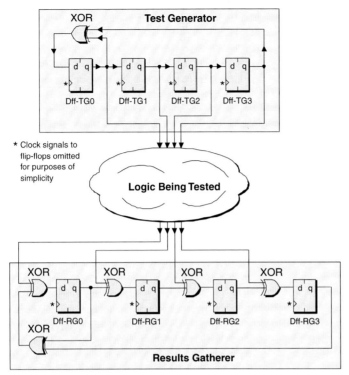

Figure C-13. BIST.

gatherer is used to capture the results. Observe that the results-gathering LFSR features modifications that allow it to accept parallel data.

Additional circuitry would be required to provide a way to load new seed values into the test generator and to access the final values in the results gatherer. This logic is not shown here for purposes of simplicity.

Note that the two LFSRs are not obliged to contain the same number of bits because the number of inputs to the logic being tested may be different to the number of outputs coming from that logic.

Also note that all of the flip-flops in the test generator would share a common clock. Similarly, all of the flip-flops in the results gatherer would also share a common clock. These two clocks might be common or they might be distinct (in the latter case they would be synchronized in some way). The clock signals are not shown in figure C-13 so as to keep things simple.

Pseudorandom-number-generation applications

Many computer programs rely on an element of randomness. Computer games such as Space Invaders employ random events to increase the player's enjoyment. Graphics programs may exploit random numbers to generate intricate patterns. All forms of computer simulation may utilize random numbers to represent the real world more accurately. For example, digital logic simulations (see also Chapter 19) may benefit from the portrayal of random stimulus such as external interrupts. Random stimulus can result in more realistic design verification, which can uncover problems that may not be revealed by more structured tests.

Random-number generators can be constructed in both hardware and software. The majority of these generators are not truly random, but they give the appearance of being random and are therefore said to be *pseudorandom*. In reality, pseudorandom numbers have an advantage over truly random numbers because the majority of computer applications typically require repeatability. For example, a designer repeating a digital simulation would expect to receive identical answers to those from the previous run. However, designers also need the ability to modify the seed value of the pseudorandom-number generator so as to spawn different sequences of values as required.

There are a variety of methods available for generating pseudorandom numbers, one of which is to use an LFSR whose tap values have been selected so as to provide a reasonably good pseudorandom source.

Last but not least

LFSRs are simple to construct and are useful for a wide variety of applications, but be warned that choosing the optimal polynomial (which ultimately boils down to selecting the tap points) for a particular application is a task that is usually reserved for a master of the mystic arts, not to mention that the maths can be hairy enough to make a grown man break down and cry (and don't even get me started on the subject of cyclotomic polynomials,[9] which are key to the tap-selection process).

_____

[9] Mainly because I haven't got the faintest clue what a *cyclotomic polynomial* is!

Glossary

ACM (adaptive computing machine)—A revolutionary new form of digital *integrated circuit (IC)* featuring a coarse-grained algorithmic element node-based architecture that can be reconfigured (adapted) hundreds of thousands of times a second.

Adaptive computing machine—*see ACM*

Address bus—A unidirectional set of signals used by a processor (or similar device) to point to memory locations in which it is interested.

A/D (analog to digital)—The process of converting an analog value into its digital equivalent.

Analog—A continuous value that most closely resembles the real world and can be as precise as the measuring technique allows.

Analog circuit—A collection of components used to process or generate analog signals.

Analog to digital—*see A/D*

Analogue—The way they spell "analog" in England.

Antifuse technology—A technology used to create programmable *integrated circuits (ICs)* whose programmable elements are based on conductive links called antifuses. When an engineer purchases a programmable device based on antifuses, none of the links is initially intact. Individual links can be selectively "grown" by applying pulses of relatively high voltage and current to the device's inputs.

Application-specific integrated circuit—*see ASIC*

Application-specific standard part—see *ASSP*

ASIC (application-specific integrated circuit)—A custom-built *integrated circuit (IC)* designed to address a specific application. Such a device can contain hundreds of millions of logic gates and can be used to create incredibly large and complex functions. Similar to an ASSP, except that an ASIC is designed and built to order for use by a specific company.

ASIC cell—A logic function in the cell library defined by the manufacturer of an ASIC.

Assertions/properties—The term *property* comes from the model-checking domain and refers to a specific functional behavior of the design that you want to (formally) verify (e.g., "after a request, we expect a grant within 10 clock cycles"). By comparison, the term *assertion* stems from the simulation domain and refers to a specific functional behavior of the design that you want to monitor during simulation (and flag violations if that assertion "fires"). Today, with the use of formal tools and simulation tools in unified environments and methodologies, the terms *property* and *assertion* tend to be used interchangeably.

ASSP (application-specific standard part)—A custom-built *integrated circuit (IC)* designed to address a specific application. Such a device can contain hundreds of millions of logic gates and can be used to create incredibly large and complex functions. Similar to an *application-specific integrated circuit (ASIC)*, except that an ASSP is marketed to multiple customers for inclusion in their products.

Asynchronous—A signal whose data is acknowledged or acted upon immediately and does not depend on a clock signal.

Ball grid array—see *BGA*

Bare die—An unpackaged *integrated circuit (IC)*.

Basic cell—A predefined group of unconnected transistors and resistors. This group is replicated across the surface of a gate-array form of ASIC.

Bebop—A form of music characterized by fast tempos and agitated rhythms that became highly popular in the decade following World War II.

BGA (ball grid array)—A packaging technology similar to a *pad grid array (PGA)*, in which a device's external connections are arranged as an array of conducting pads on the base of the package. In the case of a ball grid array, however, small balls of solder are attached to the conducting pads.

BiCMOS (bipolar-CMOS)—(1) A technology in which the logical function of each logic gate is implemented using low-power CMOS, while the output stage of each logic gate is implemented using high-drive bipolar transistors. (2) A device whose internal logic gates are implemented using low-power CMOS, but whose output pins are driven by high-drive bipolar transistors.

Binary digit—A numeral in the binary scale of notation. A binary digit (typically abbreviated to "bit") can adopt one of two values: 0 or 1.

Binary encoding—A form of state assignment for state machines that requires the minimum number of state variables.

Binary logic—Digital logic gates based on two distinct voltage levels. The two voltages are used to represent the binary values 0 and 1 along with their logical equivalents False and True.

Bipolar junction transistor—see *BJT*

BIST (built-in self-test)—A test strategy in which additional logic is built into a component, thereby allowing it to test itself.

Bit—Abbreviation *of binary digit*. A binary digit can adopt one of two values: 0 or 1.

Bit file—see *Configuration file*

BJTs (bipolar junction transistors)—A family of transistors.

Bobble—A small circle used on the inputs to a logic-gate symbol to indicate an active low input or control or on the outputs to indicate a negation (inversion) or complementary signal. Some engineers prefer to use the term *bubble*.

Boolean algebra—A mathematical way of representing logical expressions.

Built-in self-test—see *BIST*

Bus—A set of signals performing a common function and carrying similar data. Typically represented using vector notation: for example, an 8-bit database might be named data[7:0].

Byte—A group of eight binary digits, or bits.

Cache memory—A small, high-speed memory (usually implemented in SRAM) used to buffer the central processing unit from any slower, lower-cost memory devices such as DRAM. The high-speed cache memory is used to store the active instructions and data[1] associated with a program, while the bulk of the instructions and data resides in the slower memory.

Capacitance—A measure of the ability of two adjacent conductors separated by an insulator to hold a charge when a voltage differential is applied between them. Capacitance is measured in units of farads.

Cell—see *ASIC cell, Basic cell, Cell library, and Memory cell*

Cell library—The collective name for the set of logic functions defined by the manufacturer of an *application-specific integrated circuit (ASIC)*. The designer decides which types of cells should be realized and connected together to make the device perform its desired function.

---

[1] In this context, "active" refers to data or instructions that a program is currently using, or which the operating system believes that the program will want to use in the immediate future.

Central processing unit—see *CPU*

Ceramic—An inorganic, nonmetallic material, such as alumina, beryllia, steatite, or forsterite, which is fired at a high temperature and is often used in electronics as a substrate (base layer) or to create component packages.

CGA (column grid array)—A packaging technology similar to a *pad grid array (PGA)*, in which a device's external connections are arranged as an array of conducting pads on the base of the package. In the case of a column grid array, however, small columns of solder are attached to the conducting pads.

Channel—(1) The area between two arrays of basic cells in a channeled gate array. (2) The gap between the source and drain regions in a MOSFET transistor.

Channeled gate array—An *application-specific integrated circuit (ASIC)* organized as arrays of basic cells. The areas between the arrays are known as channels.

Channelless gate array—An *application-specific integrated circuit (ASIC)* organized as a single large array of basic cells. May also be referred to as a "sea-of-cells" or a "sea-of-gates" device.

Checksum—The final *cyclic-redundancy check (CRC)* value stored in a *linear feedback shift register (LFSR)* (or software equivalent). Also known as a "signature" in the guided-probe variant of a functional test.

Chemical mechanical polishing—see *CMP*

Chip—Popular name for an *integrated circuit (IC)*.

Chip scale package—see *CSP*

Circuit board—The generic name for a wide variety of interconnection techniques, which include rigid, flexible, and rigid-flex boards in single-sided, double-sided, multilayer, and discrete wired configurations.

CLB (configurable logic block)—The Xilinx term for the next logical partition/entity above a slice. Some Xilinx

FPGAs have two slices in each CLB, while others have four. See also *LAB, LC, LE,* and *Slice.*

Clock tree—This refers to the way in which a clock signal is routed throughout a chip. This is called a "clock tree" because the main clock signal branches again and again (register elements like flip-flops can be considered the "leaves" on the end of the branches). This structure is used to ensure that all of the flip-flops see the clock signal as close together as possible.

CMOS (complementary metal oxide semiconductor)—Logic gates constructed using a mixture of NMOS and PMOS transistors connected together in a complementary manner.

CMP (chemical mechanical polishing)—A process used to replanarize a wafer—smoothing and flattening the surface by polishing out the "bumps" caused by adding a metalization (tracking) layer.

Column grid array—see *CGA*

Combinatorial logic—*see Combinational logic*

Combinational logic—A digital logic function formed from a collection of primitive logic gates (AND, OR, NAND, NOR, etc.), where any output values from the function are directly related to the current combination of values on its inputs. That is, any changes to the signals being applied to the inputs to the function will immediately start to propagate (ripple) through the gates forming the function until their effects appear at the outputs from the function. Some folks prefer to say "combinatorial logic." See also *Sequential logic*.

Complementary output—Refers to a function with two outputs carrying complementary (opposite) logical values. One output is referred to as the *true output* and the other as the *complementary output*.

Complex programmable logic device—see *CPLD*

Conditioning—*see Signal conditioning*

Configurable logic block—see *CLB*

Configuration commands—Instructions in a configuration file that tell the device what to do with the associated configuration data. See also *Configuration data* and *Configuration file*.

Configuration data—Bits in a configuration file that are used to define the state of programmable logic elements directly. See also *Configuration commands* and *Configuration file*.

Configuration file—A file containing the information that will be uploaded into the FPGA in order to program (configure) it to perform a specific function. In the case of SRAM-based FPGAs, the configuration file contains a mixture of *configuration data* and *configuration commands*. When the configuration file is in the process of being loaded into the device, the information being transferred is referred to as the *configuration bitstream*. See also *Configuration commands* and *Configuration data*.

Constraints—In the context of formal verification, the term *constraint* derives from the model-checking space. Formal model checkers consider all possible allowed input combinations when performing their magic and working on a proof. Thus, there is often a need to constrain the inputs to their legal behavior; otherwise, the tool would report false negatives, which are property violations that would not normally occur in the actual design.

Core—see *Hard core* and *Soft core*

Corner condition—see *Corner case*

Corner case—A hard-to-exercise or hard-to-reach functional condition associated with the design.

CPLD (complex PLD)—A device that contains a number of SPLD (typically PAL) functions sharing a common programmable interconnection matrix.

CPU (central processing unit)—The brain of a computer where all of the decision making and number crunching are performed.

CRC (cyclic redundancy check)—A calculation used to detect errors in data communications, typically performed using a *linear feedback shift register (LFSR)*. Similar calculations may be used for a variety of other purposes such as data compression.

CSP (chip scale package)—An *integrated circuit (IC)* packaging technique in which the package is only fractionally larger than the silicon die.

Cyclic redundancy check—see *CRC*

D/A (digital to analog)—The process of converting a digital value into its analog equivalent.

Data bus—A bidirectional set of signals used by a computer to convey information from a memory location to the central processing unit and vice versa. More generally, a set of signals used to convey data between digital functions.

Data-path function—A well-defined function such as an adder, counter, or multiplier used to process digital data.

DCM (digital clock manager)—Some FPGA clock managers are based on *phase-locked loops (PLLs)*, while others are based on *digital delay-locked loops (DLLs)*. The term *DCM* is used by Xilinx to refer to an advanced clock manager that is a superset of a DLL. See also *DLL* and *PLL*.

Declarative—In the context of formal verification, the term *declarative* refers to an assertion/property/event/constraint that exists within the structural context of the design and is evaluated along with all of the other structural elements in the design (for example, a module that takes the form of a structural instantiation). Another way to view this is that a declarative assertion/property is *always* "on/active," unlike its procedural counterpart that is only "on/active" when a specific path is taken/executed through the HDL code.

Deep submicron—see *DSM*

Delay-locked loop—see *DLL*

DeMorgan transformation—The transformation of a Boolean expression into an alternate, and often more convenient, form.

Die—An unpackaged *integrated circuit (IC)*. In this case, the plural of die is also die (in much the same way that "a shoal of herring" is the plural of "herring").

Digital—A value represented as being in one of a finite number of discrete states called *quanta*. The accuracy of a digital value is dependent on the number of quanta used to represent it.

Digital circuit—A collection of logic gates used to process or generate digital signals.

Digital clock manager—see *DCM*

Digital delay-locked loop—see *DLL*

Digital signal processing/processor—see *DSP*

Digital to analog—see *D/A*

Diode—A two-terminal device that conducts electricity in only one direction; in the other direction it behaves like an open switch. These days the term *diode* is almost invariably taken to refer to a semiconductor device, although alternative implementations such as vacuum tubes are available.

Discrete device—Typically taken to refer to an electronic component such as a resistor, capacitor, diode, or transistor that is presented in an individual package. More rarely, the term may be used in connection with a simple *integrated circuit (IC)* containing a small number of primitive logic gates.

DLL (digital delay-locked loop)—Some FPGA clock managers are based on *phase-locked loops (PLLs)*, while others are based on *digital delay-locked loops (DLLs)*. DLLs are, by definition, digital in nature. The proponents of DLLs say that they offer advantages in terms of precision, stability, power management, noise insensitivity, and jitter performance. I have no clue as to why these aren't called *DDLLs*. See also *PLL*.

DSM (deep submicron)—Typically taken to refer to *integrated circuit (ICs)* containing structures that are smaller than 0.5 microns (one half of one millionth of a meter).

DSP (1) (digital signal processing)—The branch of electronics concerned with the representation and manipulation of signals in digital form. This form of processing includes compression, decompression, modulation, error correction, filtering, and otherwise manipulating audio (voice, music, etc.), video, image, and other such data for such applications like telecommunications, radar, and image processing (including medical imaging). **(2) (digital signal processor)**—A special form of microprocessor that has been designed to perform a specific processing task on a specific type of digital data much faster and more efficiently than can be achieved using a general-purpose microprocessor.

Dynamic formal verification—Some portions of a design are going to be difficult to verify via simulation because they are deeply buried in the design, making them difficult to control from the primary inputs. In order to address this, some verification solutions use simulation to reach a corner case and then automatically pause the simulator and invoke a static formal verification engine to evaluate that corner case exhaustively. This combination of simulation and traditional static formal verification is referred to as *dynamic formal verification*. See also *Corner case*, *Formal verification*, and *Static formal verification*.

Dynamic RAM—see *DRAM*

ECL (emitter-coupled logic)—Logic gates implemented using particular configurations of *Bipolar junction transistors (BJTs)*.

Edge sensitive—An input to a logic function that only affects the function when it transitions from one logic value to another.

EEPROM or E^{2}PROM (electrically erasable programmable read-only memory)—A memory *integrated circuit (IC)* whose contents can be electrically programmed by the designer. Additionally, the contents can be electrically erased, allowing the device to be reprogrammed.

Electrically erasable programmable read-only memory—see *EEPROM*

Emitter-coupled logic—see *ECL*

EPROM (erasable programmable read-only memory)—A memory *integrated circuit (IC)* whose contents can be electrically programmed by the designer. Additionally, the contents can be erased by exposing the die to *ultraviolet (UV)* light through a quartz window mounted in the top of the component's package.

Equivalency checking—see *Formal verification*

Equivalent gate—An ASIC-based concept in which each type of logic function is assigned an equivalent gate value for the purposes of comparing functions and devices. However, the definition of an equivalent gate varies depending on whom you're talking to.

Erasable programmable read-only memory—see *EPROM*

Event—In the context of formal verification, an event is similar to an assertion/property, and in general events may be considered a subset of assertions/properties. However, while assertions/properties are typically used to trap undesirable behavior, events may be used to specify desirable behavior for the purposes of functional coverage analysis.

Falling edge—see *Negative edge*

FET (field-effect transistor)—A transistor whose control (or "gate") signal is used to create an electromagnetic field that turns the transistor on or off.

Field-effect transistor—see *FET*

Field-programmable gate array—see *FPGA*

Field-programmable interconnect chip—see *FPIC[2]*

Field-programmable interconnect device—see *FPID*

FIFO (first in first out)—A special memory device or function in which data is read out in the same order that it was written in.

Finite state machine—see *FSM*

Firm IP—In the context of an FPGA, the term *firm IP* refers to a library of high-level functions. Unlike their soft IP equivalents, however, these functions have already been optimally mapped, placed, and routed into a group of programmable logic blocks (possibly combined with some hard IP blocks like multipliers, etc.). One or more copies of each predefined firm IP block can be instantiated (called up) into the design as required. See also *Hard IP* and *Soft IP*.

Firmware—Refers to programs or sequences of instructions that are loaded into nonvolatile memory devices.

First in first out—see *FIFO*

FLASH memory—An evolutionary technology that combines the best features of the EPROM and E^{2}PROM technologies. The name FLASH is derived from the technology's fast reprogramming time compared to EPROM.

Formal verification—In the not-so-distant past, the term *formal verification* was considered synonymous with *equivalency checking* for the majority of design engineers. In this context, an equivalency checker is a tool that uses formal (rigorous mathematical) techniques to compare two different representations of a design—say an RTL description with a gate-level netlist—to determine whether or not they have the same input-to-output functionality. In fact,

---

[2] FPIC is a trademark of Aptix Corporation.

equivalency checking may be considered to be a subclass of formal verification called *model checking*, which refers to techniques used to explore the state space of a system to test whether or not certain properties, typically specified as "assertions," are true. See also *Static formal verification* and *Dynamic formal verification*.

FPGA (field-programmable gate array)—A type of digital *integrated circuit (IC)* that contains configurable (programmable) blocks of logic along with configurable interconnect between these blocks. Such a device can be configured (programmed) by design engineers to perform a tremendous variety of different tasks.

FPIC (field-programmable interconnect chip)[3]—An alternate, proprietary name for a *field-programmable interconnect device (FPID)*.

FPID (field-programmable interconnect device)—A device used to connect logic devices together that can be dynamically reconfigured in the same way as standard SRAM-based FPGAs. Because each FPID may have around 1,000 pins, only a few such devices are typically required on a circuit board.

FR4—The most commonly used insulating base material for circuit boards. FR4 is made from woven glass fibers that are bonded together with an epoxy. The board is cured using a combination of temperature and pressure, which causes the glass fibers to melt and bond together, thereby giving the board strength and rigidity. The first two characters stand for "flame retardant," and you can count the number of people who know what the "4" stands for on the fingers of one hand. FR4 is technically a form of fiberglass, and some people do refer to these composites as *fiberglass boards* or *fiberglass substrates*, but not often.

Full custom—An *application-specific integrated circuit (ASIC)* in which the design engineers have complete control over

---

[3] FPIC is a trademark of Aptix Corporation.

every mask layer used to fabricate the device. The ASIC vendor does not provide a cell library or prefabricate any components on the substrate.

Functional latency—Refers to the fact that, at any given time, only a portion of the logic functions in a device or system are typically active (doing anything useful).

Fuse—see *Fusible link technology* and *Antifuse technology*

Fusible-link technology—A technology used to create programmable *integrated circuits (ICs)* whose programmable elements are based on microscopically small fusible links. When an engineer purchases a programmable device based on fusible links, all of the fuses are initially intact. Individual fuses can be selectively removed by applying pulses of relatively high voltage and current to the device's inputs.

FSM (finite state machine)—The actual implementation (in hardware or software) of a function that can be considered to consist of a finite set of states through which it sequences.

GAL (generic array logic)—A variation on a PAL device from a company called Lattice Semiconductor Corporation.[4]

Garbage in garbage out—see *GIGO*

Gate array—An *application-specific integrated circuit (ASIC)* in which the manufacturer prefabricates devices containing arrays of unconnected components (transistors and resistors) organized in groups called *basic cells*. The designer specifies the function of the device in terms of cells from the cell library and the connections between them, and the manufacturer then generates the masks used to create the metalization layers.

Generic array logic—see *GAL*

Geometry—Refers to the size of structures created on an *integrated circuit (IC)*. The structures typically referenced are

---

[4] GAL is a registered trademark of Lattice Semiconductor Corporation.

the width of the tracks and the length of the transistor's channels; the dimensions of other features are derived as ratios of these structures.

Giga—Unit qualifier (symbol = G) representing one thousand million, or 10^9. For example, 3 GHz stands for 3×10^9 hertz.

GIGO (garbage in garbage out)—An electronic engineer's joke, also familiar to the writers of computer programs.

Glue logic—The relatively small amounts of simple logic that are used to connect ("glue") together—and interface between—larger logical blocks, functions, or devices.

Gray code—A sequence of binary values in which each pair of adjacent values differs by only a single bit: for example, 00, 01, 11, 10.

Ground plane—A conducting layer in, or on, a substrate providing a grounding, or reference, point for components. There may be several ground planes separated by insulating layers.

Guard condition—A Boolean expression associated with a transition between two states in a state machine. Such an expression must be satisfied for that transition to be executed.

Guided probe—A form of functional test in which the operator is guided in the probing of a circuit board to isolate a faulty component or track.

Hard core—In the context of digital electronics, the term *core* is typically used to refer to a relatively large, general-purpose logic function that may be used as a building block forming a portion of a much larger chip design. For example, if an ASIC contains an embedded microprocessor, that microprocessor would be referred to as a "microprocessor core." Other functions that might fall into this category are microcontroller cores, *digital signal processor (DSP)* cores, communication function cores (e.g., a UART), and so forth. Such cores may be developed internally by the design team, but they are typically purchased from third-party *intellectual property (IP)* vendors.

There is some difference in how the term *hard core* is perceived depending on the target implementation technology: ASIC or FPGA. In the case of an ASIC, the hard core will be presented as a block of logic gates whose physical locations (relative to each other) and interconnections have already been defined (that is, hard-wired and set in stone). This block will be treated as a black box by the place-and-route software that is used to process the rest of the design; that is, the location of the block as a whole may be determined by the place-and-route software, but it's internal contents are completely locked down. The output from the place-and-route software will subsequently be used to generate the photo-masks that will in turn be used to fabricate the silicon chip. By comparison, in the case of an FPGA, any hard cores have already been physically implemented as hard-wired blocks that are embedded into the FPGA's fabric.

A design may comprise one or more hard cores combined with one or more soft cores along with other blocks of user-defined logic. See also *Soft core*.

Hardware—Generally understood to refer to any of the physical portions constituting an electronic system, including components, circuit boards, power supplies, cabinets, and monitors.

Hard IP—In the context of an FPGA, the term *hard IP* refers to preimplemented blocks, such as microprocessor cores, gigabit interfaces, multipliers, adders, MAC functions, and the like. These blocks are designed to be as efficient as possible in terms of power consumption, silicon real estate requirements, and performance. Each FPGA family will feature different combinations of such blocks together with various quantities of programmable logic blocks. See also *Soft IP* and *Firm IP*.

Hardware description language—see *HDL*

HDL (hardware description language)—Today's digital *integrated circuits (ICs)* can end up containing hundreds of millions of logic gates, and it simply isn't possible to capture and manage designs of this complexity at the schematic (circuit-diagram) level. Thus, as opposed to using schematics, the functionality of a high-end IC is now captured in textual form using an HDL. Popular HDLs are Verilog, SystemVerilog, VHDL, and SystemC.

HDL synthesis—A more recent name for logic synthesis. See also *Logic synthesis* and *Physically aware synthesis*.

Hertz—see *Hz*

High-impedance state—The state associated with a signal that is not currently being driven by anything. A high-impedance state is typically indicated by means of the "Z" character.

Hz (hertz)—Unit of frequency. One hertz equals one cycle, or one oscillation, per second.

IC (integrated circuit)—A device in which components such as resistors, diodes, and transistors are formed on the surface of a single piece of semiconducting material.

ICR (in-circuit reconfigurable)—An SRAM-based or similar component that can be dynamically reprogrammed on the fly while remaining resident in the system.

Impedance—The resistance to the flow of current caused by resistive, capacitive, and/or inductive devices (or undesired parasitic elements) in a circuit.

Implementation-based verification coverage—This measures verification activity with respect to microarchitecture details of the actual implementation. This refers to design decisions that are embedded in the RTL that result in implementation-specific corner cases, for example, the depth of a FIFO buffer and the corner cases for its "high-water mark" and "full" conditions. Such implementation details are rarely visible at the specification level. See also *Macroarchitecture definition*, *Microarchitecture definition*, and

Specification-level coverage.

In-circuit reconfigurable—see *ICR*

Inductance—A property of a conductor that allows it to store energy in a magnetic field which is induced by a current flowing through it. The base unit of inductance is the *henry*.

In-system programmable—see *ISP*

Integrated circuit—see *IC*

Intellectual property—see *IP*

IP (intellectual property)—When a team of electronics engineers is tasked with designing a complex *integrated circuit (IC)*, rather than reinvent the wheel, they may decide to purchase the plans for one or more functional blocks that have already been created by someone else. The plans for these functional blocks are known as intellectual property, or IP. IP blocks can range all the way up to sophisticated communications functions and microprocessors. The more complex functions, like microprocessors, may be referred to as "cores." See also *Hard IP*, *Soft IP*, and *Firm IP*.

ISP (in-system programmable)—An E^2-based, FLASH-based, SRAM-based, or similar *integrated circuit (IC)* that can be reprogrammed while remaining resident on the circuit board.

JEDEC (Joint Electronic Device Engineering Council)—A council that creates, approves, arbitrates, and oversees industry standards for electronic devices. In programmable logic, the term *JEDEC* refers to a textual file containing information used to program a device. The file format is a JEDEC-approved standard and is commonly referred to as a "JEDEC file."

Jelly-bean logic—Small *integrated circuits (ICs)* containing a few simple, fixed logical functions, for example, four 2-input AND gates.

Joint Electronic Device Engineering Council—see *JEDEC*

Kilo—Unit qualifier (symbol = K) representing one thousand, or 10^3. For example, 3 KHz stands for 3×10^3 hertz.

LAB (logic array block)—The Altera name for a programmable logic block containing a number of *logic elements (LEs)*. See also *CLB*, *LC*, *LE*, and *Slice*.

LC (logic cell)—The core building block in a modern FPGA from Xilinx is called a *logic cell (LC)*. Among other things, an LC comprises a 4-input LUT, a multiplexer, and a register. See also *CLB*, *LAB*, *LE*, and *Slice*.

LE (logic element)—The core building block in a modern FPGA from Altera is called a *logic element (LE)*. Among other things, an LE comprises a 4-input LUT, a multiplexer and a register. See also *CLB*, *LAB*, *LC*, and *Slice*.

Least-significant bit—see *LSB*

Least-significant byte—see *LSB*

Level sensitive—An input to a logic function whose effect on the function depends only on its current logic value or level and is not directly related to its transitioning from one logic value to another.

LFSR (linear feedback shift register)—A shift register whose data input is generated as an XOR or XNOR of two or more elements in the register chain

Linear feedback shift register—see *LFSR*

Literal—A variable (either true or inverted) in a Boolean equation.

Logic function—A mathematical function that performs a digital operation on digital data and returns a digital value.

Logic array block—see *LAB*

Logic cell—see *LC*

Logic element—see *LE*

Logic gate—The physical implementation of a simple or primitive logic function.

Logic synthesis—A process in which a program is used to automatically convert a high-level textual representation of a design (specified using a *hardware description language (HDL)* at the *register transfer level (RTL)* of abstraction) into equivalent registers and Boolean equations. The synthesis tool automatically performs simplifications and minimizations and eventually outputs a gate-level netlist. See also *HDL synthesis* and *Physically aware synthesis*.

Lookup table—see *LUT*

LSB—**(1) (least-significant bit)** The binary digit, or bit, in a binary number that represents the least-significant value (typically the right-hand bit). **(2) (least-significant byte)**—The byte in a multibyte word that represents the least-significant values (typically the right-hand byte).

LUT (lookup table)—There are two fundamental incarnations of the programmable logic blocks used to form the medium-grained architectures featured in FPGAs: MUX (multiplexer) based and LUT (lookup table) based. In the case of a LUT, a group of input signals is used as an index (pointer) into a lookup table. See also *CLB, LAB, LC, LE,* and *Slice*.

Macroarchitecture definition—A design commences with an original concept, whose high-level definition is determined by system architects and system designers. It is at this stage that *macroarchitecture* decisions are made, such as partitioning the design into hardware and software components, selecting a particular microprocessor core and bus structure, and so forth. The resulting specification is then handed over to the hardware design engineers, who commence their portion of the development process by performing *microarchitecture definition* tasks. See also *Microarchitecture definition*.

Magnetic random-access memory—see *MRAM*

Magnetic tunnel junction—see *MTJ*

Mask—see *Photo-mask*

Mask programmable—A device such as a *read-only memory (ROM)* that is programmed during its construction using a unique set of photo-masks.

Maximal displacement—A *linear feedback shift register (LFSR)* whose taps are selected such that changing a single bit in the input data stream will cause the maximum possible disruption to the register's contents.

Maximal length—A *linear feedback shift register (LFSR)* with n bits that sequences through $2^n - 1$ states before returning to its original value.

Maxterm—The logical OR of the inverted variables associated with an input combination to a logical function.

MCM (multichip module)—A generic name for a group of advanced interconnection and packaging technologies featuring unpackaged *integrated circuits (ICs)* mounted directly onto a common substrate.

Mega—Unit qualifier (symbol = M) representing one million, or 10^6. For example, 3 MHz stands for 3×10^6 hertz.

Memory cell—A unit of memory used to store a single binary digit, or bit, of data.

Memory word—A number of memory cells logically and physically grouped together. All the cells in a word are typically written to, or read from, at the same time.

Metalization layer—A layer of conducting material on an *integrated circuit (IC)* that is selectively deposited or etched to form connections between logic gates. There may be several metalization layers separated by dielectric (insulating) layers.

Metal-oxide semiconductor field-effect transistor—see *MOSFET*

Microarchitecture definition—A design commences with an original concept, whose high-level definition is determined by system architects and system designers. The resulting specification is then handed over to the hardware design engineers, who commence their portion of the development process by performing *microarchitecture definition* tasks such as detailing control structures, bus structures, and primary datapath elements. A simple example would be an element such as a FIFO, to which one would assign attributes like *width* and *depth* and characteristics like *blocking write*, *nonblocking read*, and how to behave when empty or full. Microarchitecture definitions, which are often performed in brainstorming sessions on a whiteboard, may include performing certain operations in parallel verses sequentially, pipelining portions of the design versus nonpipelining, sharing common resources—for example, two operations sharing a single multiplier—versus using dedicated resources, and so forth.

Micro—Unit qualifier (symbol = µ) representing one millionth, or 10^{-6}. For example, 3 µS stands for 3×10^{-6} seconds.

Microcontroller—see *µC*

Microprocessor—see *µP*

Milli—Unit qualifier (symbol = m) representing one thousandth, or 10^{-3}. For example, 3 mS stands for 3×10^{-3} seconds.

Minimization—The process of reducing the complexity of a Boolean expression.

Minterm—The logical AND of the variables associated with an input combination to a logical function.

Mixed signal—Typically refers to an *integrated circuit (IC)* that contains both analog and digital elements.

Model checking—see *Formal verification*

Moore's law—In 1965, Gordon Moore (who was to cofound Intel Corporation in 1968) noted that new generations of memory devices were released approximately every 18 months and that each new generation of devices contained roughly twice the capacity of its predecessor. This observation subsequently became known as *Moore's Law,* and it has been applied to a wide variety of electronics trends.

MOSFET (metal-oxide semiconductor field-effect transistor) —A family of transistors.

Most-significant bit—see *MSB*

Most-significant byte—see *MSB*

MRAM (magnetic RAM)—A form of memory expected to come online circa 2005 that has the potential to combine the high speed of SRAM, the storage capacity of DRAM, and the nonvolatility of FLASH, while consuming very little power.

MSB—(1) (**most-significant bit**) The binary digit, or bit, in a binary number that represents the most-significant value (typically the left-hand bit). (2) (**most-significant byte**) The byte in a multibyte word that represents the most-significant values (typically the left-hand byte).

MTJ (magnetic tunnel junction)—A sandwich of two ferromagnetic layers separated by a thin insulating layer. An MRAM memory cell is created by the intersection of two wires (say, a "row" line and a "column" line) with an MJT sandwiched between them.

Multichip module—see *MCM*

Multiplexer (digital)—A logic function that uses a binary value, or address, to select between a number of inputs and conveys the data from the selected input to the output.

Nano—Unit qualifier (symbol = n) representing one thousandth of one millionth, or 10^{-9}. For example, 3 nS stands for 3×10^{-9} seconds.

Negative edge—A signal transition from a logic 1 to a logic 0.

Nibble—see *Nybble*

NMOS (N-channel MOS)—Refers to the order in which the semiconductor is doped in a MOSFET device, that is, which structures are constructed as N-type versus P-type material.

Noise—The miscellaneous rubbish that gets added to an electronic signal on its journey through a circuit. Noise can be caused by capacitive or inductive coupling or by externally generated electromagnetic interference.

Nonrecurring engineering—see *NRE*

Nonvolatile—A memory device that does not lose its data when power is removed from the system.

NPN (N-type–P-type–N-type)—Refers to the order in which the semiconductor is doped in a *bipolar junction transistor (BJT)*.

NRE (nonrecurring engineering)—In the context of this book, this refers to the costs associated with developing an ASIC, ASSP, or FPGA design.

N-type—A piece of semiconductor doped with impurities that make it amenable to donating electrons.

Nybble—A group of four binary digits, or bits.

Ohm—Unit of resistance. The Greek letter omega, Ω, is often used to represent ohms; for example, 1 MΩ indicates one million ohms.

One-hot encoding—A form of state assignment for state machines in which each state is represented by an individual state variable, and only one such variable may be "on/active" ("hot") at any particular time.

One-time programmable—see *OTP*

OpenVera Assertions—see *OVA*

Open Verification Library—see *OVL*

Operating system—The collective name for the set of master programs that control the core operation and the base-level user interface of a computer.

OTP (one-time programmable)—A programmable device, such as an SPLD, CPLD, or FPGA, that can be configured (programmed) only a single time.

OVA (OpenVera Assertions)—A formal verification language that has been specially constructed for the purpose of specifying assertions/properties with maximum efficiency. OVA is very powerful in creating complex regular and temporal expressions, and it allows complex behavior to be specified with very little code. This language was donated to Accellera's SystemVerilog committee, which is controlled by the Accellera organization (www.accellera.org), and is based on IBM's Sugar language. See also *PSL*, *Sugar*, and *SVA*.

OVL (Open Verification Library)—A library of assertion/property models available in both VHDL and Verilog 2K1 that is managed under the auspices of the Accellera organization (www.accellera.com).

Pad grid array—see *PGA*

PAL (programmable array logic)[5]—A programmable logic device in which the AND array is programmable, but the OR array is predefined (see also *PLA, PLD, and PROM*).

Parasitic effects—The effects caused by undesired resistive, capacitive, or inductive elements inherent in the material or topology of a track or component.

PCB (printed circuit board)—A type of circuit board that has conducting tracks superimposed, or "printed," on one or both sides and may also contain internal signal layers and power and ground planes. An alternative name—*printed wire board (PWB)*—is commonly used in America.

Peta—Unit qualifier (symbol = P) representing one thousand million million, or 10^{15}. For example, 3 PHz stands for 3×10^{15} hertz.

PGA (1) (pad grid array)—A packaging technology in which a device's external connections are arranged as an array of conducting pads on the base of the package. **(2) (pin grid array)**—A packaging technology in which a device's external connections are arranged as an array of conducting leads, or pins, on the base of the package.

---

[5] PAL is a registered trademark of Monolithic Memories

Phase-locked loop—see *PLL*

Physically aware synthesis—For most folks, physically aware synthesis means taking actual placement information associated with the various logical elements in the design, using this information to estimate accurate track delays, and using these delays to fine-tune the placement and perform other optimizations. Interestingly enough, physically aware synthesis commences with a first-pass run using a relatively traditional logic/HDL synthesis engine. See also *logic synthesis*.

Photo-mask—A sheet of material carrying patterns that are either transparent or opaque to the *ultraviolet (UV)* light used to create structures on the surface of an *integrated circuit (IC)*.

Pico—Unit qualifier (symbol = p) representing one millionth of one millionth, or 10^{-12}. For example, 3 pS stands for 3×10^{-12} seconds.

Pin grid array —see *PGA*

PLA (programmable logic array)—The most user configurable of the traditional programmable logic devices because both the AND and OR arrays are programmable (see also *PAL, PLD, and PROM*).

PLD (programmable logic device)—An *integrated circuit (IC)* whose internal architecture is predetermined by the manufacturer, but which is created in such a way that it can be configured (programmed) by engineers in the field to perform a variety of different functions. For the purpose of this book, the term PLD is assumed to encompass both *simple PLDs (SPLDs)* and *complex PLDs (CPLDs)*. In comparison to an FPGA, these devices contain a relatively limited number of logic gates, and the functions they can be used to implement are much smaller and simpler.

PLI (programming-language interface)—One very cool concept that accompanied Verilog (the language) and Verilog-XL (the simulator) was the Verilog programming-language interface, or PLI. The more generic name for this sort of thing is *application programming interface (API)*. An API is a library of software functions that allow external software programs to pass data into an application and access data from that application. Thus, the Verilog PLI is an API that allows users to extend the functionality of the Verilog language and simulator.

PLL (phase-locked loop)—Some FPGA clock managers are based on *phase-locked loops (PLLs)*. PLLs have been used since the 1940s in analog implementations, but recent emphasis on digital methods has made it desirable to process signals digitally. Today's PLLs can be implemented using either analog or digital techniques. See also *DLL*.

PMOS (P-channel MOS)—Refers to the order in which the semiconductor is doped in a MOSFET device, that is, which structures are constructed as P-type versus N-type material.

PNP (P-type–N-type–P-type)—Refers to the order in which the semiconductor is doped in a *bipolar junction transistor (BJT)*.

Positive edge—A signal transition from a logic 0 to a logic 1.

Power plane—A conducting layer in or on the substrate providing power to the components. There may be several power planes separated by insulating layers.

Pragma—An abbreviation for "pragmatic information" that refers to special pseudocomment directives inserted in source code (including C/C++ and HDL code) that can be interpreted and used by parsers/compilers and other tools. (Note that this is a general-purpose term, and pragma-based techniques are used by a variety of tools in addition to formal verification technology.)

Primitives—Simple logic functions such as BUF, NOT, AND, NAND, OR, NOR, XOR, and XNOR. These may also be referred to as *primitive logic gates*.

Printed circuit board—see *PCB*

Printed wire board—see *PWB*

Procedural: In the context of formal verification, the term *procedural* refers to an assertion/property/event/constraint that is described within the context of an executing process or set of sequential statements such as a VHDL process or a Verilog "always" block (thus, these are sometimes called "in-context" assertions/properties). In this case, the assertion/property is built into the logic of the design and will be evaluated based on the path taken through a set of sequential statements.

Product-of-sums—A Boolean equation in which all of the *maxterms* corresponding to the lines in the truth table for which the output is a logic 0 are combined using AND operators.

Product term—A set of literals linked by an AND operator.

Programmable array logic—see *PAL*

Programmable logic array—see *PLA*

Programmable logic device—see *PLD*

Programmable read-only memory—see *PROM*

Programming-language interface—see *PLI*

PROM (programmable read-only memory)—A programmable logic device in which the OR array is programmable, but the AND array is predefined. Usually considered to be a memory device whose contents can be electrically programmed (once) by the designer (see also *PAL, PLA, and PLD*).

Properties/assertions—see *Assertions/properties*

Property-specification language—see *PSL*

Pseudorandom—An artificial sequence of values that give the appearance of being random, but which are also repeatable.

PSL (property-specification language)—A formal verification language that has been specially constructed for the purpose of specifying assertions/properties with maximum efficiency. PSL is very powerful in creating complex regular and temporal expressions, and it allows complex behavior to be specified with very little code. This industry standard language, which is controlled by the Accellera organization (www.accellera.org), is based on IBM's Sugar language. See also *OVA*, *Sugar*, and *SVA*.

P-type—A piece of semiconductor doped with impurities that make it amenable to accepting electrons.

PWB (printed wire board)—A type of circuit board that has conducting tracks superimposed, or "printed," on one or both sides and may also contain internal signal layers and power and ground planes. An alternative name—*printed circuit board (PCB)*—is predominantly used in Europe and Asia.

QFP (quad flat pack)—The most commonly used package in surface mount technology to achieve a high lead count in a small area. Leads are presented on all four sides of a thin square package.

Quad flat pack—see *QFP*

Quantization—(1) Part of the process by which an analog signal is converted into a series of digital values. First of all the analog signal is sampled at specific times. For each sample, the complete range of values that the analog signal can assume is divided into a set of discrete bands or quanta. Quantization refers to the process of determining which band the current sample falls into. See also *Sampling*. (2) The process of changing floating-point representations into their fixed-point equivalents.

RAM (random-access memory)—A data-storage device from which data can be read out and into which new data can be written. Unless otherwise indicated, the term *RAM* is typically taken to refer to a semiconductor device in the form of an *integrated circuit (IC)*.

Random-access memory—see *RAM*

Read-only memory—see *ROM*

Read-write memory—see *RWM*

Real estate—Refers to the amount of area available on a substrate.

Register transfer level—see *RTL*

Rising edge—see *Positive edge*

ROM (read-only memory)—A data storage device from which data can be read out, but into which new data cannot be written. Unless otherwise indicated, the term *ROM* is typically taken to refer to a semiconductor device in the form of an *integrated circuit (IC)*.

RTL (register transfer level)—A *hardware description language (HDL)* is a special language that is used to capture (describe) the functionality of an electronic circuit. In the case of an HDL intended to represent digital circuits, such a language may be used to describe the functionality of the circuit at a variety of different levels of abstraction. The simplest level of abstraction is that of a gate-level netlist, in which the functionality of the digital circuit is described as a collection of primitive logic gates (AND, OR, NAND, NOR, etc.) and the connections between them. A more sophisticated (higher) level of abstraction is referred to as *register transfer level (RTL)*. In this case, the circuit is described as a collection of storage elements (registers), Boolean equations, control logic such as *if-then-else* statements, and complex sequences of events (e.g., "If the clock signal goes from 0 to 1, then load register A with the contents of register B plus register C"). The most popular languages used for capturing designs in RTL are VHDL and Verilog (with SystemVerilog starting to gain a larger following).

RWM (read-write memory)—An alternative (and possibly more appropriate) name for a *random-access memory (RAM)*.

Sampling—Part of the process by which an analog signal is converted into a series of digital values. Sampling refers to observing the value of the analog signal at specific times. *See also Quantization.*

Schematic—Common name for a circuit diagram.

Sea of cells—Popular name for a channelless gate array.

Sea of gates—Popular name for a channelless gate array.

Seed value—An initial value loaded into a *linear feedback shift register (LFSR)* or random-number generator.

Semiconductor—A special class of material that can exhibit both conducting and insulating properties.

Sequential logic—A digital function whose output values depend not only on its current input values, but also on previous input values. That is, the output value depends on a "sequence" of input values. See also *Combinational logic.*

Signal conditioning—Amplifying, filtering, or otherwise processing a (typically analog) signal.

Signature—Refers to the *checksum* value from a *cyclic redundancy check (CRC)* when used in the guided-probe form of functional test.

Signature analysis—A guided-probe functional-test technique based on signatures.

Silicon chip—Although a variety of semiconductor materials are available, the most commonly used is silicon, and *integrated circuits (ICs)* are popularly known as "silicon chips," or simply "chips."

Simple PLD—see *SPLD*

Single sided—A *printed circuit board (PCB)* with tracks on one side only.

Skin effect—The phenomenon where, in the case of high-frequency signals, electrons only propogate on the outer surface (the "skin") of a conductor.

Slice—The Xilinx term for an intermediate logical partition/entity between a *logic cell (LC)* and a *configurable logic block (CLB)*. Why "slice"? Well, they had to call it something, and—whichever way you look at it—the term *slice* is "something." At the time of this writing, a slice contains two LCs. See also *CLB*, *LAB*, *LC*, and *LE*.

SoC (system on chip)—As a general rule of thumb, a SoC is considered to refer to an *integrated circuit (IC)* that contains both hardware and embedded software elements. In the not-so-distant past, an electronic system was typically composed of a number of ICs, each with its own particular function (say a microprocessor, a communications function, some memory devices, etc.). For many of today's high-end applications, however, all of these functions may be combined on a single device, such as an ASIC or FPGA, which may therefore be referred to as a *system on chip*.

Soft core—In the context of digital electronics, the term *core* is typically used to refer to a relatively large, general-purpose logic function that may be used as a building block forming a portion of a much larger chip design. For example, if an ASIC contains an embedded microprocessor, that microprocessor would be referred to as a "microprocessor core." Other functions that might fall into this category are microcontroller cores, *digital signal processor (DSP)* cores, communication function cores (e.g., a UART), and so forth. Such cores may be developed internally by the design team, but they are often purchased from third-party *intellectual property (IP)* vendors.

In the case of a soft core, the logical functionality of the core is often provided as RTL VHDL/Verilog. In this case, the core will be synthesized and then placed-and-routed along with the other blocks forming the design. (In some cases the core might be provided in the form of a gate-level netlist or as a schematic, but these options are rare and extremely rare, respectively). One advantage of a soft core is that it may be customizable by the end user; for example, it may be possible to remove or modify certain subfunctions if required.

There is some difference in how the term *soft core* is perceived, depending on the target implementation technology: ASIC or FPGA. In the case of an ASIC, and assuming that the soft core is provided in RTL, the core is synthesized into a gate-level netlist along with the other RTL associated with the design. The logic gates forming the resulting gate-level netlist are then placed-and-routed, the results being used to generate the photo-masks that will, in turn, be used to fabricate the silicon chip. This means that the ultimate physical realization of the core will be in the form of hard-wired logic gates (themselves formed from transistors) and the connections between them. By comparison, in the case of an FPGA, the resulting netlist will be used to generate a *configuration file* that will be used to program the lookup tables and configurable logic blocks inside the device.

A design may comprise one or more soft cores combined with one or more hard cores, along with other blocks of user-defined logic. See also *Hard core*.

Soft IP—In the context of a FPGA, the term *soft IP* refers to a source-level library of high-level functions that can be included in users' designs. These functions are typically represented using a *hardware description language (HDL)* such as Verilog or VHDL at the *register transfer level (RTL)* of abstraction. Any soft IP functions the design engineers decide to use are incorporated into the main body of the design, which is also specified in RTL, and subsequently synthesized down into a group of programmable logic blocks (possibly combined with some hard IP blocks like multipliers, etc.). See also *Hard IP* and *Firm IP*.

Software—Refers to programs, or sequences of instructions, that are executed by *hardware*.

Solder—An alloy of tin and lead with a comparatively low melting point used to join less fusible metals. Typical solder contains 60 percent tin and 40 percent lead; increasing the proportion of lead results in a softer solder with a lower melting point, while decreasing the proportion of lead results in a harder solder with a higher melting point.

Specification-based verification coverage—This measures verification activity with respect to items in the high-level functional or macroarchitecture definition. This includes the I/O behaviors of the design, the types of transactions that can be processed (including the relationships of different transaction types to each other), and the data transformations that must occur. See also *Macroarchitecture definition*, *Microarchitecture definition*, and *Implementation-level coverage*.

SPLD (simple PLD)—Originally all PLDs contained a modest number of equivalent logic gates and were fairly simple. These devices include PALs, PLAs, PROMs, and GALs. As more *complex PLDs (CPLDs)* arrived on the scene, however, it became common to refer to their simpler cousins as *simple PLDs (SPLDs)*.

SRAM (static RAM)—A memory device in which the core of each cell is formed from four or six transistors configured as a latch or a flip-flop. The term *static* is used because, once a value has been loaded into an SRAM cell, it will remain unchanged until it is explicitly altered or until power is removed from the device.

Standard cell—A form of *application-specific integrated circuit (ASIC)*, which, unlike a gate array, does not use the concept of a *basic cell* and does not have any prefabricated components. The ASIC vendor creates custom photomasks for every stage of the device's fabrication, allowing each logic function to be created using the minimum number of transistors.

State diagram—A graphical representation of the operation of a *state machine*.

State machine—see *FSM*

State variable—One of a set of registers whose values represent the current state occupied by a state machine.

Static formal verification—Formal verification tools that examine 100 percent of the state space without having to simulate anything. Their disadvantage is that they can typically be used for small portions of the design only because the state space increases exponentially with complex properties and one can quickly run into "state space explosion" problems. See also *Formal verification* and *dynamic formal verification*.

Static RAM—see *SRAM*

Structured ASIC—A form of *application-specific integrated circuit (ASIC)* in which an array of identical modules (or tiles) is prefabricated across the surface of the device. These modules may contain a mixture of generic logic (implemented either as gates, multiplexers, or lookup tables), one or more registers, and possibly a little local RAM. Due to the level of sophistication of the modules, the majority of the metallization layers are also predefined. Thus, many structured ASIC architectures require the customization of only two or three metallization layers (in one case, it is necessary to customize only a single via layer). This dramatically reduces the time and cost associated with creating the remaining photo-masks used to complete the device.

Sum-of-products—A Boolean equation in which all of the *minterms* corresponding to the lines in the truth table for which the output is a logic 1 are combined using OR operators.

SVA (SystemVerilog Assertions)—The original Verilog did not include an *assert* statement, but SystemVerilog has been augmented to include this capability. Furthermore, in 2002, Synopsys donated its *OpenVera Assertions (OVA)* to the Accellera committee in charge of SystemVerilog. The SystemVerilog folks are taking what they want from OVA and mangling the syntax and semantics a tad. The result of this activity may be referred to as SystemVerilog Assertions, or SVA.

Synchronous—(1) A signal whose data is not acknowledged or acted upon until the next active edge of a clock signal. (2) A system whose operation is synchronized by a clock signal.

Synthesis—see *Logic synthesis* and *Physically aware synthesis*.

Synthesizable subset—When *hardware description languages (HDLs)* such as Verilog and VHDL were first conceived, it was with tasks like simulation and documentation in mind. One slight glitch was that logic simulators could work with designs specified at high levels of abstraction that included behavioral constructs, but early synthesis tools could only accept functional representations up to the level of RTL. Thus, design engineers are obliged to work with a *synthesizable subset* of their HDL of choice. See also *HDL* and *RTL*.

System gate—One of the problems FPGA vendors run into occurs when they are trying to establish a basis for comparison between their devices and ASICs. For example, if someone has an existing ASIC design that contains 500,000 equivalent gates, and they wish to migrate this design into an FPGA implementation, how can they tell if their design will "fit" into a particular FPGA. In order to address this issue, FPGA vendors started talking about "system gates" in the early 1990s. Some folks say that this was a noble attempt to use terminology that ASIC designers could relate to, while others say that it was purely a marketing ploy that doesn't do anyone any favors.

System on chip—see *SoC*

SystemVerilog—A *hardware description language (HDL)* that, at the time of this writing, is an open standard managed by the Accellera organization (www.accellera.com).

SystemVerilog Assertions—see *SVA*

Tap—A register output used to generate the next data input to a *linear feedback shift register (LFSR)*.

Tera—Unit qualifier (symbol = T) representing one million million, or 10^{12}. For example, 3 THz stands for 3×10^{12} hertz.

Tertiary—Base-3 numbering system.

Tertiary digit—A numeral in the tertiary scale of notation. Often abbreviated to "trit," a tertiary digit can adopt one of three states: 0, 1, or 2.

Tertiary logic—An experimental technology in which logic gates are based on three distinct voltage levels. The three voltages are used to represent the tertiary digits 0, 1, and 2 and their logical equivalents False, True, and Maybe.

Time of flight—The time taken for a signal to propagate from one logic gate, *integrated circuit (IC)*, or optoelectronic component to another.

Toggle—Refers to the contents or outputs of a logic function switching to the inverse of their previous logic values.

Trace—see *Track*

Track—A conducting connection between electronic components. May also be called a *trace* or a *signal*. In the case of *integrated circuits (ICs)*, such interconnections are often referred to collectively as *metallization*.

Transistor—A three-terminal semiconductor device that, in the digital world, can be considered to operate like a switch.

Tri-state function—A function whose output can adopt three states: 0, 1, and Z (high impedance). The function does not drive any value in the Z state and, when in this state, the function may be considered to be disconnected from the rest of the circuit.

Trit—Abbreviation of *tertiary digit*. A tertiary digit can adopt one of three values: 0, 1, or 2.

Truth table—A convenient way to represent the operation of a digital circuit as columns of input values and their corresponding output responses.

TTL (transistor-transistor logic)—Logic gates implemented using particular configurations of *bipolar junction transistors (BJTs)*.

Transistor-transistor logic—see *TTL*

UDL/I—In the case of the popular HDLs, Verilog was originally designed with simulation in mind, while VHDL was created as a design documentation and specification language with simulation being taken into account. The end result is that one can use both of these languages to describe constructs that can be simulated, but not synthesized. In order to address these problems, the *Japan Electronic Industry Development Association (JEIDA)* introduced its own HDL called the *Unified Design Language for Integrated Circuits (UDL/I)* in 1990. The key advantage of UDL/I was that it was designed from the ground up with both simulation and synthesis in mind. The UDL/I environment includes a simulator and a synthesis tool and is available for free (including the source code). However, by the time UDL/I arrived on the scene, Verilog and VHDL already held the high ground, and this language never really managed to attract much interest outside of Japan.

μC (microcontroller)—A microprocessor augmented with special-purpose inputs, outputs, and control logic like counter timers.

μP (microprocessor)—A general-purpose computer implemented on a single *integrated circuit (IC)* (or sometimes on a group of related chips called a *chipset*).

ULA (uncommitted logic array)—One of the original names used to refer to gate-array devices. This term has largely fallen into disuse.

Uncommitted logic array—see *ULA*

Vaporware—Refers to either hardware or software that exists only in the minds of the people who are trying to sell it to you.

Verilog—A *hardware description language (HDL)* that was originally proprietary, but which has evolved into an open standard under the auspices of the IEEE.

VHDL—A *hardware description language (HDL)* that came out of the American *Department of Defense (DoD)* and has evolved into an open standard. VHDL is an acronym for *VHSIC HDL* (where VHSIC is itself an acronym for "very high-speed integrated circuit").

Via—A hole filled or lined with a conducting material, which is used to link two or more conducting layers in a substrate.

VITAL—The VHDL language is great at modeling digital circuits at a high level of abstraction, but it has insufficient timing accuracy to be used in sign-off simulation. For this reason, the VITAL initiative was launched at the *Design Automation Conference (DAC)* in 1992. Standing for *VHDL Initiative toward ASIC Libraries*, VITAL was an effort to enhance VHDL's abilities for modeling timing in ASIC and FPGA design environments. The end result encompassed both a library of ASIC/FPGA primitive functions and an associated method for back-annotating delay information into these library models.

Volatile—Refers to a memory device that loses any data it contains when power is removed from the system, for example, random-access memory in the form of SRAM or DRAM.

Word—A group of signals or logic functions performing a common task and carrying or storing similar data; for example, a value on a computer's data bus can be referred to as a "data word" or "a word of data."

About the Author

Clive "Max" Maxfield is 6'1" tall, outrageously handsome, English, and proud of it. In addition to being a hero, trendsetter, and leader of fashion, he is widely regarded as an expert in all aspects of electronics (at least by his mother).

After receiving his B.Sc. in control engineering in 1980 from Sheffield Polytechnic (now Sheffield Hallam University), England, Max began his career as a designer of central processing units for mainframe computers. To cut a long story short, Max now finds himself president of TechBites Interactive (www.techbites.com). A marketing consultancy, TechBites specializes in communicating the value of technical products and services to nontechnical audiences through such mediums as Web sites, advertising, technical documents, brochures, collaterals, books, and multimedia.

In his spare time (Ha!), Max is coeditor and copublisher of the Web-delivered electronics and computing hobbyist magazine *EPE Online* (www.epemag.com) and a contributing editor to www.eedesign.com. In addition to writing numerous technical articles and papers that have appeared in magazines and at conferences around the world, Max is also the author of *Bebop to the Boolean Boogie (An Unconventional Guide to Electronics)* and *Designus Maximus Unleashed (Banned in Alabama)* and coauthor of *Bebop BYTES Back (An Unconventional Guide to Computers)* and *EDA: Where Electronics Begins.*

On the off-chance that you're still not impressed, Max was once referred to as an "industry notable" and a "semiconductor design expert" by someone famous, who wasn't prompted, coerced, or remunerated in any way!

Index

Edwards Brothers Malloy
Ann Arbor MI. USA
September 25, 2012